NUREG-1055
For Comment

Improving Quality and the Assurance of Quality in the Design and Construction of Nuclear Power Plants

I0488191

A Report to Congress

Manuscript Completed: April 1984
Date Published: May 1984

W. Altman, T. Ankrum, W. Brach

Division of Quality Assurance, Safeguards, and Inspection Programs
Office of Inspection and Enforcement
U.S. Nuclear Regulatory Commission
Washington, D.C. 20555

ABSTRACT

At the request of Congress, NRC conducted a study of existing and alternative programs for improving quality and the assurance of quality in the design and construction of commercial nuclear power plants. A primary focus of the study was to determine the underlying causes of major quality-related problems in the construction of some nuclear power plants and the untimely detection and correction of these problems. The study concluded that the root cause for major quality-related problems was the failure or inability of some utility managements to effectively implement a management system that ensured adequate control over all aspects of the project. These management shortcoming arose in part from inexperience on the part of some project teams in the construction of nuclear power plants. NRC's past licensing and inspection practices did not adequately screen construction permit applicants for overall capability to manage or provide effective management oversight over the construction project.

The study recommends a number of improvements in industry and NRC programs. For industry, the study recommends self-imposed rising standards of excellence, treatment of quality assurance as a management tool, not a substitute for management, improved trend analysis and identification of root causes of quality problems, and a program of comprehensive third party audits of present and future construction projects. To improve NRC programs, the study recommends a heavier emphasis on team inspections and resident inspectors, an enhanced review of new applicant's capabilities to construct commercial nuclear power plants, more attention to management issues, improved diagnostic and trending capabilities, improved quality and quality assurance for operating reactors, and development of guidance to facilitate the prioritization of quality assurance measures commensurate with the importance of plant structures, systems, and components to the achievement of safety.

April 20, 1984

The Honorable George Bush
President of the United
States Senate
Washington, DC   20515

Dear Mr. Speaker:

The NRC Authorization Act for fiscal years 1982-83 (P. L.
97-415) directed that the NRC "shall conduct a study of
existing and alternative programs for improving quality
assurance and quality control in the construction of
commercial nuclear power plants." Section 13 of that Act
contained specific study requirements, including
requirements to analyze five alternative approaches to
improving the assurance of quality in the nuclear industry
and to describe any administrative actions or legislative
proposals that the Commission has taken or plans to
undertake for improving quality assurance in construction.

In response, the NRC staff recently completed its report of
the required study. The Commissioners received a briefing
concerning that report on April 4, 1984. A brief overview
of the staff's report is attached (Enclosure 1) along with
a copy of the report itself (Enclosure 2).

The staff's report is complex and contains a large number
of interrelated actions recommended to be undertaken by the
NRC. Due to the complexity of the report and the need for
the Commission to fully understand the plans, schedules,
and resource implications if the recommendations are
implemented, we believe it necessary to take considerably
more time to study the matter before informing the Congress
of our final recommendations. While we are considering the
details of the report, we also believe it desirable to
request comments from the public on the staff's report.

The above deliberations by the Commission will likely take
several months. At the end of that time, we will forward
the Commission's final recommendations to the Congress.

Sincerely,

Nunzio J. Palladino

Enclosures: As stated

Brief Overview of NRC Staff Report on Improving
Quality and the Assurance of Quality in the
Design and Construction of Commercial Nuclear
Power Plants

The staff's report focuses heavily on improvements to the
NRC program. Improvements to NRC's programs are necessary,
but not sufficient, to achieve significant improvements in
quality in the nuclear industry. Significant improvements
can come only from the industry. We view the industry's
Institute of Nuclear Power Operations as a positive step in
that direction. The staff expresses the hope that NRC's
initiatives regarding the importance of excellence in
management to the achievement and assurance of quality will
act as a catalyst for such change.

A primary focus of the required study was to determine the
underlying causes, of (1) the occurrence of major
quality-related problems in the construction of some
nuclear power plants, and (2) the untimely detection and
correction of these problems. The answers to these
questions provided the staff with a foundation for
evaluating the specific alternatives proposed by Congress
in the Act and for recommending improvements to NRC's and
the nuclear industry's approach to and programs for both
achieving quality and assuring quality.

The staff concluded that the root cause for the major
quality-related problems in design and construction was the
failure or inability of some utility management to
effectively implement a management system that ensured
adequate control over all aspects of the project. These
management shortcomings arose in part from inadequate
nuclear design and construction experience on the part of
one or more of the key participants in the nuclear
construction project: the owner utility, architect-
engineer, nuclear steam supply system manufacturer,
construction manager, or the constructor, and the
assumption by some participants of a project role which was
not commensurate with their level of experience. As a
corollary, NRC's past licensing and inspection practices
did not adequately screen construction permit applicants
for overall capability to manage or provide effective
management oversight over the construction project.

The staff found a number of reasons why the utilities and
the NRC were slow to detect or recognize the extent of
major problems in quality or quality assurance. The
reasons include an inability on the part of either to
recognize the underlying programmatic and managerial
deficiencies that caused individual quality problems, an

attenuation in the flow of essential project information from the working level to top management, and a tendency on the part of NRC to set the threshold for taking action for construction problems higher than for operational problems because of the lack of an immediate threat to public health and safety.

The staff's conclusions with respect to the five specific alternative approaches to quality assurance described in the Act were as follows:

(1) Making architectural and engineering criteria more prescriptive would not have a substantial impact on quality; however, reducing the number of design changes during construction would. More complete designs at initiation of construction would enhance quality.

(2) Construction permits (CP) for future CP applicants should be conditioned on post-CP demonstration by the applicant of its capability and effectiveness in managing a nuclear construction project, including the quality assurance program. NRC's pre-CP screening should be modified to evaluate the management competence and prior nuclear experience of applicants, and a special advisory board should be established to provide further advice to the NRC on the qualifications of new applicants.

(3) Audits by certain associations of professionals including the American Society of Mechanical Engineers and the National Board of Boiler and Pressure Vessel Inspectors, cover certain narrow technical areas in more depth than NRC's inspection program but are not sufficiently comprehensive in scope to substitute for NRC inspections. The new construction evaluation program of the Institute of Nuclear Power Operations (INPO) provides the most comprehensive construction audit of any professional association, and it represents a positive industry initiated step toward helping the nuclear industry raise its own standards of performance. This INPO program should not be construed as a substitute for NRC oversight of construction quality, however. The roles of the NRC and INPO are necessarily different, and INPO serves the government, the industry and the public best in its present role. Although the roles of the NRC and INPO must remain separate, they are not fixed, and NRC

needs to be alert to industry improvements resulting from INPO programs and adjust its programs accordingly.

(4) There are a number of ways in which the NRC program has improved in the past several years and can be improved further. The resident inspector program has become the foundation of the NRC inspection program, and it may be expanded. Team inspections such as the new Construction Appraisal Team (CAT) inspections offer significant detection and diagnostic capability for quality problems, and their use should be expanded. NRC's past quality assurance efforts have focused on form and paper at the expense of implementation and evaluating quality of completed work, and they should be reoriented to emphasize performance and effectiveness. The inspection program should address the issue of management capability and effectiveness on a routine basis, not just when the need for remedial action has become apparent.

(5) Comprehensive periodic audits by independent (third-party) inspectors should be required of plants currently under construction as well as future CP applicants. In the interim until such a program can be established by regulation, the CAT program should be expanded to cover more plants for an operating license for additional assurance that their plant's design complies with licensing commitments and NRC regulations.

Administrative actions underway and planned to address these conclusions and others are found in the report and are summarized in tables 2.1, 2.2, and 2.3 of Chapter 2. Chapter 2 provides a comprehensive summary of the study, its conclusions and its recommendations.

The staff's report concludes that at this time there are no legislative changes required. Each of the recommended staff actions could be implemented within NRC's current statutory authority. However, the staff identifies several issues that after subsequent analysis may result in legislative proposals.

The staff notes that the actions which have been identified and recommended by the study are extremely comprehensive and several of them could consume all of NRC's current budget and manpower allocated to development of the quality assurance program. It will be necessary to establish priorities for the quality assurance issues within the other issues faced by the NRC and make resource allocations. As a result, some of the recommended actions may necessarily be deferred until the higher priority actions are completed.

TABLE OF CONTENTS

LIST OF TABLES

ACKNOWLEDGEMENTS

A number of individuals other than the immediate project staff and management made important contributions to this report. Mr. Richard C. DeYoung, Director of NRC's Office of Inspection and Enforcement (IE), selected the key project participants and provided continuing support and suggestions throughout the project. In addition to providing continuous senior management oversight of the study, Mr. James M. Taylor, Deputy Director of IE, was actively involved in the performance of the study. Among his many contributions for the project, he set up and managed the pilot program study and activities.

Mr. Brian Grimes, former Deputy Director for the Division of Quality Assurance, Safeguards, and Inspection Programs, IE, provided valuable advice on several complex issues and was largely responsible for establishment of the special review group and selection of its members. Mr. Tom Dorian of the Office of the Executive Legal Director provided continuing legal review and advice on the project as well as comprehensive editorial advice.

The manuscript was typed by Ms. Donna Zetts, and edited by Ms. Vicki Lee, both of whom made personal and family sacrifices to contribute their considerable talents to the project to ensure that the report was completed on time.

The insights gained in this study would not have been possible without the information obtained through the case studies and the pilot programs. The study staff is indebted to the following utilities, each of whom agreed to participate in the case studies and/or the pilot program:

> Public Service of Indiana (both)
> Arizona Public Service Company (both)
> Houston Lighting and Power Company (both)
> Georgia Power Company (Case Studies)
> Pacific Gas and Electric Company (Case Studies)
> Florida Power and Light Company (Case Studies)
> Consumers Power Company (Pilot Program)

The authors of the report are solely responsible for any awkwardness of phrase, redundancy (of which there is much) and lack of logical thought.

PROJECT ORGANIZATION

FORD AMENDMENT STUDY

Richard C. DeYoung, Director, Office of Inspection and Enforcement (IE)
James M. Taylor, Deputy Director, IE
J. Nelson Grace, Director, Division of Quality Assurance, Safeguards, and
  Inspection Programs, (DQASIP), IE
James G. Partlow, Deputy Director, DQASIP, IE
G. Ted Ankrum, Chief, Quality Assurance Branch, DQASIP, IE
W. D. Altman, Study Project Manager, DQASIP, IE

Project Staff

E. W. Brach   Edward T. Baker
Paul Keshishian   Ronald M. Langstaff
George Gower   James E. Kennedy

Administrative Staff

Donna Zetts
Vicki Lee

Other Contributors

J. M. Sniezek, Former Deputy Director, IE
E. L. Jordan, Former Director, Division of Engineering and Quality Assurance, IE
B. K. Grimes, Former Deputy Director, DQASIP, IE
T. L. Harpster, Former Chief, Quality Assurance Branch, DQASIP, IE
Tom Dorian, Office of Executive Legal Director
Chip Cameron, Office of Executive Legal Director
John Fair, IE
Bill Beach, IE

Contractors

Pacific Northwest Laboratory

Harold Harty   W. J. Apley
Miles Patrick   J. F. Nesbit
James Christensen   B. W. Smith
Robert Sorenson   P. L. Hendrickson

N. C. Kist and Associates, Inc.

John L. Heidenreich
Richard M. Kleckner

Battelle Human Affairs Research Center

Marilyn Walsh
Mary McGuire
B. L. Hansen

EG&G Idaho

Ken Carroll   Earl Bradford
Larry Kubicek   Dave Ross
Harley Kirschenmann

Dellon Associates, Inc.

Robert M. Dellon

REVIEW GROUP

FORD AMENDMENT STUDY

Fred Albaugh, Chairman
Independent Consultant

John Amaral
Bechtel Power Corporation

Spencer Bush
Review and Synthesis Associates

Thomas Cochran
National Resources Defense Council

George Coulbourn
Boeing Engineering Company

John Gray
International Energy Associates, Ltd.

John Hansel
System Development Corporation

Robert Laney
Independent Consultant

Leland Bohl
General Electric

1.0 INTRODUCTION

1.1 PURPOSE OF THE REPORT

In recent years, major problems relating to the quality of design and/or construction have arisen at several nuclear power plant construction projects. Projects having received widespread attention in this regard include Marble Hill, Midland, Zimmer, South Texas, and Diablo Canyon. Because of these quality-related problems and others in the U.S. nuclear industry, many in the public and in Congress have questioned (1) the nuclear industry's ability to design, construct and operate reactors in a manner consistent with maintaining public health and safety, and (2) the Nuclear Regulatory Commission's (NRC's) ability to provide effective regulatory oversight of these activities. As a result of these Congressional concerns, the NRC was directed by Congress in Section 13(b)* of Public Law 97-415 (the NRC Authorization Act for fiscal years 1982 and 1983) to conduct a study of existing and alternative programs for improving quality assurance and quality control in the construction of nuclear power plants. The study requirements of that law are as follows:

> Sec. 13(b) The Commission shall conduct a study of existing and alternative programs for improving quality assurance and quality control in the construction of commercial nuclear powerplants. In conducting the study, the Commission shall obtain the comments of the public, licensees of nuclear powerplants, the Advisory Committee on Reactor Safeguards, and organizations comprised of professionals having expertise in appropriate fields. The study shall include an analysis of the following:**
>
> > (1) providing a basis for quality assurance and quality control, inspection, and enforcement actions through the adoption of an approach which is more prescriptive than that currently in practice for defining principal architectural and engineering criteria for the construction of commercial nuclear powerplants;
> >
> > (2) conditioning the issuance of construction permits for commercial nuclear powerplants on a demonstration by the licensee that the licensee is capable of independently managing the effective performance of all quality assurance and quality control responsibilities for the powerplant;

---

*This amendment to the NRC Authorization Act was introduced by Senator Wendell Ford of Kentucky and was co-sponsored by Senators Simpson, Mitchell, Levin, and Hart. It was called the "Ford Amendment" by its sponsors and this term is adopted in this report.

**These five alternatives will frequently be referred to as "alternatives b(1)-b(5)" in the remainder of this report.

(3) evaluations, inspections, or audits of commercial
nuclear powerplant construction by organizations comprised
of professionals having expertise in appropriate fields
which evaluations, inspections, or audits are more effective
than those under current practice;

(4) improvement of the Commission's organization, methods,
and programs for quality assurance development, review, and
inspection; and

(5) conditioning the issuance of construction permits for
commercial nuclear powerplants on the permittee entering
into contracts or other arrangements with an independent
inspector to audit the quality assurance program to verify
quality assurance performance.

For purposes of paragraph (5), the term "independent inspector"
means a person or other entity having no responsibility for the
design or construction of the plant involved. The study shall
also include an analysis of quality assurance and quality control
programs at representative sites at which such programs are
operating satisfactorily and an assessment of the reasons therefor.

(c) For purposes of --

(1) determining the best means of assuring that commercial
nuclear power plants are constructed in accordance with the
applicable safety requirements in effect pursuant to the
Atomic Energy Act of 1954; and

(2) assessing the feasibility and benefits of the various
means listed in subsection (b);

the Commission shall undertake a pilot program to review and
evaluate programs that include one or more of the alternative
concepts identified in subsection (b) for the purposes of assessing
the feasibility and benefits of their implementation. The pilot
program shall include programs that use independent inspectors for
auditing quality assurance responsibilities of the licensee for
the construction of commercial nuclear powerplants, as described
in paragraph (5) of subsection (b). The pilot program shall include
at least three sites at which commercial nuclear powerplants are
under construction. The Commission shall select at least one site
at which quality assurance and quality control programs have operated
satisfactorily, and at least two sites with remedial programs underway at which major construction, quality assurance, or quality
control deficiencies (or any combination thereof) have been identified
in the past. The Commission may require any changes in existing
quality assurance and quality control organizations and relationships
that may be necessary at the selected sites to implement the pilot
program.

(d) Not later than fifteen months after the date of the enactment of this Act, the Commission shall complete the study required under subsection (b) and submit to the United States Senate and House of Representatives a report setting forth the results of the study. The report shall include a brief summary of the information received from the public and from other persons referred to in subsection (b) and a statement of the Commission's response to the significant comments received. The report shall also set forth an analysis of the results of the pilot program required under subsection (c). The report shall be accompanied by the recommendations of the Commission, including any legislative recommendations, and a description of any administrative actions that the Commission has undertaken or intends to undertake, for improving quality assurance and quality control programs that are applicable during the construction of nuclear powerplants.

This report describes the activities and results of the special study of quality assurance required by the Ford Amendment. Congress' action to elevate concern for quality in construction of commercial nuclear power plants to the national level will be of continuing help to the NRC in attaining its goals for quality in the nuclear industry.

In its 1984 "Policy and Planning Guidance" to the NRC staff, the Commission states its policy for raising the quality of nuclear plants as follows:

Policy:

1. The NRC must improve its activities that affect quality in the nuclear industry. NRC's goal is to assure a high level of quality in management of reactor design, construction, operations, and maintenance.

2. For both construction activities and operating facilities the NRC needs to understand the causal factors leading to problems and to develop a modified institutional and legislative framework for future nuclear plants which will decrease the probability of repetition of past mistakes. The theme of "do it right the first time" should be adopted to ensure plants are built properly and can operate safely.

3. In order to reduce operational problems including mainten-ance and modification activities, the NRC needs to pursue more aggressively efforts (1) to assure utilities provide the appropriate management framework and capability for safe operation and maintenance of nuclear power plants; (2) to improve quality in utility operations and in procedures, systems, and components used in operations; and (3) to develop better guidance for the treatment of plant systems, components, and equipment that can adversely affect safe operation.

4. NRC should highlight the necessity for highly trained and qualified professionals for licensees, contractors and vendors to manage those functions that relate to safety.

This study reflects the above Commission policy statement. It is a look to the future--an opportunity for a mid-course correction that builds upon past experience to chart a future course for assuring quality in nuclear power plant design and construction. While this study has looked at the past, it has been from the perspective of what should be done in the future.

In any complex endeavor, some errors will be made. The more complex the endeavor, the greater the chance of errors. If some risk is associated with the endeavor, measures must be taken to provide assurance that errors are found, corrected, and do not pose an undue threat to public health and safety. Construction of nuclear power plants is a very complex endeavor, and uncorrected errors in construction may seriously threaten public health and safety when operation begins. The primary measure used by the nuclear industry to provide assurance that construction errors are found and corrected is a quality assurance (QA) program. As used by the NRC, "quality assurance" comprises all those planned and systematic actions necessary to provide adequate confidence that a structure, system or component will perform satisfactorily in service. Quality assurance includes "quality control, which comprises those quality assurance actions related to the physical characteristics of a material, structure, component or system which provide a means to control the quality of the material, structure, component or system to predetermined requirements."*

Congress has posed several very specific questions, and this study undertakes to answer those questions. However, to provide a foundation for the answers to those specific questions, the study sought also to answer the following underlying questions:

1.   Why have certain nuclear construction projects experienced significant quality-related problems while others have not?

2.   Why have the NRC and the utilities failed or been slow to detect and/or respond to these quality-related problems?

The answers to these underlying questions provide a foundation for answering the following question which, in the NRC's opinion, summarizes the thrust of the Ford Amendment:

3.   What changes should be made to the current policies, practices, and procedures governing commercial nuclear power plant design, construction and regulation to prevent major quality problems in the future or to provide more timely detection and correction of problems?

These questions helped to focus the study activities and approach, and their answers provided the central themes for this report to Congress.

Perhaps equally important to stating what questions this study did answer is to state what questions it did not answer. Primary among questions that this study did not answer are the following:

---

*Code of Federal Regulations, Title 10, Part 50 (10 CFR 50), Appendix B.

(1) This study did not attempt to quantify the relationship between quality and quality assurance and safety, nor did the study develop a quantifiable relationship between risk and quality assurance. In particular, this study did not address the question of the extent to which the quality or quality assurance problems that occurred at plants such as Marble Hill, Midland, Zimmer, South Texas, or Diablo Canyon may have affected the safety of those plants.

(2) This study did not address the issue of quality and quality assurance for operating plants.

(3) This study did not develop a methodology to measure the effectiveness of quality assurance programs. In particular, this study did not attempt to evaluate the effectiveness of various non-NRC QA programs covered in the study, including those of other government agencies, other industries, or other countries, but rather sought to identify individual features of those programs that should be considered for adoption in NRC's program.

(4) The study took as a given that NRC's statutory role is not to ensure the survival of the nuclear option but rather to ensure that if nuclear power is used in the U.S., such use is consistent with maintaining the common defense and security and public health and safety. Consistent with this premise, the study (1) did not consider the appropriate role of nuclear power in the U.S.'s national energy policy, (2) did not attempt to determine whether NRC's present statutory role should be changed, and (3) did not attempt to assess the future of nuclear power in the U.S. or the effect of quality assurance programs on that future. Exploration of such questions is beyond the statutory purview of the NRC. In this regard, the Congressional Office of Technology Assessment (OTA) has recently published a major study that deals with these issues: Nuclear Power in an Age of Uncertainty.* The OTA report and this study complement each other in many ways and, while dealing with overlapping issues from different perspectives, each reinforces findings of the other (e.g., the critical role of utility management in constructing and operating nuclear power plants, and predictability in the licensing process).

Each of these questions was considered outside the scope of this study, which was tailored to be as responsive as possible to the specific questions asked by the Ford Amendment.

This report focused on developing an understanding of the quality or quality assurance problems that have occurred in plants currently under construction. Some of these projects have experienced problems in plant quality--parts of the plants were built incorrectly. Some of these projects experienced problems in

_____
*U.S. Congress, Office of Technology Assessment. February 1984. OTA-E-216, Washington, D.C.

the assurance of quality--the utility was unable to demonstrate whether its plant was built correctly. Some projects experienced problems in both quality and the assurance of quality. To acknowledge this overlap, the report throughout will refer to problems in quality and/or quality assurance or quality and the assurance of quality, etc. For simplicity of writing, problems generally falling under this umbrella will sometimes be referred to as "quality-related" problems.

1.2 ROLES OF THE NRC AND UTILITIES IN NUCLEAR CONSTRUCTION

Before describing the study activities and results, the statutory role of the NRC in nuclear construction, quality and quality assurance should be made clear. The NRC is not directly responsible for nuclear power plant quality. The public policy of the United States, established in the Atomic Energy Act of 1954, is that ownership and operation of commercial nuclear power plants rest in the hands of the public and privately owned utilities of the United States, but only to the extent their use is consistent with the common defense and security and the public's health and safety. The Act directs the NRC to issue licenses only to persons "who are equipped to observe and who agree to observe such safety standards to protect health and to minimize danger to life or property as the Commission may by rule establish."*

It is the owner/licensee who is responsible for achieving and assuring the quality and reliability of a nuclear power plant. The designers, the constructors, labor contractors, and component vendors are responsible to the licensee to the extent that the owner/licensee delegated responsibility. However, ultimate responsibility, even though delegated, is retained by the licensee.** The NRC is responsible for the health and safety of the public, not the quality or lack of quality of the nuclear power plant. If the licensee has not fulfilled its responsibility for building a safe plant, the NRC can still fulfill its responsibility by denying an operating license.

However, neither the interests of the public (who may also happen to be the owners, stockholders and/or customers of the utilities) nor the utilities are well served by a regulatory system that introduces uncertainty about the ultimate acceptability of an expensive and long-in-the-making facility until its completion date. All parties are best served by a regulatory process that establishes relevant standards, exercises due process in the change of those standards, screens out at the beginning those organizations that are not equipped to attain those standards, provides inspections that effectively measure the attainment of those standards in a time frame that permits corrective action as early as possible and takes enforcement action in all cases where corrective action is not adequate, and finally provides reasonable confidence that a project has demonstrably met all requirements and can be operated safely. Many of this study's recommendations, when implemented, should improve NRC's ability to provide such a regulatory process.

---

*Atomic Energy Act of 1954, as amended, Section 103(b)(2).

**10 CFR 50, Appendix B.

1.3 EVOLUTION OF NRC PROGRAMS FOR QUALITY ASSURANCE

It is important to understand the evolution of the regulatory framework within which the major quality-related problems have occurred. The regulatory framework governing the nuclear industry has developed and changed along with the nuclear industry over the years, often in response to specific events. The major quality-related problems at the five nuclear projects cited previously provide a new set of events and programmatic failures that will lead to further evolution of the regulatory framework. The purpose of this study is to provide direction for that evolution and also to identify any factors that may be beyond NRC's regulatory purview but that may have contributed to those major quality-related problems.

The following sections describe the evolution of quality assurance requirements and guidance, quality assurance licensing programs, and quality assurance inspection programs.

1.3.1 Quality Assurance Requirements and Guidance

In July 1967, the Atomic Energy Commission (AEC) published for public and industry comment Appendix A to 10 CFR Part 50, "General Design Criteria for Nuclear Power Plants." Among the 55 criteria in Appendix A covering plant design, one criterion required a quality assurance program for certain structures, systems and components. Following review, public comments, and subsequent revisions, Appendix A was issued as an effective regulation in February 1971. Although its criterion for the QA program was very general, the July 1967 draft of Appendix A was the first AEC proposal that would require nuclear power plant licensees to have a quality assurance program.

The lack of AEC requirements and criteria for quality assurance was a key issue raised by the Atomic Safety and Licensing Board (ASLB) in the operating license hearings for the Zion plant in 1968. The board ruled that until the licensee presented a program to assure quality and until the AEC developed criteria by which to evaluate such a QA program, the hearings would be halted. Following the board's rulings, the AEC developed requirements and criteria for quality assurance programs and prepared a proposed new regulation, Appendix B to 10 CFR Part 50, which would require licensees to develop programs to assure the quality of nuclear power plant design, construction, and operation.

Appendix B contains 18 criteria that must be a part of the quality assurance program for safety-related systems and components. Experience from military, the National Aeronautics and Space Administration (NASA), and commercial nuclear projects, as well as the AEC's own nuclear reactor experience, was used in developing the criteria. Appendix B clearly places the burden of responsibility for quality assurance on the licensee. Although the licensee may delegate to others the work of establishing and executing part or all of the quality assurance program, the licensee retains responsibility for the program. Visible QA documentation is required for all activities affecting the quality of safety-related systems. Appendix B was published for comment in April 1969 and implemented in June 1970.

In addition to establishing QA regulations (i.e., Appendices A and B) in the early 1970s, the AEC and the industry began issuing guidance that provided acceptable ways of meeting the intent and requirements of the specific regulations. In October 1971, the American National Standards Institute (ANSI) issued N45.2, "Quality Assurance Program Requirements for Nuclear Power Plants." This standard was subsequently endorsed by the AEC in Safety Guide 28 (now Regulatory Guide 1.28) in June 1972. In 1973-1974, the AEC issued three guidance documents for quality assurance in design and procurement, construction, and operation to help licensees establish QA programs. In July 1973, two AEC Commissioners and senior AEC staff participated in a series of regional conferences with utilities to explain the role of quality assurance in designing, constructing, and operating nuclear power plants and the NRC's role in licensing, inspecting, and implementing licensee's quality assurance programs. Since 1970, as the nuclear industry grew, as experience was gained in nuclear regulation, and as the need for such guidance was recognized, many consensus standards and AEC/NRC regulatory guides have been developed and published to address various aspects of quality and quality programs.

1.3.2 Quality Assurance Licensing Programs

Appendices A and B of 10 CFR 50 set quality assurance requirements but left open the issue of how to meet them. Industry standards were subsequently, developed, and AEC guidance documents for quality assurance were prepared and published. The standards and guidance documents helped both the AEC and the license applicants understand what quality assurance is and how the quality assurance program should function. AEC staff guidance was prepared for the licensing staff to use as criteria for evaluating licensees' applications.

In the early 1970s, the regulatory staff believed that license applications should contain additional information on the licensee's quality assurance programs. In an effort to establish standards for the licensees' description of their QA program in their construction permit applications, a proposed "Standard Format and Content of Safety Analysis Reports for Nuclear Power Plants" was issued for comment in February 1972 and later adopted. After the new standard was developed, a staff Standard Review Plan (SRP) was published in 1974 and adopted in 1975 to standardize and guide the licensing staff in its review of license applications. Licensing staff use the SRP as a benchmark in reviewing the QA programs of license applicants. Updated and revised versions of the SRP have been issued about every three years since.

1.3.3 Quality Assurance Inspection Programs

Before 1968, the AEC performed little inspection at nuclear power plants under construction. Few inspection procedures and only minimal guidance were available. As a result of quality-related problems in the construction of some nuclear power plants, including the Oyster Creek plant in New Jersey at which major problems in vendor-supplied materials were discovered, the AEC recognized the need to examine construction activities more closely and to develop more formalized programs for inspecting construction activities. The AEC reassigned inspectors from operations to construction and hired personnel with construction backgrounds.

As the number of inspectors and reactors increased, so did the need for more inspection guidance. The AEC began developing a "General Facility Under Construction Inspection Program" and began writing inspection procedures to implement the program. In late 1969, the AEC issued a directive to the regional compliance offices to implement the procedures. In 1972, a procedure entitled "QA During Design and Construction" was issued. This procedure addressed Appendix B of 10 CFR 50 and required a review of the licensee's quality assurance manual, a meeting with corporate utility management, and an initial inspection after the construction permit application was docketed but before it was issued.

In 1973, more detailed inspection procedures were issued covering pre-docketing and pre-construction permit inspections. The AEC then began preparing a more comprehensive inspection program, which greatly expanded and clarified the inspection program during the pre-construction and post-construction permit issuance period. These inspection programs have basically the same structure today as when the major revised programs were issued in 1975. However, major changes have recently been made to refine and prioritize the inspection procedures, to increase inspection coverage with resident inspectors and team inspections, and to direct more inspection effort to independently confirming the quality of hardware and completed work and less inspection to quality assurance documentation and programmatic aspects.

1.4 PROJECT TECHNICAL APPROACH

The findings, conclusions, and recommendations of this study are based on the following project activities: (1) case studies of several commercial nuclear power plant projects that have had major quality-related problems in design and construction and several that have not; (2) pilot programs to assess the feasibility and benefits of third-party inspections to evaluate QA program effectiveness; (3) evaluation of audits of nuclear power plant construction by the Institute of Nuclear Power Operations (INPO); (4) analysis of the feasibility and benefits of a more prescriptive approach for defining principal architectural and engineering criteria; (5) review and analysis of NRC's organization, methods, and programs for quality assurance; (6) analysis of project, organizational, and institutional issues associated with quality in nuclear power plant design and construction; (7) review of other selected programs for the assurance of quality, including programs of other U.S. government agencies, other industries, and foreign countries; (8) consultations and interaction with the public, licensees, the Advisory Committee on Reactor Safeguards (ACRS), associations of professionals and others to solicit their ideas and input; and (9) establishment of a group of outside senior and expert consultants to provide individual comments on study activities and findings.

Because the case studies and pilot program involved some common sites (Marble Hill, South Texas, and Palo Verde), they may be confused with each other. In the case studies, six projects were analyzed to identify the reasons for the success or lack of success of their quality assurance programs, whereas the pilot program was a test of the use of independent auditors at four sites to evaluate QA program effectiveness. Although the pilot program audits did analyze the quality assurance programs of four different licensees and over-lapped some case study activity, the desired result from the pilot program was an assessment of whether independent, third-party audits could feasibly enhance the detection capability currently provided by existing NRC and licensee programs.

1.5 PUBLIC COMMENTS

Because only fifteen months were available to complete the required analyses and to prepare this report, time was not available to publish preliminary study findings for public comment. The NRC elected to request public comments on the Ford Amendment at the beginning of the study so that the comments could be used to develop and refine the study plan. The NRC did not want to develop a study plan and discover, through a later public comment process, that a significant item had been missed and could not be added because of time. Some of the comments received were used in conducting the study, and several of the study conclusions support comments received. As a result, many of the comments that were received have been adopted within NRC's planned actions or included in issues slated for further study. The resulting study plan was presented at a public meeting of an ACRS subcommittee in July 1983.

To provide some outside review, the NRC arranged for nine persons who were independent of the NRC to examine NRC's plans and progress several times during the study. These outside professionals had expertise in nuclear power plant quality assurance, project management, engineering, and other relevant areas. The names, positions, and a summary of the comments of the reviewers are contained in Chapter 10 of this report.

1.6 ORGANIZATION OF THE REPORT

Based on the previously described project activities, the remainder of this report is organized as follows. Chapter 2 is a summary of the report and contains the study findings, conclusions, actions and recommendations. Chapter 3 describes findings from the case studies and contains an assessment of the reasons the quality programs at some nuclear projects have operated satisfactorily while others have not. The case study methodology, analysis and findings are described in more detail in Appendix A. In Chapter 4, the pilot program and the results of the pilot program analysis are described. This chapter also includes an analysis of the feasibility and benefits of conditioning construction permits on a positive post-construction permit demonstration of the applicant's QA management ability and on the applicant's entering into arrangements with third parties to audit its QA program performance. Chapter 5 is an analysis of the benefits and feasibility of audits by associations of professionals, with a focus on the INPO's Construction Project Evaluation program.

Chapter 6 is an analysis of the benefits and feasibility of adopting a more prescriptive approach to defining principal architectural and engineering criteria. Chapter 7 contains the results of an analysis of the NRC's organization, methods and programs for quality assurance. Appendix B, which is an analysis of the NRC's QA program by a management consulting firm, covers the NRC program in more detail. Chapter 8 contains the results of an analysis of contractual, organizational, and institutional issues associated with quality in nuclear power plant design and construction. The issues in this section emerged as a result of other study activities, and the results of this analysis help provide a more comprehensive understanding of indirect factors that have some effect on quality in the nuclear industry. A more detailed analysis of these issues is found in Appendix C.

Chapter 9 contains the results of a review of selected quality programs outside the U.S. commercial nuclear industry, including those of other government agencies, other industries and foreign countries. The purpose of this outside program review was to identify aspects of other programs that could be translated to the NRC program and might improve the NRC program. Appendix D contains a more detailed analysis of this review. Neither the Chapter 8 nor the Chapter 9 analyses were required by the Ford Amendment, but they were included in the study to provide a broader spectrum of information and analysis from which to draw findings and conclusions and to develop recommendations. Chapter 10 briefly summarizes information received from the public, licensees, the ACRS, associations of professionals, and the special review group established for this study, together with NRC's response to the significant public comments received.

The report has been structured so that Chapters 3 through 10 individually describe the analyses and study results summarized in Chapter 2. Each of Chapters 3 through 10 has been written as a stand-alone document so that anyone who is interested in a particular subject (e.g., more prescriptive architectural and engineering criteria) can read the chapter pertaining to that subject and understand the study's conclusions on that subject without having to read the rest of the report. The study's major results, conclusions and recommendations are summarized in Chapter 2. This organization has resulted in some necessary redundancy between Chapter 2 and the rest of the report to achieve the goals of (1) summarizing the study results in one place, and (2) covering each major topic in a self-contained, stand-alone treatise.

2.0 SUMMARY: STUDY CONCLUSIONS AND RECOMMENDATIONS

This chapter summarizes the conclusions and recommendations of the Ford
Amendment study. Section 2.1 describes the findings and conclusions stemming
from NRC's analysis of the underlying questions introduced in Chapter 1. The
study conclusions with respect to the five specific alternative approaches to
improve quality assurance and quality control described in the Ford Amendment
are presented in Section 2.2. Section 2.3 discusses conclusions and recommendations from several consultant studies that were conducted as part of the
overall study. Section 2.4 describes administrative actions already undertaken
by the NRC or recommended by the study to be undertaken or further analyzed by
the NRC as a result of the findings and conclusions in the preceding sections.
These actions are summarized in Tables 2.1 and 2.2. Section 2.5 covers actions
that the study found to be appropriate for consideration by the nuclear
industry. Table 2.3 summarizes the differences among the former (pre-1980),
the present (1982-83) and the recommended future NRC and industry programs for
the assurance of quality in designing and constructing nuclear power plants.
Section 2.6 describes an issue that was identified in the study that requires
further analysis before any legislative recommendations can be made.

As with the report as a whole, individual sections of this chapter have been
written as stand-alone treatises so that the reader may develop a quick understanding of the study's conclusions or recommendations on a particular topic
without reading the whole chapter. This has resulted in some redundancy
between sections of the chapter. To the extent possible, the text has been
annotated to refer the reader to other similar material in the report.

Most of the actions recommended by this study are directed toward revising
NRC's program for the assurance of quality in nuclear power plant design and
construction. The recommended actions are intended to improve the capabilities
of the NRC and the nuclear industry to better achieve the overall quality
assurance (QA) program goals of prevention, detection, and assurance. Although
most of the recommended actions are directed at changes in NRC's performance of
its QA activities, they will also influence the way the nuclear industry
conducts its QA activities. The industry's activities are ultimately the more
important of the two, because the actual work activities that result in whether
a nuclear power plant is built and operated safely remain where they have
always been--with the owner/licensee.

2.1 CONCLUSIONS STEMMING FROM UNDERLYING QUESTIONS

While conducting this study, it became apparent that the root causes of quality
assurance breakdowns went well beyond the purview of the formal QA program
itself and that the solution of the QA problem went beyond how to devise new or
better quality assurance programs. To provide a foundation for the answers to
the specific questions asked by the Ford Amendment, there were two underlying
questions that needed to be answered first. The answers to these underlying
questions also form the foundation for the actions proposed by this study and
the conclusions formed concerning the five specific approaches Congress
prescribed for study. The following subsections discuss each of these underlying concerns.

2.1.1. Why Have Several Nuclear Construction Projects Experienced Significant Quality-Related Problems While Others Have Not?

The principal conclusion of this study is that nuclear construction projects having significant quality-related problems in their design or construction were characterized by the inability or failure of utility management to effectively implement a management system that ensured adequate control over all aspects of the project. Each of the major quality-related problems cited in Chapter 1 was related to breakdowns or shortcomings in the implementation of the project's quality assurance programs; however, the quality assurance program's deficiencies had as their root cause shortcomings in corporate and project management. At several projects, breakdowns in the quality assurance program were part of larger breakdowns in overall project management, including planning, scheduling, procurement, and oversight of contractors.

There are two major corollary findings associated with management capability and effectiveness. First, in today's environment, prior nuclear design and construction experience of the collective project team (defined as the architect-engineer (A/E), nuclear steam supply system (NSSS) manufacturer, construction manager (CM), constructor, and owner) is essential, and inexperience of some members of the project team must be offset and compensated for by experience of other members of the team. Each member of the project team should assume a project role consistent with its prior nuclear experience and not overstep its capabilities. A false sense of security growing out of prior success in fossil plant construction led several first-time utilities into underestimating the complexity of nuclear design and construction. This miscalculation resulted in the assembly of a project team that lacked the requisite experience, background, and management capability, individually or collectively, to successfully design and construct a commercial nuclear power plant without the development of significant quality problems. Although prior nuclear design construction experience of the collective project team appears necessary for future plants, it is not sufficient to assure the completed construction of a quality nuclear plant.

The second corollary finding is that in the past, the NRC has not adequately assessed the factors of management capability and prior nuclear experience in its pre-construction permit reviews and inspections. The substantial changes the NRC has required of some licensees' projects to bring them up to minimum standards are evidence that some utilities that were not adequately prepared to undertake a nuclear construction project were granted construction permits (CPs). It is clear in retrospect that some utilities granted CPs under previous standards would not, based on the same qualifications, be granted a CP in today's regulatory environment without substantial personnel and organizational improvements in experience levels and management approach. Besides not performing a searching evaluation of licensee management capability before issuing the CP, the NRC also did not foresee that even an otherwise adequate management could be overwhelmed and demoralized by increasingly numerous regulatory, design, and hardware changes mandated during the design and construction process.

Other factors that contributed to major construction quality problems in the past include the changing regulatory, political, and economic environment surrounding nuclear power over the past several years and some licensees' inability to recognize and adjust to the changes as they occurred; the NRC's and licensees' inability to manage change well; some licensees' failure to treat quality assurance as a management tool, rather than as a paperwork exercise or, conversely, as a substitute for their own management involvement; and NRC's inability to convince some licensees of the necessity for implementing their quality assurance program.

The major quality problems that have arisen in design were related to short-comings in management oversight of the design process, including failure to implement quality assurance controls over the design process that were adequate to prevent or detect mistakes in an environment of many design changes.

An essential characteristic of a successful nuclear construction project is prior nuclear construction experience of the project team (utility owner, A/E, NSSS manufacturer, CM, and constructor) collectively, with individual team members assuming roles consistent with their prior level of nuclear experience and capabilities. Prior nuclear design and construction experience is necessary for key project personnel for each of the organizations comprising the project team.

Although it is necessary that each team member assume a project role commen-surate with its capability and prior experience for project success, it is not sufficient. Prior nuclear construction experience of the utility owner is particularly helpful, although not mandatory if the corporate entities com-prising the rest of the project team are sufficiently experienced and if the utility and the other members of the project team assume project roles con-sistent with their respective levels of nuclear experience. However, the utility is ultimately responsible for the project, and it cannot delegate its management and oversight responsibilities to others. This thought was summar-ized well by the Deputy Administrator of one of the NRC regional offices:

> It is essential that a utility undertaking the construction and operation of a power reactor facility have strong project manage-ment capability within its own organization to enable independent owner direction and assessment of overall management and assurance of quality of the project.

Another essential characteristic of a successful nuclear construction project is an understanding and appreciation of the complexities and difficulties of nuclear construction by top corporate management that manifests itself in a project management approach that includes adequate financial, organizational, and staffing support for the project; good planning and scheduling; and close management oversight of the project and the project contractors. Other factors contributing to project success include strong management commitment to quality and support for the quality program that starts at the top of the corporate structure and flows down through project-level management to first-line super-visors and foremen; involvement of top corporate management in the project; commitment of resources sufficient to complete the project in a quality manner; careful selection of key project staff; an atmosphere that encourages looking for problems and solving them; an openness to ideas for improvements; effective project communications vertically and across project interfaces; an under-standing of the symptoms of poor management practices; use of the quality assurance program as a management tool, rather than as a substitute for manage-ment; and an understanding of the role, mission, and constraints of the NRC.

Nuclear construction is sufficiently different from and more complex than fossil construction that fundamental changes to a utility's corporate structure and project approach may be necessary to successfully complete the project.

Finally, of several projects studied, there tended to be a direct correlation between the project's success and the utility's view of NRC requirements. More successful utilities tended to view NRC requirements as minimum, not maximum, levels of performance, and they strove to establish and meet increasingly higher, self-imposed goals. This rising standard of excellence theme was an important part of the study's analysis of industry initiatives for self-improvement, such as industry establishment and support of the Institute of Nuclear Power Operations (INPO) (Chapter 5).

The case studies (Chapter 3) of nuclear construction projects having various levels of quality success confirmed, through the analysis of actual cases, several widely held opinions about the cause of major quality-related problems. These opinions include shortcomings in management oversight of the project, lack of management commitment to quality, insufficient prior nuclear experience, and use of a fossil approach to nuclear construction. The case studies also confirmed the phenomenon of top corporate management setting the tone for a project and affecting the emphasis of its subordinates, both managers and workers. In this regard, management's actions have much more influence than their words.

The case studies were also useful in understanding what the principal causes of the quality-related problems were not, e.g., craftsmanship. The case studies found that while poor craftsmanship played a role in some of the major quality-related problems, it was an effect, not the cause, of the underlying problems. The principal underlying cause of poor craftsmanship in constructing nuclear power plants, as well as the quality problem, was found to be poor utility and project management.

This discussion is not meant to minimize the importance of craftsmanship in achieving quality. Clearly, it is craftsmen who build or fail to build quality into a nuclear plant, and quality craftsmanship is necessary for achieving quality in nuclear construction. However, good craftsmanship is not a sufficient condition to achieve quality. Good craftsmanship can be defeated in its attempts to build a quality plant by conditions out of its control. Such conditions include unavailability of tools or materials, rework due to excessive design changes, design completion not sufficiently ahead of construction activity, untimely scheduling of quality of work inspection activities, unqualified or uninformed supervisors and foremen, a project environment that emphasizes production to the detriment of quality, and a project environment that takes away the craftsman's sense of pride and accomplishment in his work. Each of these conditions is within the control of management, not the craftsman, and until project management is improved to minimize these conditions, the effect of improved craft skills alone on nuclear plant quality will be minimal.

For further discussion of these findings and conclusions, refer to Chapter 3 and Appendix A (Case Studies) and Chapter 5 (Audits by Associations of Professionals).

2.1.2. Why Have the NRC and the Utilities Failed or Been Slow To Detect and/or Respond To These Quality-Related Problems?

The utilities, which have primary responsibility for the safe construction and operation of nuclear power plants, have been slow in detecting or responding to quality-related problems for several reasons. The reasons include abdication of project oversight responsibilities to contractors or to subcontractors, inadequate implementation of quality assurance programs, cost and schedule pressures, inadequate QA/QC staffing, and attenuation of vital project information flowing from the working level to top management. Each of these reasons was found to have its roots in shortcomings of project and corporate management; many of these shortcomings were caused or exacerbated by inexperience in constructing nuclear power plants. In some cases, the licensees did not have effective management control of their project as a whole, and the quality problems were symptomatic of a much broader malaise that affected the entire project.

At some projects there was a tacit delegation by management of its responsibility for the achievement of quality to the NRC-required organization (the QA organization) whose mission is the assurance of quality. Inappropriate delegation of responsibility for quality, along with top management not knowing what their quality assurance programs were discovering, either through lack of interest or understanding or through attenuation of information as it passed through layers of intermediate management, contributed in no small part to the untimely detection of and response to some quality problems. Licensee QA managers and their programs have not been without fault, but they can be only as effective as top utility management permits. As with the improvement of craftsmanship, substantial improvements in quality assurance programs must start at the very top of the corporate structures of those organizations involved in the nuclear industry.

The NRC was slow to detect and/or take strong action in the major quality-related problems cited previously for several reasons. These reasons include, but are not limited to the following. The NRC made a tacit but incorrect assumption that there was a uniform level of industry and licensee competence. NRC inspection presence at construction sites was sporadic (before the NRC resident inspector program was implemented). The NRC inspection program was slow to synthesize scattered quality-related inspection findings coming in over a period of time into a comprehensive picture of a project-wide breakdown. Limited NRC inspection resources were so prioritized to address operations first, construction second, and design last, that inadequate inspection of the design process resulted. The threshold for reacting to construction-related problems was set higher than for operational problems because of (1) no immediate threat to public health and safety posed by construction deficiencies, (2) an attitude that construction problems would be found during an intensive period of startup testing prior to issuance of an operating license, and (3) an attitude that required a project-wide pervasive breakdown to be demonstrated before strong enforcement action would be taken for construction quality problems. The inspection program was oriented to focus heavily on paperwork at the expense of examining either actual work in progress or QA program implementation. The inspection program focused on detail rather than on whether the overall management process for the project was working.

Finally, the NRC was reluctant to address the issue of capability of utility management until the need for a massive remedial program for a particular licensee became evident.

2.2 CONCLUSIONS FROM NRC'S ANALYSES OF FORD AMENDMENT ALTERNATIVES b(1)-b(5)

The following conclusions summarize NRC's analyses of the specific alternatives proposed for study by Congress. Collectively, the study conclusions on these five Ford Amendment alternatives answer study question 3, which was introduced in Section 1.1:

> What changes should be made to the current policies, practices, and procedures governing commercial nuclear power plant design, construction and regulation to prevent major quality problems in the future or to provide more timely detection and correction of problems?

Later parts of this report will provide additional detail on the analyses and on the specific actions that NRC has undertaken or that are recommended. In this section, each alternative is first reprinted and then is followed by the major conclusions resulting from this study's analysis of that alternative.

Alternative b(1)

> Providing a basis for quality assurance and quality control, inspection, and enforcement actions through the adoption of an approach which is more prescriptive than that currently in practice for defining principal architectural and engineering criteria for the construction of commercial nuclear powerplants.

Conclusions:

The study concluded that while more prescriptive architectural and engineering (A&E) (i.e., design) criteria would provide a stronger basis for inspection and enforcement action, neither the degree of prescriptiveness of principal A&E criteria nor the enforcement of such criteria were factors in the major quality-related problems that led to the Congressional mandate to perform this study. None of the five plants having quality-related problems would have found their problems lessened if more prescriptive A&E criteria during the plant's design and construction had been required.

Quality problems in design were directly attributable to changes in the design basis and to inadequate management oversight of the design process, including implementation of quality assurance controls over the design process, rather than to the degree of prescriptiveness of A&E criteria. Historically, neither the industry nor the NRC has done a good job in managing change, whether the changes be technical, regulatory, or procedural. Recent NRC action to control the rate of regulatory change and to prevent unnecessary change by establishing the Committee to Review Generic Requirements has been a positive force in reducing the impact of regulatory change on the industry.

Two other considerations argue against more prescriptive design criteria. First, there is usually more than one satisfactory way to accomplish design activity and more prescription would unnecessarily limit the designer's choices. Second, too much prescription by the NRC tends to shift the licensee's responsibility for safety to the NRC.

The study did find that a more complete design early in the construction process would enhance several project activities, including planning, scheduling, and procurement, and would facilitate readiness reviews (to evaluate readiness to proceed to a new project phase of activity), thereby improving the prospects for greater project quality. Current NRC initiatives concerning standardized designs address this point.

The study also found that current practice does not provide a strong basis for inspection against Preliminary Safety Analysis Report (PSAR) commitments. The study concluded that an effective way of providing a stronger basis for inspection (and subsequent enforcement, if necessary) would be to provide more definitive procedures for management of changes to principal A&E design criteria. One way to accomplish this would be to make licensee commitments to certain A&E design criteria contained in the PSAR conditions of the CP.

No new administrative action is recommended under this alternative other than to revise future staff review practices to accommodate the above conclusions and to further evaluate the impact of changes on the collective NRC-industry regulatory and project management structure in order to develop further guidelines for controlling unnecessary change and for better managing necessary changes. The NRC has several actions currently under way, including a legislative proposal, which address the issue of standardized designs.

Alternative b(2)

> Conditioning the issuance of construction permits for commercial nuclear powerplants on a demonstration by the licensee that the licensee is capable of independently managing the effective performance of all quality assurance and quality control responsibilities for the powerplant.

Conclusions:

The study concluded that this alternative would offer significant advantages over current and past NRC practice. In the past, CPs have been issued to some applicants who would not have met this criterion. Past NRC reviews of CP applicants did not deal substantively with management experience or capability either in an overall sense or in the context of QA program effectiveness. The study found that deficiencies in utility and project management were root causes of the major quality-related problems experienced and that in such projects, problems in the quality program were often accompanied by deficiencies in other management aspects, including planning, scheduling, procurement, and oversight over contractors. The study established a strong correlation between the effectiveness of the QA program and the effectiveness of overall project management. Therefore, any future assessment of the effectiveness of the licensee's management and oversight of its QA/QC responsibilities should cover other management aspects of the project as well.

This study recommends that future CP applicants be required to meet this criterion. While the licensee could use contractors to manage the project or parts of it, the licensee would retain ultimate responsibility for the effective management of the project, including its quality aspects. Demonstrations of management capability and effectiveness would be required both before CP issuance and throughout the construction process, at about two-year intervals. The CP would be conditioned on the applicant's successful performance on each of these post CP-audits. Poor performance on any single audit would not necessarily result in license suspension but could lead to other enforcement action. Poor performance repeated in a subsequent audit would lead to more extensive enforcement action, including the possibility of license suspension. To perform these audits, NRC staff should develop a better capability to assess, prospectively, project management and quality program management capability.

In addition to this prospective staff review of an applicant's management capability, the NRC should also establish an advisory board that would be similar in function to the Advisory Committee on Reactor Safeguards (ACRS) but whose members would have appropriate background and experience to review the management qualifications, experience, and capability of future CP applicants. This board would advise the NRC of their findings and recommendations regarding the applicant's capability and competence to construct a nuclear power plant.

Comprehensive third-party audits such as those envisioned by alternative b(5) could be used to periodically confirm management and QA/QC program effectiveness after NRC's initial prospective finding of adequacy. Therefore, the third-party audits that were examined in conjunction with alternative b(5) would represent an acceptable method for meeting the post-CP demonstration requirements of this alternative.

Alternative b(3)

Evaluations, inspections, or audits of commercial nuclear powerplant construction by organizations comprised of professionals having expertise in appropriate fields, which evaluations, inspections, or audits are more effective than those under current practice.

Conclusions:

The study concluded that audits conducted by the American Society of Mechanical Engineers (ASME) for ASME code work and by the National Board of Boiler and Pressure Vessel Inspectors (NB) provide detection capability in certain specific areas beyond that provided by the NRC. Those audits therefore provide a valuable and continuing contribution that complements the NRC inspection program.

The new INPO Construction Project Evaluation (CPE) program fits the alternative b(3) criteria of "evaluations...by organizations comprised of professionals having expertise in appropriate fields, which evaluations... are more effective than those under current practice." INPO implemented its CPE program after Public Law 97-415 was enacted, and this program represents a significant enhancement of efforts by the nuclear industry to improve quality assurance and quality control in design and construction.

Of all audit or evaluation activities by associations of professionals having appropriate expertise, only the CPE is comprehensive enough to be considered as a potential surrogate for NRC inspections. However, the INPO construction evaluations do not attempt to cover all of the areas that a regulatory inspection must cover and do not evaluate the quality of installed hardware to the extent that NRC's Construction Appraisal Team (CAT) inspections do. The study concluded that INPO's current mission of assisting nuclear utilities in raising their levels of performance and standards of excellence will do more to improve industry performance and to prevent future problems than any attempt to transpose INPO's activities into a quasi-regulatory role. Consequently, the study concludes that little change should be sought in INPO's current mission, which is to help the nuclear industry improve itself by establishing standards of industry performance and excellence, and evaluation against those standards.

Although the study concludes that NRC's and INPO's roles presently are separate, INPO's potential is not yet fully realized. Therefore, the NRC should remain alert to future changes in INPO's program that would justify NRC's placing greater reliance on it and that would lessen the combined impact of NRC and INPO evaluation programs on individual licensees. The NRC should find ways to reinforce the INPO concept of improving levels of performance in all areas of nuclear power, including operations, design and construction. The goal should be to ensure that licensees who do not choose to strive for standards of excellence do not find the alternative path any easier.

Currently, none of the designated organizations of professionals have the NRC's technical inspection depth, breadth, and experience. Moreover, no other organization has the statutory strength of the NRC. Effectiveness is not only measured by technical competence, but also by the ability to assure that identified problems are fixed. Only the NRC has the statutory ability to provide such incentives.

Alternative b(4)

> Improvement of the Commission's organization, methods, and programs for quality assurance development, review, and inspection.

Conclusions:

The study found that the NRC shares responsibility with the utilities for the occurrence and magnitude of the major quality-related problems that stimulated this study. The major findings and conclusions relating to NRC's organization, methods, and programs for quality are summarized below. Improvements to NRC's organization, methods and programs for quality are discussed in Section 2.4 (NRC Administrative Actions) and in Chapters 4 and 7. Each of these conclusions are conclusions of the study and any related recommended regulatory actions are only proposed for implementation at this time. Those recommendations that would result in new regulatory requirements will be subject to the Administrative Procedures Act and established NRC procedures, including review by the Committee to Review Generic Requirements, by public comment, and by the NRC Commissioners before being enacted.

NRC's program for the assurance of quality in design and construction in the nuclear industry has several primary objectives that are achieved through a hierarchy of organizational oversight arrangements involving the licensee, its contractors, independent auditors, the ASME and NB, INPO and the NRC. The three primary objectives of this total program for the assurance of quality are (1) to prevent major quality-related problems such as those cited in the introduction from occurring, (2) to detect, in a timely fashion, developing quality problems and to take corrective action before isolated problems multiply into a programmatic breakdown, and (3) to provide assurance to the NRC, the public, and the Congress that plants that are licensed to operate have met applicable legal requirements and are designed and built in a manner consistent with public safety. The NRC is not primarily responsible for accomplishing any of these three activities, but the NRC is the architect and monitor of the total system for assurance of quality and must share in the blame when the system does not work. This NRC-required system has, on occasion, missed its goals in some or all of the three objectives: prevention, detection, and assurance. The study's conclusions on each of those objectives are discussed below.

Prevention

(1) NRC CP licensing reviews and pre-CP inspections should deal more substantively with prior nuclear construction experience within the project team and the capability of the licensee's management to carry out its intended role within the project team. The NRC should review the aggregate capability, prior nuclear experience, and project roles proposed of each corporate entity within the project team.

To execute these new reviews, the NRC needs to develop methods to assess project and utility management capability and effectiveness prospectively. The capability for effective management should be a criterion for license issuance and retention. The NRC should develop evaluation criteria or characteristics, based on this study and refined through further research, for the elements of successful and unsuccessful organization and management practices of commercial nuclear power plant construction projects. These criteria should be codified as part of NRC's pre-CP issuance inspection guidelines.

(2) The NRC should revise its quality assurance programmatic requirements to emphasize performance rather than form and to establish QA principles as an integral part of licensee construction management philosophy. As an NRC Regional Administrator observed, NRC quality assurance efforts to date have, unfortunately, succeeded in establishing licensee QA organizations that are short on technical

expertise, long on bureaucratic paperwork and essentially isolated from the safety-related licensee programs they were designed to improve. This has resulted from a licensing process that has emphasized organizational and programmatic form while failing to impress licensees with the need to be effective in the day-to-day management of engineering and construction activities. Similarly, the requirement to establish QA functional independence has, in many cases, convinced construction managers that QA is someone else's job. NRC's failure is in not effectively communicating to licensees that the 18 quality assurance program criteria of 10 CFR 50, Appendix B, describe a comprehensive closed-loop management control system that is worthy of adoption as an overall construction management system. Other knowledgeable officials have suggested that those 18 criteria should probably be given a new name in an effort to take them out of the province of the QA department and establish them as the provenance of the corporate boardroom.

(3) The NRC and industry need to improve their capability to manage change. A key step in improving the management of change is reducing change. The NRC and industry should continue and expand their efforts to control procedural, technical, and regulatory change and to stabilize design requirements.

Detection

(1) The NRC and industry need to focus more on the implementation of quality assurance programs including the quality of completed hardware, and less on the details of the programs (e.g., program description, organization chart, independence of reporting chain, etc.).

(2) The NRC should continue current efforts to match its inspection program to its resources so that areas of greatest safety significance are inspected more heavily. The inspection program should focus more on licensee management performance and effectiveness than it has in the past.

(3) The NRC should continue its newly established integrated design inspections.

(4) The NRC needs to do a better job of synthesizing and analyzing findings from individual inspections and other sources to lower its threshold for taking action on construction quality problems. Team inspections have been found to be one way to address this problem. The NRC should continue and expand current efforts to include more team inspection activity in the inspection program.

(5) Comprehensive third-party inspections are a viable supplement to the NRC inspection program and should be required of future and current CP holders. The third-party audits should assess the effectiveness of both QA program implementation and project management as well as a verification of achieved quality in construction.

Assurance

Assurance exists on at least two levels: the level of the total NRC program and the nuclear industry as a whole and the level of an individual project. Each time some part of the total NRC QA program for the assurance of quality fails to prevent or provide timely detection of a major quality-related problem, such as those cited previously, the level of assurance that the total system provides to the public is lowered, no matter which party (e.g., NRC, licensee, contractor) is primarily to blame. Collectively, the five major quality-related problems cited previously so lowered the level of assurance provided by the total program that Congress directed that this study be conducted to find ways to redesign the system and to restore public confidence in it.

The recent decision by the owners of the Zimmer project to convert their nuclear project to coal underscores the importance of assurance at the individual project level. The NRC had halted safety-related construction on the project because of deficiencies in the system that was intended to provide assurance that the Zimmer project had been constructed in compliance with NRC regulations. It appears that the high cost of a remedial program designed to provide such assurance resulted in termination of the nuclear portions of the project.

Alternative b(5)

> Conditioning the issuance of construction permits for commercial nuclear powerplants on the permittee entering into contracts or other arrangements with an independent inspector to audit the quality assurance program to verify quality assurance performance.

Conclusions:

This study concluded that comprehensive audits of nuclear construction projects by qualified third parties (independent inspectors) can provide significant additional preventive and detection capability as well as enhanced assurance that nuclear plants are built according to their design and licensing commitments. This study found that this alternative, including its provision for conditioning the CP, offers significant benefits over current and past practice. Just as periodic independent audits are conducted of publicly held corporations to determine their financial condition, periodic independent audits of a licensee's construction project would provide the public, regulators and utility stockholders greater assurance that the project's design and construction were of high quality and according to applicable safety requirements. The independent auditor would be required to meet independence criteria to be established by the NRC, and the audits would be reviewed and monitored by the NRC. The NRC also would establish criteria for audit coverage and completeness. An audit frequency of approximately once every two years appears most appropriate. The study concluded that a program of comprehensive periodic audits by qualified third parties should be implemented both for plants currently under construction and for future plants.

2.3. OTHER CONCLUSIONS

While preparing the analyses required by Congress, it became apparent that the study should be expanded beyond Congress' specific questions to the previously described underlying questions that seemed to go to the root of public concerns. Expanding the study revealed several topics that affected the underlying concerns but that required additional study before specific action could be recommended. These topics and the additional study performed on them are summarized below.

2.3.1. The Kist Report on Improvements to NRC's Programs

When it became apparent that NRC's past policies and practices contributed to the development of quality-related problems in design and construction, the NRC arranged for an independent contractor to assess NRC's activities and requirements for quality and quality assurance during design and construction. This assessment was conducted by a management consulting firm, N. C. Kist and Associates, which specializes in nuclear industry QA program audits and reviews. The Kist Report comprises Appendix B of this report. Not all of its conclusions and recommendations have yet been evaluated for adoption. The Kist Report includes the following recommendations:

(1) The regulatory process should be stabilized through more preventive action and planning.

(2) The NRC should make the required elements of control more definitive in guidance documents without specifying how those elements must be implemented.

(3) The NRC should define the applicability of quality program requirements for items considered important to safety.

(4) The NRC should focus QA licensing reviews more on the licensee's QA manual itself and less on pro forma commitments in the PSAR application.

(5) The NRC should evaluate licensees' and contractors' experience, attitude and management capability before authorizations and permits are issued. The NRC should establish acceptance criteria for that evaluation.

(6) The NRC should require the licensee to demonstrate its capability to implement the QA program before authorizations or permits are issued.

(7) The NRC should devote greater attention to design activities.

(8) The NRC should develop programs based on what must be done to assure safety and then obtain necessary resources to implement the programs.

(9) The NRC should require a master Inspection Plan from licensees and contractors, showing planned QA/QC inspection activity.

(10) The NRC should change regulations to permit industry organizations to evaluate vendors instead of requiring individual licensees to evaluate vendors.

(11) The NRC should take stronger, more expeditious enforcement action for quality problems in design or construction, including determining the magnitude of problems and correcting their root causes.

(12) The NRC should perform or require detailed periodic audits of each licensee's implementation of its QA program.

(13) The NRC should increase the training of NRC inspectors in quality assurance, auditing, and implementation of inspection modules.

(14) The NRC should establish an audit program of NRC activities, using qualified personnel not having responsibility in the areas audited.

(15) The NRC should establish a quality assurance program within the NRC.

A number of the Kist Report's recommendations coincide with this study's recommendations. The remainder are being evaluated by the NRC staff for possible followup action.

2.3.2. Battelle Reports on Contractual and Institutional Issues and on QA Programs of Other Industries

This study found that major quality problems were caused by breakdowns or inadequate implementation of quality programs, which invariably stemmed from problems with project management and/or with the project team's inexperience in their assumed roles. Many factors indirectly influence these primary causal factors. Battelle Human Affairs Research Center (HARC) and Pacific Northwest Laboratory (PNL) (operated by Battelle) conducted analyses to identify or better understand some of these less obvious factors. This section describes the results of two special substudies undertaken to develop a broader perspective on which to base study conclusions and recommendations. As with the Kist Report, not all of these conclusions and recommendations have yet been fully evaluated for adoption.

Chapter 8 and Appendix C of this report examine some of the contractual, organizational, and institutional issues associated with designing and constructing nuclear power plants. HARC performed this analysis, with the following results:

(1) Previous nuclear experience appears to provide a significant advantage in a nuclear construction effort. Utilities not possessing such experience initially should consider hiring either a project staff or contractors who can provide such expertise.

(2) A nuclear construction project appears to benefit when its procurement entity is large and experienced enough to exert "marketplace presence". A large procurement entity offers the advantage of market familiarity and commercial leverage as well as the "clout" needed to secure satisfactory performance on procurements.

(3) Without substantially more complete designs before construction is begun and stabilization of technical requirements, fixed-price contracting does not appear to be justified for most aspects of nuclear power plant construction.

(4) Achieving quality objectives includes attention to detail in procurement documents and specifications, careful evaluation of a bidder's capability before a contract is issued, and followup to evaluate contractors' performance after a contract is issued.

(5) The NRC should focus more attention on how a licensee proposes to ensure quality work is performed rather than on written descriptions of QA/QC programs.

(6) Along with the NRC, state Public Utility Commissions (PUCs) provide a major source of regulatory oversight for nuclear construction projects. Historically, state PUCs do not appear to have been active in disallowing construction costs that may have resulted from lapses in quality assurance or project management. Recent developments suggest that this practice is changing with unknown implications for the course of nuclear projects currently under construction.

Chapter 9 and Appendix D describe a second analysis that was undertaken to give this report additional perspective--an analysis of the existing programs for assurance of quality of other U.S. government agencies, other industries, and other countries. The analysis focused on identifying aspects of alternative QA programs that might be transferred to NRC's program and improve it. This analysis was performed in conjunction with NRC staff by PNL. Major insights from this analysis and related work include the following:

(1) Plant designs should be well advanced before construction activities begin.

(2) The NRC should consider establishing a QA system that prioritizes quality efforts commensurate with the relative importance of equipment, components, and systems to safety, reliability and availability.

(3) The NRC should consider adopting "readiness reviews" during nuclear plant construction similar to those used by the Department of Energy (DOE) and the National Aeronautics and Space Administration (NASA). In some industries, readiness reviews are conducted before embarking on a major new phase of a project to ensure that appropriate planning, coordination and design work have been completed and that the project team is "ready" to proceed. These would not be regulatory "hold points" but rather a requirement for licensees to perform a self-assessment at critical points of the construction process.

(4) The NRC should study ways to better integrate NRC inspection functions with system design reviews, test program reviews, and test program evaluations.

(5) The NRC should look at alternative ways of improving its vendor inspection program.

(6) The NRC should emphasize that achieving quality is the responsibility of licensee management, not the QA organization. Several alternative programs studied emphasized the responsibility for quality of line management from top executives down to first-level supervisors and foremen. Several examples demonstrated that if this responsibility is fulfilled, a large contingent of QC inspectors is not needed.

4 NRC ADMINISTRATIVE ACTIONS

This section describes the administrative actions that the NRC has undertaken
or that are recommended by this study for improving quality assurance and
quality control programs. Each action may address several of the study's
findings and conclusions and is grouped according to the QA program objective
it most strongly supports: prevention/improved management; detection/lowered
threshold; assurance/increased public confidence. For convenience these
actions are summarized in tabular form at the end of this section in Tables 2.1
and 2.2. The tables make it easier to understand the actions under way and
actions recommended, applying to future plants and to plants currently under
construction, and actions requiring more analysis.

Although some of the requirements of the Ford Amendment were futuristic (e.g.,
two of the five alternatives spoke of conditioning future CPs on certain requirements), several of this study's results are immediately applicable for plants
presently under construction. The actions described in the remainder of this
chapter collectively define both a framework for future CPs and a framework
within which existing plants under construction can be completed safety,
according to NRC requirements, and with high assurance of the quality of
construction necessary for licensing and safe operation.

2.4.1 NRC Administrative Actions To Support the Prevention Objective and To Improve Management

This section is divided into discussions of actions already undertaken and
actions recommended for consideration by the NRC.

Actions Already Undertaken

(1) Systematic Assessment of Licensee Performance

   The study found that historically the NRC inspection program has not
   focused on the quality, capability and effectiveness of licensee management. Following the accident at Three Mile Island, the NRC initiated an
   effort to better address the issue of management performance through the
   Systematic Assessment of Licensee Performance (SALP) program. Under the
   SALP program, the overall performance of each nuclear power plant licensee
   (both CP and operating license holders) is reviewed periodically (approximately every 9 to 18 months). Evaluation results are discussed with senior
   licensee management and help prioritize the level of NRC inspection for the
   coming period for each licensee. The SALP program is discussed in more
   detail in Chapter 7.

(2) Committee to Review Generic Requirements

   The study found that historically neither the NRC nor the industry
   had managed changes well, whether they were technical, procedural,
   or regulatory. The most direct way to improve management's capability
   to handle change is to reduce the rate of change itself. In 1981, the NRC
   established the Committee to Review Generic Requirements (CRGR) for the
   NRC to exercise better management control over the flow of new regulatory
   requirements and to carefully examine the feasibility and benefits of
   proposed NRC staff actions having generic implications. The CRGR is

generally credited with bringing order to the promulgation of new
regulatory requirements and thereby giving more stability to the
regulatory process.

Recommended Actions

(1) Enhanced Pre-CP Review of Applicants' Experience and Managerial
    Qualifications

Past NRC reviews of CP applications have not dealt substantively with
management experience and capability or prior nuclear experience. The
Commission has no CP applications at this time nor does it expect any in
the near future. This hiatus presents an excellent opportunity to review
and revise Commission practice in this area without impacting any current
applications. This study has concluded that this issue should be
addressed in two ways: (1) enhancing NRC staff review, and (2)
establishing an advisory board.

As a result of this study, the NRC staff has improved its understanding of
the management factors that have resulted in both satisfactory and less
than satisfactory quality in construction. Based on this improved understanding and further analysis in this area, the study recommends that
the NRC staff revise portions of the Standard Review Plan (SRP) and the
inspection program to greater emphasize reviews of the applicant's management capability, quality assurance program, project team experience and
management's prior nuclear experience before CP issuance. The revised
SRP and inspection program are intended to provide substantial additional
guidance to the staff for its review of the applicant's ability to
effectively implement a quality program and manage a nuclear construction
project. The staff's efforts are anticipated to be augmented with expert
consultants in conducting these management reviews.

In addition to this enhanced staff review of management capability, the
study has concluded that independent advice on this subject is needed from
persons having expert knowledge of and experience in various aspects of
the management of a commercial nuclear power plant construction project.
One alternative is to establish an advisory board that is similar in some
regards to the Advisory Committee on Reactor Safeguards (ACRS) but whose
charter is to address management, organizational, experience, and qualification issues associated with constructing a commercial nuclear power
plant. In particular, the board would independently advise the NRC on the
applicant's capability to effectively manage all aspects of a nuclear construction project, including its quality assurance program. The duties of
this board might also be expanded later to include advice on the applicant's capability to manage the plant's operation.

The Commission is authorized to establish advisory boards by Section 161a.
of the Atomic Energy Act. The creation and operation of such boards and
committees are subject to the requirements of the Federal Advisory
Committee Act and 10 CFR 7 of the Commission's regulations. The
proposed board would be a balanced body of persons having direct experience and knowledge of managing the design and construction of a large
commercial nuclear power plant. Board membership would be formed on

an ad-hoc basis from a slate of experienced persons from such organizations as other nuclear utilities, investment banking firms that arrange
financing for nuclear projects, state PUCs, nuclear insurance firms,
nuclear-experienced A/E firms, NSSS manufacturers, legal firms with an
extensive nuclear practice, and perhaps management consulting firms. In
creating such a board (whose membership would be voluntary), procedural
safeguards would have to be carefully structured to avoid conflicts of
interest.

An alternative to the proposed construction advisory panel would be to
expand the duties of the ACRS to advise the NRC on the managerial qualification of CP applicants. Such an expansion in scope of ACRS purview would
represent a significant change from the highly technical reviews ACRS now
performs. Moreover, the type of background and experience envisioned for
the proposed advisory board historically has not been available on the
ACRS. This proposed administrative action directly addresses Congressional Alternative b(2).

(2) Post-CP Demonstration of Managerial Competence and Effectiveness

The study concluded that future CPs for commercial nuclear power plants
should be conditioned on a licensee's post-CP demonstration that it is
capable of managing or providing effective management oversight over the
construction project. This would include a demonstration that the
licensee is capable of independently managing or overseeing the management
of the effective performance of all quality assurance and quality control
responsibilities for the power plant. Although the licensee could delegate
some project responsibility, it would retain responsibility for the
effectiveness of project management, including the effectiveness of the
quality program.

In some cases in the past, the NRC has been slow to conclude that a major
breakdown has occurred in a licensee's quality assurance program, although
the symptoms of and practices leading to the breakdown were, in hindsight,
evident early in the project. In such cases, neither the interests of the
public nor the licensee have been well served by the delays inherent in
the NRC accumulating sufficient foundation for a Show Cause Order or other
enforcement action.

The study has concluded that a post-CP demonstration of management
capability and effectiveness, as a condition of the license, is the most
effective way to impress upon an applicant the importance the Congress and
the Commission attach to proper implemention of the applicant's QA program.
Such a requirement would provide a substantial incentive for the licensee,
its reactor manufacturer, its A/E, and all its contractors to demonstrate that
the QA program committed to in the licensing process has been implemented
and is being effectively managed. Public confidence in the quality of the
project's design and construction would also be enhanced. The system of
independent third-party audits proposed by Congressional Alternative b(5)
could be one method for verifying such demonstration.

The first of the periodic independent third-party audits, proposed by Congressional Alternative b(5) and recommended by this study in Section 2.4.2, could appropriately evaluate this demonstration and could assure the NRC and the public that the licensee is properly implementing its QA/QC program and building a high-quality plant. If the performance in this first audit were unsuccessful, the CP could be suspended or other enforcement action could be taken.

NRC's past practice has not been to comprehensively assess, at an early stage, a licensee's implementation of the QA/QC program. The Commission's adoption of the requirement to demonstrate such implementation as a condition of the CP would correct that shortcoming. A regulatory analysis should be performed to assess the feasibility and benefits of alternative approaches for implementing this proposed action. Alternatives include promulgating a new rule requiring that the CP be conditioned on a post-CP demonstration of management capability. This proposed administrative action directly addresses Congressional Alternatives b(2) and b(5). See Chapter 4 for further discussion of this recommendation.

(3)   Performance Objectives for QA Programs

The study found that the regulatory basis for QA in the nuclear industry, i.e., 10 CFR 50, Appendix B, was sound. The only significant change the study envisions is that Appendix B should be viewed by the NRC and industry as a "comprehensive, closed-loop management system", not just a program for the assurance of quality. While the study found the management practices advocated by Appendix B to be sound and not needing improvement, NRC's methods for implementing Appendix B emphasize form and paper at the expense of substance, and program implementation and effectiveness. As one member of the ACRS noted, any new QA initiatives will not have the effect of improving quality unless steps are taken to motivate people, both in design, construction and vendor operations. The current methods of quality assurance alienate professional and technically oriented people, as well as craftsmen and foremen. He said a way must be found to make these people feel that they can make an important contribution to design, construction and safe operation.

The study concluded that NRC's methods to get licensees to implement the management practices of Appendix B need to be changed so that licensees and their employees are motivated to achieve results rather than merely comply with regulations. The study recommends that this be done by re-examination of NRC's method of ensuring that Appendix B is implemented. Both Appendix B and NQA-1-1983, the voluntary consensus code and the standard, describe performance standards. The NRC must translate these performance standards into performance objectives; implementing Appendix B by establishing performance objectives would define what a licensee's QA program is expected to accomplish. NRC inspections would then measure the effectiveness of licensee management and the QA program in meeting the performance objectives.

The study recognizes that successfully achieving this fundamental shift in program emphasis from compliance to performance will not be easy. However, such a shift in NRC (and industry) emphasis is necessary if substantial improvements in quality and quality assurance are going to be made. The following paragraphs describe how such a program could be structured.

NRC currently establishes very prescriptive requirements for a "QA program" in Chapter 17 of NRC's Standard Review Plan. Once NRC has approved a QA program, the licensee develops a set of detailed implementing procedures in the form of a "QA manual". The licensee's employees use the QA manual to guide their actions.

The "QA program" reviews conducted by the NRC have emphasized description of the QA program and provide reasonable certainty that any NRC-approved QA program will have met all of the requirements of the Standard Review Plan Chapter 17 guidelines. However, major difficulties have arisen at some projects in implementing the written QA programs approved by the NRC. NRC inspection experience suggests, and this study has confirmed, that the major problems with QA programs are in their implementation, not in their description.

The study concluded that an alternative to the current approach should be developed in which performance objectives or criteria govern a licensee QA program rather than its written description. These performance objectives would establish what the NRC wants the licensee's QA activities to actually accomplish. The licensee would then develop a QA manual that establishes detailed procedures designed to meet NRC's performance objectives. The intermediate step of a "QA Program Description", which is currently reviewed and approved by NRC, would be eliminated. The performance objectives would be based upon 10 CFR 50, Appendix B, and would be a substitute for the current Chapter 17 guidelines. A licensee could elect to establish procedures that exceed NRC's minimums. However, a licensee's actual performance would be evaluated against NRC's minimum performance criteria rather than the procedures described in the licensee's QA manual, which could exceed NRC's minimums.

To implement this study conclusion on a trial basis, the NRC staff should begin developing a set of performance objectives for an operations QA program and implement it on a voluntary trial basis with one or more licensees who are currently constructing a plant and approaching the operating license stage. Currently, no CP applicants are pending, so the program would have to be tested on an operating license applicant. Because all CP licensees are required to prepare a new QA program for the operating phase of their project, this approach should allow an opportunity to test performance QA objectives in parallel with the existing program. If the proposed program is successful, the NRC should consider adopting performance objectives for all QA activities and should evaluate the benefits and costs of backfit of these performance objectives to all licensees. Although staff action to test the approach in a limited way has begun, this action cannot be considered to be a short-term action in terms of its effect on the assurance of quality. This proposed administrative action directly addresses Congressional Alternative b(4).

(4) Management Appraisals as an Adjunct to the CAT Inspections

The case studies conducted for this study produced a set of project and management characteristics evidenced by more successful projects, as well as a set of characteristics that tended to be shared by projects experiencing major quality-related problems. The empirical lessons learned about the quality, capability, and effectiveness of management should be applied in future Construction Appraisal Team (CAT) inspections. (See Section 2.4.2 for a discussion of the CAT program.) Current CAT methodology emphasizes hardware inspection and indirectly draws inferences about the quality and effectiveness of project and quality management by assessing the finished project's quality. Management problems are thus identified indirectly and inferentially. The proposed adjunct to the CAT methodology would complement the existing methodology by viewing project and quality performance from the top down as well as from the bottom up. It is believed that potential or actual problems in the management of the project will be more quickly identified and better characterized through this augmentation of the CAT inspection approach.

This recommendation differs from the previously described recommended activities in that it can be implemented immediately and applied to plants currently under construction. This activity, coupled with the recommended interim expansion of the CAT program to cover plants currently under construction pending action on a third-party audit rule (see description of interim expanded CAT program in the next section) would provide a significant near-term enhancement in NRC's oversight of utility and project management. As one Regional Administrator noted, "The solution of the short-term effective management problem must be based on observed results and proper use of governmental authority." This proposed administrative action directly addresses Congressional Alternative b(4).

(5) Application of Ford Study Lessons to Plants Currently Under Construction/Inspection Prioritization

The NRC should apply lessons learned from this study regarding the elements of successful and unsuccessful commercial nuclear power plant construction experience, project organization, and management to projects currently under construction. This retrospective look would be used to identify any plants that might be more susceptible than others to problems during design and construction. An enhanced inspection effort should be undertaken to ensure that any such problems are detected as early as possible. This administrative action directly addresses Congressional Alternative b(4). This recommendation is discussed in more detail in Chapter 7.

(6) Improved Diagnostic Capability Including Trend Analysis

NRC inspection program management recognizes and this study confirms the need for NRC management and staff to recognize and treat NRC inspection findings and licensee event reports as symptoms of potential utility management shortcomings and to pursue them accordingly. In several of the major construction quality problems, the NRC was slow to diagnose the programmatic illnesses underlying the symptomatic information trickling into the NRC via the inspection program and licensee reports.

To address this problem, the study concluded that NRC inspection staff and management should (1) make a conscious effort to analyze each inspection finding to determine its root cause, (2) based on inspection experience, the results of this study, and other information, develop a set of construction performance indicators to be monitored, trended and evaluated by each licensee for his own performance and by the NRC. These activities are discussed in more detail in Chapter 7. Such indicators should be oriented toward measuring the effectiveness of activities that contribute to, control, and verify construction quality. The trending program would be an extension of some present SALP activities and would provide input for future SALP evaluations. A goal of this "trending" program would be for the licensee and NRC to more quickly detect and correct quality problems. QA problems at any one site should be clearly and accurately identified, including root causes, and that information should be provided to all sites immediately. Strong results-oriented management of this activity is needed to ensure adequate followup and problem resolution.

As a corollary to developing this trending program, the NRC should revise its training program to instruct inspectors, supervisors, and managers in the use of the system and followup of findings. Also, as the inspection program is further revised from a compliance-based orientation to a performance-based orientation, inspector, supervisor, and management training must be revised to reflect the change in emphasis and to help develop the skills needed for effective evaluation of performance. This proposed administrative action directly addresses Congressional Alternative b(4).

(7) NRC/Utility Senior Management Meetings

The NRC should expand the existing practice of conducting senior-level meetings between NRC and utility management to discuss the status, progress, and problems of ongoing construction activities, particularly those relating to quality and quality assurance. In such meetings both top NRC and utility management have to focus on the problems of construction, including its quality. Such meetings require that top management of both the regulator and regulatee become personally aware of specific details of construction projects, including quality problems, and help to combat the attenuation of information that contributed to the quality-related problems at some projects and that is inherent to some degree in most organizational structures. This concept is strongly supported by one NRC Regional Administrator, who writes:

> Frequent planned meetings must be held between Regional Administrators, cognizant Office Directors, and high level licensee management for projects under construction. In addition, periodic meetings with the Commission that involve both a licensee and the staff should be held to assure Commission support, advice and project familiarity. Such meetings will serve to ensure direct involvement at the highest levels of licensee and NRC management in QA-related matters such as the adequacy of resources; the clear recognition of significant problems at licensee and other sites; and the acceptance (or non-acceptance) of corrective measures, including root causes and timeliness, by the NRC.

This administrative action directly addresses Congressional Alternative b(4).

(8) Enhanced Vendor and Supplier Inspection Program

The NRC is in the process of modifying its vendor and supplier inspection program to better prioritize its effort according to the significance of safety concerns. However, this NRC inspection program, like the construction inspection program, fulfills an oversight role only. The responsibility for the quality of a vendor's or supplier's product, like the construction quality of a nuclear power plant, lies with the licensee. With the decline in nuclear plant orders, the entire supplier/vendor/licensee infrastructure is changing, with unknown implications for safety and quality in the future. While this issue needs more study, within the present structure enhanced NRC enforcement is clearly appropriate against some licensees for failing to provide effective quality assurance oversight over their vendors, including in some cases failure to audit vendors and/or to detect work of unacceptable quality.

Although not the focus of this study, there are many examples of poor quality products supplied by vendors for use at nuclear power plants, which makes the vendor issue of considerable importance to the NRC. Three of the five NRC Regional Administrators provided comments on the vendor issue:

> I think the NRC should take a strong stand on unacceptable vendor performance, including enforcement action and "blackballing", as appropriate.

> I agree with (the above) comment concerning the role of the vendors. We need to take a much stronger stand on unacceptable vendor performance. As I have stated many times over the past 3 years, we need to have a strong enforcement policy for vendors, including AEs, NSSS and component suppliers, and equipment qualification facilities. In addition, we need to review our inspection programs to address the utilities vendor surveillance programs. Too many utilities sit back and expect the NRC to do their work with regard to vendors. We need to reverse this role and place the responsibility directly on the shoulders of the utility.

> Heavy emphasis must be placed on the identification of generic and QA weaknesses in the following organizations: Nuclear Steam Supply System Manufacturers, Architect Engineers, and Vendors supplying safety equipment. The recommendations relating to High Level Meetings with licensees are directly applicable to meetings with these organizations - including the Commission. This area must be aggressively pursued by the NRC to assure formal and prompt feedback to licensees.

The NRC vendor program is in the process of being restructured, reoriented, reprioritized, and relocated. While it is too early to characterize all effects of this transformation, the following is clear for the near-term:

° The licensee will continue to be held responsible for the quality of work performed for it by vendors.

2-23

° The NRC vendor inspection program in no way substitutes for or relieves the licensee of its responsibility for vendor oversight; the NRC vendor inspection program is NRC's QA check of the effectiveness of licensee oversight programs.

° Stronger enforcement action than in the past can be expected against licensees whose vendor and supplier oversight is demonstrably inadequate.

Special note should be taken here about the first bullet above. Many comments have been received on the desirability of licensing vendors, and in particular, the major vendors such as the A/E and NSSS manufacturers. This study has concluded that the current organizational environment that requires that the utility take all or most of the price risk for the nuclear power plant virtually demands that only the utility be licensed. The licensing of vendors would inevitably reduce some of the control utilities currently have over licensing-driven actions while still requiring the utility to pay for those actions. However, there are circumstances under which it may be desirable to license vendors, and this is discussed in Section 2.4.5 under the heading, "Project Ownership and Management Arrangements". This administrative action directly addresses Congressional Alternative b(4).

2.4.2 NRC Administrative Actions To Support the Detection Objective and To Lower the Threshold for Taking Action for Construction Quality Problems

This section is divided into discussion of actions already undertaken and actions recommended for consideration by the NRC.

Actions Already Under Way

(1) Resident Inspector Program

As directed by the Ford Amendment [Section 13(a)], the NRC has assigned at least one resident inspector to all sites under active construction where construction is more than 15% complete. The study found that the resident inspector program is the backbone of the present NRC inspection program and provides the NRC with a better awareness and understanding of the status of a construction project as well as a more continuous inspection presence than previously. Each of the five major quality-related problems that stimulated this study began or occurred before the resident inspector program was implemented. The day-to-day presence of the resident at a site allows him to better understand the project and improves the NRC's capability to determine the extent and magnitude of quality or quality assurance problems and to require corrective action in a more timely fashion.

While it cannot be conclusively demonstrated that major quality-related problems in construction would not have occurred if the resident program been in place earlier, the study found that several of the major quality-related problems would have been detected sooner and would not have been as serious if the program been implemented sooner. For future applicants, the study concluded that the NRC should assign resident inspectors to the construction site as early as CP issuance and possibly as early as the start of any construction begun under a Limited Work Authorization before CP issuance. The exact timing would be determined on a case-by-case basis and such factors as prior nuclear construction experience would be considered. This administrative action directly addresses Congressional Alternative b(4).

(2) Team Inspections

One reason that NRC was slow to detect or realize the extent of some of the quality problems in design and construction is the difficulty in integrating and synthesizing, into a comprehensive picture, site-specific inspection results determined at different times by different inspectors in different disciplines. For several of the projects having significant quality-related problems, the extent and magnitude of the problem was eventually established by a comprehensive team inspection involving several inspectors in different disciplines and several weeks of concurrent field work. With such comprehensive team inspections, information can be interchanged frequently and quickly among inspectors looking at different areas, and synthesizing and integrating findings and developing project-wide conclusions are made easier.

Team inspections have also been shown to effectively overcome the problem of reaching the "threshold" for taking action in response to quality problems in construction. The NRC is establishing a pilot program in one of its five regional offices to test the feasibility and benefits of reorienting the present routine inspection program. The present inspection program generally supplements the resident program with inspections by individual specialists from the regional office and uses few team inspections. The reoriented program would (1) provide for more residents at each site where special circumstances apply, and (2) use team inspections as the primary inspection activity of the regional office. This trial program is consistent with this report's findings, and pending the results of the pilot inspection program, the NRC inspection program for all regions may be reoriented to place more residents at sites and place region-based inspection emphasis on team inspections. This administrative action directly addresses Congressional Alternative b(4).

(3) Construction Appraisal Team (CAT) Inspections

The team inspection approach for reactor construction projects has been tested by the NRC regions and instituted by NRC headquarters. A regional trial Construction Appraisal Team (CAT) inspection program was conducted in 1981, with eight trial inspections being performed by region-based inspectors. These inspections were effective in identifying hardware and construction quality problems not identified by the routine inspection program. However, the manpower demand of these team inspections caused the Regional Administrators to defer routine performance of this type of inspection. Although some regions have conducted subsequent CAT-type inspections on an as-needed basis (the inspection program encourages the regions to perform CAT-type inspections), they are not mandatory. The previously described pilot program was a test of whether they should be made mandatory. A headquarters CAT program was instituted by the NRC Headquarters Office of Inspection and Enforcement (IE) in 1982. These headquarters-based CAT inspections serve as both an audit of the licensee's performance and the NRC's resident and regional-based inspection program. The primary emphasis of the CAT is to concentrate on examining safety-related hardware after installation and after the licensee's own quality control inspection process has been completed. The study recommends that future CAT inspections be modified to more directly address management issues through the addition of a management appraisal. See Section 2.4.1.

Each CAT inspection involves about ten professionals in various specialties who spend four to five weeks and 1,600 to 2,000 manhours on site. Counting preparation time, analysis, and report writing, each CAT inspection takes about three months to complete. As of February 1984, six headquarters-based CAT inspections had been conducted and further CAT inspections had been planned at a frequency of four per year. This frequency is not sufficient to provide CAT inspection coverage of the current population of plants under construction. Consequently, this study recommends an expansion of the CAT program to ensure that plants presently under construction are subject to either a CAT inspection or a comprehensive third-party audit. This recommendation is discussed later in this chapter. The CAT program is discussed also in Chapter 7. This administrative action directly addresses Congressional Alternative b(4).

(4)   Integrated Design Inspection (IDI)

The NRC has also developed a special design inspection program whose object is to assess the quality of design activities. The design area received little inspection attention in the past, and recent experience, including some of the major quality-related problems that stimulated this study, indicated that NRC should increase its design inspection efforts. Like the CAT program, the Integrated Design Inspection (IDI) program uses the team approach and is conducted by the NRC Headquarters Office of Inspection and Enforcement.

The IDI inspection supplements a core group of NRC staff members with contractors or consultants having specific design expertise and experience. This design inspection program encompasses the total design process on a selected plant system, from formulating design and A&E criteria through developing and translating the design and its reviews to actual site construction. The inspection staff evaluates and confirms certain basic design information previously submitted in connection with license applications. Inspections are conducted at the A/E design organization and the site to verify that proper design control programs are in place. This program examines the adequacy and consistency of the integration of all the design details within a selected sample area. It is believed that conclusions about the adequacy of the overall design process can be drawn from this very detailed audit of a selected sample.

Each IDI requires about twelve persons and four months to complete. As of December 1983, three IDIs had been performed and current plans are to conduct three IDIs per year. This frequency is based on staffing limitations and is not sufficient to provide coverage of every plant under construction. For the forseeable future, IDI inspections will concentrate on plants nearing completion of the construction process and for which the design is essentially complete. Among this group of plants, candidates for the IDI inspection are selected based on a review of all pertinent data, including such things as whether any other form of independent design review has been performed (such as an Independent Design Verification Program, see Section 2.4.3), the nuclear experience of the licensee and the A-E, results of other inspections, and advice from the NRC Regional Administrator. This administrative action directly addresses Congressional Alternative b(4). The IDI program is discussed further in Chapter 7.

(5) Contractor Support to the NRC Inspection Program

An increase in direct NRC inspection of licensee-sponsored design and construction would increase confidence that licensee commitments are being met. This is particularly true when special circumstances require added inspection attention (e.g., oversight of a project with a remedial program under way or one with many allegations of safety-related deficiencies).

On a trial basis, the CAT and IDI inspections have used substantial contractor support as one method for increasing the expert technical resources available to the NRC for carrying out its inspection responsibilities. Such contractor augmentations have proven to be extremely helpful for these headquarters-based inspection efforts. Like all NRC team inspections, contractor-supported team inspections are led by an NRC team leader having inspection authority and responsibility. There is no delegation of NRC inspection authority or responsibilities to a contractor. The use of contractor assistance for NRC inspections is being expanded in both headquarters and the region-based inspection programs, including regional team inspections. Other appropriate uses for contractor support are being sought. This administrative action directly addresses Congressional Alternative b(4).

(6) Revised Construction Inspection Program

The construction inspection program was recently revised for two reasons: (1) a recognition that procedures in NRC's inspection program manual exceeded inspection manpower resources; and (2) review of the licensee's written QA program and QA program documentation was being emphasized at the expense of observing work and inspecting hardware. The NRC staff is presently revising the individual inspection procedures in the construction inspection manual to better match the budgeted resources and to better focus the inspection effort to improve effectiveness.

The main goals of the revisions are as follows: (1) to shift emphasis of inspection from reviewing records to observing work; (2) to facilitate performance of certain procedures by resident inspectors; (3) to re-examine the scope and frequency of some inspections based on limitations of inspector resources; and (4) to eliminate redundancies in the procedures. Current plans will substantially consolidate procedures. It is too early to determine the full effect of these revisions of the written inspection program on the effectiveness of the implementation of the NRC inspection program. This administrative action directly addresses Congressional Alternative b(4) and is discussed further in Chapter 7.

A word of caution: Improvements resulting from the revised procedures are limited, as are any other improvements to the inspection program, by the following two considerations. First, NRC's inspection program is an oversight program only. It does not perform direct first-line QC inspection. It is not sufficiently staffed to perform a 100% oversight function and performs direct inspections of at most 1-2% of the safety-related work at a construction site, on a sampling basis. Second, only about 1.5 manyears per year of direct NRC inspection effort are budgeted for each reactor under construction.

Recommended Actions

(1) Independent Third-Party Audits

As indicated in Section 2.2, this study found that a program of periodic independent third-party evaluations, inspections, or audits of commercial nuclear power plant construction by qualified individuals would represent a significant improvement over current practice and would complement the Commission's own inspection program. Such independent audits would bring an additional measure of confidence that licensing commitments are being met and increase the probability that any major systematic quality deficiencies will be identified earlier than in the past. Current NRC direct inspection resources of about 1.5 staff years per reactor under construction per year have not been adequate to provide timely detection of all major problems. The added use of qualified, independent auditors would increase the probability of more timely detection of major problems.

The study recommends that for future CP applicants, CP issuance be conditioned on the applicant's entering into contracts or other arrangements with independent inspectors to periodically verify the adequacy of its achieved construction quality, quality assurance program performance, and ability to independently manage the effective performance of all QA and QC responsibilities. That is, the study recommends that the proposed third-party audit program meet the performance criteria implicit in both Congressional Alternatives b(2) and b(5).

The study recommends that current CP holders also be subject to a program of periodic independent third-party audits. Until the third-party audit program is established as a requirement, the NRC should continue with the current voluntary Independent Design Verification Program (IDVP) on a case-by-case basis and implement an expanded CAT program. These recommended actions are discussed below.

The recommended independent audits would be conducted for each plant under construction about every two years, with the scope and nature of the audit being adjusted to the construction schedule and level of completion. For example, the first audit should occur within the first 12 to 20 months of construction and would concentrate on civil and structural work and the design control process in addition to its primary objective of verifying management capability to successfully implement an effective QA program. Later audits would cover electrical work, piping, instrumentation and control, etc. The last audit would cover completed design verification as well as review proposed technical specifications against the plant design and serve the purpose, among others, of the current voluntary IDVP program. Each audit would be designed to meet the requirements of Congressional Alternatives b(2) and b(5), i.e., to verify that the licensee had demonstrated the capability to independently manage or oversee the management of the effective performance of all QA and QC responsibilities for the project over the previous two years.

Criteria for the third-party audits, including independence criteria similar to those now used in the IDVP efforts, should be developed by the NRC staff in consultation with appropriate professionals and other interested groups. Those criteria should incorporate lessons learned from the NRC's evaluation of the third-party audits reviewed as part of the pilot program (Chapter 4), the case studies (Chapter 3), and the current IDVP, CAT, and IDI programs.

A regulatory analysis will have to be performed before this proposed action can be implemented as a new regulatory requirement. This proposed administrative action is also discussed in Chapter 4. This action directly addresses Congressional Alternatives b(2), b(4), and b(5).

(2) Interim Expanded CAT Program

Implementing a program for third-party audits for plants under construction would probably take two years or more from the date of initiation of action before it could become effective, if it were approved by the Commission. This time delay stems from the procedural safeguards that are a part of the rulemaking process. According to current estimates, many of the plants currently under construction will be completed within this time frame, and the third-party audit requirement would not apply to over half of the plants presently being constructed. Therefore, in the interim, pending the approval and implementation of a third-party audit rule, the study recommends that the NRC expand its CAT program to ensure that as many plants under construction as possible are subjected to either an intensive audit by a qualified third party or an NRC CAT inspection. Thereafter, CATs would be required on a sampling basis (to check third-party audit effectiveness). The management appraisal recommended in the preceding section as an adjunct to the CAT program should apply to the expanded CAT program as well. This proposed administrative action directly addresses Congressional Alternative b(4) and indirectly addresses Alternative b(5).

(3) Regional Team Inspections

The use of contractor support to assist headquarters-based team inspections has been successful. The study recommends that the regional inspection program be supplemented with additional use of contractor support for the routine regional inspection program. This will allow more NRC staff time for reactive inspections such as allegation followup, remedial program inspections, and regional team inspections. As indicated previously, increased use of regional team inspections is being tested in one NRC regional office. Depending on its results, the NRC inspection program in all regions may be reoriented to emphasize team inspections. This administrative action addresses Congressional Alternative b(4) and is also discussed in Chapter 7.

(4) Resident Inspectors

The study found that for new applicants or for the restart of construction at projects presently in suspension, resident inspectors should be assigned to the site as early as possible, preferably before CP issuance and the start of safety-related construction activities. This study recommends that this finding become part of NRC's future policy on placing residents at construction sites. As indicated previously, the NRC is also establishing a pilot program in one of its regional offices which will place more resident inspectors at plant sites where special circumstances dictate. Depending on the outcome of this trial program, the NRC inspection program may be reoriented to an even heavier emphasis on resident inspectors. This proposed administrative action directly addresses Congressional Alternative b(4) and is discussed further in Chapter 7.

(5) Improved Licensee Detection Capability

In licensee QA programs, additional emphasis must be placed on identifying problems and trends, including the processing of nonconformance reports and design changes. The NRC should develop more definitive guidance to be followed by utilities for determining root causes of nonconformances, timeliness of corrective action, and evaluation of generic implications of nonconformances found both in the design and construction process. While the NRC needs to improve its own capability in these areas, the NRC sees, on a nation-wide basis, both good and bad practices and is in the most logical position to develop and share such information and generic guidance with the utilities. This proposed administrative action addresses Congressional Alternative b(4).

2.4.3 NRC Administrative Actions To Support the Assurance Objective and To Increase Public Confidence

This section is divided into discussions of actions already undertaken and recommended actions for consideration by the NRC.

Actions Already Under Way

(1) Independent Design Verification Program (IDVP)

On a case-by-case basis, the NRC staff has requested an applicant for an operating license to provide additional assurance that the design process used in constructing the plant has fully complied with NRC regulations and licensing commitments.

Many licensees have responded to this request by initiating a design review through an independent third-party contractor. This review has been termed the Independent Design Verification Program (IDVP). This program has been mentioned several times previously in conjunction with other actions under way or proposed, and is also discussed in Chapter 7. Reviews conducted under this program have provided an evaluation of the quality of design based on a detailed examination of a small sample. The independent review has also addressed programmatic areas, e.g., classification of systems and components, design and verification records, interface control and interdisciplinary review, consistency with the Final Safety Analysis Report (FSAR), nonconformances and corrective actions, and audit findings and resolutions. The review includes verifying specific design features by independent calculations and comparing installations against as-built drawings. The NRC staff reviews the selection of the independent review organization and the audit plan before they are implemented, reviews the completed report, and assesses the applicant's response to the audit findings. In all cases to date, the NRC staff has concluded that the applicant has complied with NRC regulations and licensing commitments.

The usefulness of these audits has varied from site to site because of the variability between each audit's scope and methodology. With the recent transfer of IDVP responsibility to the same NRC program office (IE) responsible for the IDI program, future IDVPs will be modeled somewhat like an IDI, and the degree of variability should decrease.

Recommended Actions

(1) Interim IDVP/Third-Party Audit

This study has concluded that a series of comprehensive third-party audits required by regulation with a clearly established set of audit criteria will better enable the NRC to meet its responsibilities than the current IDVP practice. Until this requirement has been established, however, the NRC should continue to encourage licensees to perform independent design reviews on a case-by-case basis.

The recommended third-party audit program was listed in Section 2.4.2 under the detection objective. However, it also strongly supports the assurance objective. The independent oversight brought to the nuclear construction process by the third-party audit concept should increase public confidence in the construction process. This administrative action directly addresses Congressional Alternatives b(4) and b(5).

(2) Audit Program for the NRC

One of the findings of the Kist Report was that the NRC should have a QA program for its own activities. While the CAT, IDI, and PAT (Performance Appraisal Team inspections) programs, as well as NRC Headquarters audits of regional performance, provide some degree of quality assurance over NRC regional activities, there is no formal NRC program for QA of NRC QA activities. In view of the study findings that shortcomings in the NRC QA program contributed partly to the quality problems that led to this study, both the overall assurance of quality for nuclear power and the public's confidence in NRC's oversight of it would be enhanced if NRC had a formal QA program covering its own QA activites. The study recommends that such a program be established and that it include an audit program for NRC QA activities that provides for periodic independent audits.

2.4.4 Summary of NRC Actions Under Way and Actions To Be Taken

Table 2.1 summarizes the NRC actions under way and proposed actions to be taken.

Note: The NRC actions that have been identified and recommended by the study are extremely comprehensive, and several of them could consume all of NRC's current budget and manpower allocated to development of the quality assurance program. It will be necessary to prioritize the quality assurance issues within the other issues faced by the NRC and make resource allocations. As a result, some of the recommended actions may necessarily be deferred until the higher priority actions are completed.

TABLE 2.1. NRC Administrative Actions Under Way and Recommended for Nuclear Plants Under Construction to Support the NRC QA Program Objectives of Prevention, Detection, and Assurance

Objective	Applies To	
	Current Plants	Future Plants
I. Prevention/Improved Management		
Under Way		
1. Systematic Assessment of Licensee Peformance	X	X
2. Committee to Review Generic Requirements	X	X
Recommended		
1. Enhanced Pre-CP Review of Experience and Managerial Qualif./Advisory Board		X
2. Post-CP Demonstration of Management Effectiveness		X
3. QA Program Performance Objectives*	X	X
4. Management Appraisals/CAT Adjunct*	X	X
5. Inspection Prioritization of Plants Currently Under Construction*	X	X
6. Improved Diagnostic Capability/Trend Analysis	X	X
7. Senior Management Meetings	X	X
8. Enhanced Vendor Program*	X	X
II. Detection/Lowered Threshold		
Under Way		
1. Resident Inspector Program	X	X
2. Team Inspections	X	X
3. CAT Program	X	X
4. IDI Program	X	X
5. Contractor Support to the NRC Inspection Program	X	X
6. Revised Inspection Program	X	X
Recommended		
1. Third-Party Audit/Interim CAT*/Interim IDVP*	X	X
2. Regional Team Inspections*	X	X
3. Expanded Resident Program*	X	X
4. Improved Licensee Detection Capability	X	X
III. Assurance/Public Confidence		
Under Way		
1. IDVP	X	
Recommended		
1. Interim IDVP*/Third-Party Audit	X	X
2. QA of NRC	X	X

* Action on recommendation already begun.

2.4.5 Actions Requiring Further Analysis

During the course of this study, several possible actions were identified that unfortunately could not be sufficiently analyzed in the time frame of this report to be included as study recommendations. These possible actions are described below. In some cases further study is needed to determine the feasibility and benefits of further changes to NRC's programs. In other cases, further study is required to better understand certain issues that may have an impact on quality and the assurance of quality in the nuclear industry.

(1) Ford Amendment Study to Improve QA for Plants in Operation

The Ford Amendment directed the NRC to conduct a study designed to improve quality and the assurance of quality in the design and construction of nuclear power plants. An effort of similar magnitude and scope should be undertaken for plants in operation. Many more nuclear plants are in operation today in the U.S. (about 80) than are under active construction (about 40), and operating plants represent a more immediate threat to public health and safety than do plants under construction. The 1983 ATWS (anticipated transient without scram) event at the Salem nuclear station is a recent example of the importance of quality and quality assurance in nuclear power plant operations and maintenance. The near-term future focus of U.S. nuclear power will be in operations and maintenance, not design and construction and serious, though less publicized, operational problems with safety implications have occurred because of poor QA.

(2) Prioritization of QA Measures

The NRC needs to establish more detailed guidance for QA systems that prioritize quality-related efforts. Such a QA system is currently required by NRC regulations, but it has been unevenly implemented, partly because of a lack of appropriate NRC guidance. In some prioritized approaches, quality assurance measures are prioritized based on the safety, reliability and availability analyses such as discussed under (7), "Quality Engineering" below. The usefulness of this approach is suggested by findings of the study on the DOE, NASA and shipbuilding programs. The goal of new NRC guidance in this area would be to provide a logical foundation for applying quality measures to plant structures, systems, and components commensurate with their relative importance to achieving some system objective such as safety or reliability. This guidance should also reduce the application of deterministic engineering judgment to the lowest possible level. Although such guidance is expected to extend beyond the current "safety-related" class, it may also reduce quality program requirements for some equipment, systems or components that are presently considered to be "safety-related". This topic is discussed also in Chapters 7 and 9 and Appendices B and D.

(3) Measuring Effectiveness of QA Programs

As indicated in Chapter 1, this study did not attempt to quantify the relationships among quality, quality assurance, and safety, nor did it attempt to quantify the relationship between risk and quality assurance. It became increasingly clear during the study that clearly defined measures need to be developed to assess QA program effectiveness.

Developing such measures is crucial to meaningfully address the above unanswered questions. Moreover, without such measures, it is virtually impossible to evaluate the benefits that would accrue from adopting an alternative approach to QA (such as that of NASA, the Federal Aviation Administration (FAA), or DOE).

The NRC should set as top QA research priorities development of ways to measure QA program effectiveness and analyses to quantify the quality, quality assurance and safety relationship, and the relationship of risk to quality assurance. In particular, the effect of a QA program on plant safety should be evaluated through probabilistic and other risk analyses.

(4) Essentially Complete Design at CP Stage

The NRC should further analyze the feasibility and benefits of requiring that plant designs of future CP applicants be well advanced before construction activities begin. This analysis should also consider whether future applicants should be required to have scale models of their plants and computer-assisted drawings. (See public comment (3) in Section 10.2.1.) This research is suggested by the findings from the case studies (Chapter 3 and Appendix A), the review of outside programs (Chapter 9 and Appendix D), the study of contracts (Chapter 8 and Appendix C), and other study activities.

(5) Configuration Control/Management of Change

The NRC needs to further analyze the feasibility of applying the techniques of the aerospace industry's apparently successful configuration management approach to the nuclear industry's need for improved management of change. Change and the difficulty in managing change were found to have significant impacts on design and construction quality. This research is suggested by the results of the case studies (Chapter 3 and Appendix A), the study of outside QA programs (Chapter 9 and Appendix D), and comments from the study's special review group (Chapter 10).

As part of this effort, NRC should determine how best to revise staff review practices to provide more definitive procedures for managing changes to principal A&E design criteria. This analysis would include consideration of including licensee commitments to certain A&E design criteria contained in the PSAR as conditions of the CP. See the study conclusion on Alternative b(1) in Section 2.2.

(6) Feasibility of Readiness Reviews

The NRC should analyze the feasibility and benefits of requiring formal assessments by licensees of their readiness to proceed to the next critical phase of a project (i.e., planning to construction, construction to pre-operational testing, testing to operations). In such "readiness reviews" plant designers, construction managers, owner/operators, and (possibly) NRC staff would participate. The reviews could be required at key points in the project beginning with "design ready for construction" and could be repeated at selected key milestone points. The usefulness of this approach is suggested by the findings from the DOE, NASA and ship-building programs (see Chapter 9 and Appendix D).

(7) Quality Engineering

The NRC should analyze the degree to which NRC design requirements should include the completion of safety, reliability, and availability analyses, including failure modes and effects analyses, and fault tree and hazard or safety analyses. The usefulness of this approach is suggested by the findings from the DOE, NASA, FAA, foreign nuclear, and shipbuilding programs and the movement of the NRC toward expanded use of Probabilistic Risk Assessment. See Chapter 9 and Appendix D.

(8) Project Ownership and Management Arrangements/PUC Interface

Projects under construction appear to benefit significantly when the owners and members of the project team possess strong management experience and a strong financial position (see Chapter 8 and Appendix C). The advantage of these circumstances appears great enough to warrant NRC's examination of ways in which beneficial ownership and management arrangements can be stimulated and fostered. The specific advantages/disadvantages of various ownership and management arrangements for assuring safe and successful nuclear projects need careful study. Such a study should include determining which desirable changes are possible within the present statutory framework and which would require legislation.

Recent events affecting the nuclear industry suggest that financial considerations will be the principal determinant of any new CP applications and that a possible form of a new construction project may be the presentation to a utility of an essentially "turnkey" proposal by an NSSS manufacturer and A&E joint venture. One essential component of this proposal is likely to be assumption of a significant portion of the price risk by the joint venture. Consistent with the previous discussion (see Section 2.4.1, "Enhanced Vendor and Supplier Inspection Program") concerning the necessity for the entity having control of the funding also having responsibility for licensing, the appropriate CP licensee in this case might be the joint venture, not the utility. Further analysis must be undertaken to understand the potential implications of such "dual licensing" where the CP holder may be different from the operating licensee. For example, this process would be much simplified by using pre-approved sites whose licensing was separate from the CP process. It would also require a careful scrutiny of whether an operating license could reasonably be granted to a utility with no prior nuclear operating experience.

Further study of the NRC/PUC interaction must also be undertaken. There are indications that certain major preventive maintenance actions, such as replacing the recirculation piping in a boiling water reactor or replacing a steam generator in a pressurized water reactor, may be deferred by utilities because of concern over PUC policies. In cases like these, good engineering judgment and safety concerns indicate that the work should go forward, but it might be deferred because of a lack of confidence that PUCs will consider the "non-essential" maintenance expenses to have been prudently incurred, absent an NRC order to perform the maintenance. Other lessor examples of utilities deferring or postponing important maintenance activities because of concern over PUC policies exist. The NRC must develop a clearer understanding of its options and possible actions when faced by a new regulatory activism by state PUCs.

(9) Feasibility of Designated Representatives

One possible way to increase the resources available to carry out NRC inspections is the use of a "designated representative" (DR) program analogous to that employed by the FAA. Under the FAA's DR program, employees of an aircraft designer or manufacturer are deputized by the FAA to perform examinations, inspections, and tests on behalf of the FAA. If an analogous NRC program were established, it would place some NRC inspection responsibility and authority in the hands of employees of the licensee. This is a potentially controversial program whose advantages and disadvantages have not been fully assessed. Further analysis of this issue is needed before any conclusion can be reached. This topic is also discussed in Chapter 7 and was the subject of several NRC staff papers to the Commission (SECY 83-26 and SECY 83-499).

(10) Limiting Construction Permits

Many of the problems experienced by the nuclear industry recently were exacerbated by the surge of reactor orders and CP applications that occurred in the early and mid-1970s. This surge caused utilities to assemble project teams having key members with little or no prior nuclear experience. (See discussion in Chapter 3 and Appendix A.) Extraordinary demands were also placed on component suppliers and subcontractors, with many entities competing for increasingly scarce nuclear experienced personnel. The inevitable result was that performance declined--to sometimes unacceptable levels.

The NRC was also faced with problems caused by the earlier rapid growth of the nuclear industry: increased CP applications to be reviewed, safety evaluation reports to be prepared with practically every reactor design different from the last one reviewed, more and more construction projects to be inspected, competition with the industry for a limited pool of experienced personnel.

Consideration should be given to establishing limits on the rate of growth of any future resumption in nuclear power plant construction. Depending on when a resumption might begin and the circumstances causing such a resumption, the U.S. could be faced with problems similar to those that ocurred with the last rapid buildup. Many factors could influence a decision on the number of construction permits issued in a year. Such factors include the degree of standardization of design; the experience of the potential operators; industry capacity and residual experience, including major vendors, subcontractors and suppliers; NRC staffing levels and ability to respond to workload fluctuations; and the availability of sites.

Further analysis should be performed to identify the rapid-expansion-related problems that previously occurred and to develop guidelines for assessing whether and what future limits should be placed on issuing CPs by the NRC. These efforts should not be directed to establish such limits at this time but rather to identify the key parameters that could be used to establish such limits in the future.

Table 2.2 lists all the the actions discussed in 2.4.5 requiring further analysis.

TABLE 2.2. Actions Requiring Further Analysis

(1) Ford Amendment Study for Plants in Operation

(2) Prioritization of QA Measures:

     Guidance on "Safety-Related" vs. "Important to Safety"

(3) Measuring Effectiveness of QA Programs

(4) Essentially Complete Design at CP Stage

(5) Feasibility of Aerospace Industry's Configuration Management Approach

(6) Feasibility of Readiness Reviews

(7) Quality Engineering

(8) Alternative Project Ownership and Management Arrangements/PUC Interface

(9) Feasibility of Designated Representatives

(10) Limiting Construction Permits

Note: The NRC actions that have been identified and recommended by the study
are extremely comprehensive, and several of them could consume all of NRC's
current budget and manpower allocated to development of the quality assurance
program. It will be necessary to prioritize the quality assurance issues
within the other issues faced by the NRC and to make resource allocations. As a
result, some of the recommended actions may necessarily be deferred until the
higher priority actions are completed.

2.5 ACTIONS OF THE NUCLEAR INDUSTRY

This section discusses actions already undertaken and future actions by the nuclear industry to improve quality and the assurance of quality in the industry. The preceding section discussed in detail the framework of NRC actions under way to improve quality and the assurance of quality in the nuclear industry. NRC actions were emphasized because the Ford Amendment specified that NRC actions be highlighted. While improvements to NRC's programs, methods, and organization are necessary for improving quality in the nuclear industry, they are not sufficient. The study concluded that the primary cause of the quality-related problems in the nuclear industry was shortcomings in utility management.

Real improvements to address this root cause must come from the industry itself. The NRC cannot write a regulation that will achieve good utility management. Better utility management must come from the utilities themselves, from the boards of directors, from the stockholders, and from the ratepayers. The NRC and the PUCs can provide penalties for poor utility management, but these negative incentives are of limited value without the utilities' conscious commitment to raise their own performance standards. Quality must be built into a plant by the builder, it cannot be inspected in by QA. Similarly, achieving quality in nuclear design, construction, and operation is the responsibility of the utility and utility management, and it must be achieved by them. The NRC cannot inspect quality into a plant.

Given that the sine qua non to improved quality in the nuclear industry is improved, informed, capabable utility management, this section discusses industry actions already taken or recommended by the study to improve quality.

2.5.1 Actions Already Undertaken

In 1979, in response to the accident at Three Mile Island, the nuclear industry created the Institute of Nuclear Power Operations (INPO). INPO's chartered mission is to promote the highest level of safety and reliability in operating nuclear power plants. In carrying out this mission, INPO strives to encourage excellence in all phases of design, construction, and operation. This study performed a thorough review of INPO's new program for construction evaluation and concluded that the program was consistent with INPO's stated mission of promoting excellence in construction and design (See Chapter 5.).

Another INPO activity that bears directly on improving utility management has been the sponsorship of several management workshops for utility chief executive officers, plant managers, and others to stress the importance of quality and management responsibility for quality and to strengthen management awareness, understanding and commitment to safe operation and quality construction of nuclear facilities. NRC Commissioners and senior managers have participated in these workshops to the mutual benefit of both the industry and the NRC. The study endorses the INPO program of management workshops, which is consistent with the belief that any significant improvements in the nuclear industry must start at the top.

2.5.2 Future Action

The already undertaken and proposed NRC actions described in Section 2.4 should result in many improvements on the part of the nuclear industry in the design and construction of nuclear power plants. Many of those actions were modifications to improve the NRC inspection program. It is important to understand the limitations of any NRC inspection program, no matter how many improvements are made to it.

The NRC inspection program is a sampling program that covers at most 1% to 2% of the safety-related construction activities at a site. Presently, only 1.5 staff years/year/reactor is budgeted for direct inspection of reactors under construction. Even if the NRC spent four or five times that inspection effort, it could not keep pace with all of the activities of the several thousand workers at a nuclear construction site. Although reshaping the NRC inspection programs along the lines indicated in earlier discussion will improve the programs and the overall assurance of quality, NRC actions alone will not be enough to stop future quality problems of the type that stimulated this report. As one NRC Regional Administrator noted, "While I endorse reshaping our inspections along the line described, if the licensee doesn't do the job properly, I don't believe we can ever count on our limited inspection program alone to provide timely identification of the scope of the problems. We have to achieve the principle of the licensee building quality in from the beginning."

The study confirmed the intuitively obvious observation that quality has to be put into a product or project by the producer or builder, not by the inspector. Because the NRC does not build nuclear plants, but only inspects them, no matter how much NRC inspection effort is devoted to plants under construction, the builder (i.e., the nuclear industry: utility-owners, A/E, CM, reactor supplier and other vendors) must ultimately achieve quality in the construction. If the nuclear industry does not take positive action, this report's recommendations will do little more than assure that poorly or questionably built plants do not operate. The recommendations will not assure that plants, once started, are not stopped in mid-construction due to quality problems. Such positive industry action cannot be successfully elicited through regulation; it must come because the nuclear industry wants it to. It must come because the nuclear industry, and each of its members, believes it is the right and necessary, but not the obligatory thing to do. In this regard, three conclusions of this study require voluntary industry action to be accomplished:

(1) Industry should view NRC requirements as minimum levels of performance, not absolute goals, and should capitalize on and expand on the practice of some utilities that continually seek to improve their level of performance and seek excellence in their operations. Industry establishment and support of INPO is a positive step in this direction.

The overriding, predominant conclusion of this report is that the common cause of poor quality in nuclear power plant construction is poor management by the responsible licensees--the utilities. It follows that the solution to the problem must also lie with utility management. To the extent the utilities use INPO, their performance can be aided measurably by the programs, reviews, common knowledge, experience and peer pressure provided by INPO as an integral part of utility management. The NRC is farther removed and does not have responsibility for managing the utilities. In pursuing its statutory responsibilities for ensuring the health and safety of the public, the regulations, inspections, and penalties NRC imposes can motivate utility management, including INPO, to strive toward high quality in construction and operations through excellence in their management. However, since the problem and the ultimate solution lie with the utilities, NRC must recognize, encourage, support and nurture the efforts of utility management, including INPO, to improve their performance through their self-improvement, self-inspection, and self-developed programs and peer pressure. Their programs and practices are no substitute for NRC practices because the NRC has different responsibilities with the same goal. The NRC cannot and must not manage for them and they cannot fulfill NRC's statutory responsibilities to the public. This requires a rather critical balance: if NRC over-prescribes and over-regulates, it can stifle the efforts of utility management through INPO to do their job themselves. If this should happen, the net result would be the opposite of what was intended.

The study found that of the utilities studied, there was a strong correlation between project success in design and construction and embracement of the "rising standard of excellence" concept by the owner utility (see Section 3.4.3). INPO efforts in this direction will improve quality and safety in the nuclear industry and should contribute to increased public confidence in and acceptance of nuclear power. However, INPO alone cannot accomplish this goal. The active support and commitment of each nuclear power plant licensee to achieving excellence are needed. No regulation can achieve its full potential effect unless the regulatees comply with it because they believe in it, not just because they have to.

(2) The nuclear industry needs to treat quality assurance as a management tool, not as just another regulatory requirement, or as a substitute for active management oversight of a project.

The words of one NRC Regional Administrator are particularly appropriate on this point and merit repeating. He wrote:

> NRC's failure is in not effectively communicating to licensees that 10 CFR 50, Appendix B, describes a comprehensive closed loop management control system that is worthy of adoption as an overall construction management system. Consequently, managers often rely on inspecting quality into a plant rather than doing it right the first time. We believe additional NRC effort is warranted in establishing QA principles as an integral part of licensee construction management philosophy.

Quality assurance as a discipline cannot achieve or assure quality. In some organizations, management views QA as being responsible for quality and fires the QA manager if quality is not achieved. This study concluded that too often top utility management assessed blame in the wrong place and fired the wrong person(s). Top management, and through them, intermediate management and the workers, are primarily responsible for quality. Quality assurance is a management tool to provide feedback on how well quality objectives are being attained. Achieving quality requires effective management of the design and construction process and placing quality as a high priority. The 18 criteria of Appendix B could just as easily be entitled "elements for effective management of a project" as "quality assurance criteria." Because they really are elements of effective management, they must be implemented; similarly, they will not serve as substitutes for active line management involvement in their implementation.

(3) Additional emphasis must be placed on aspects of licensee QA programs that identify problems and trends, including the processing of noncompliance reports and design changes.

In the past, neither the utilities nor the NRC have done well in analyzing trends and recognizing the root causes of quality problems. Several activities to improve NRC's capability in this regard are described in Section 2.2 and Chapter 7. Management of ongoing construction projects should develop trend analysis capabilities of their own, improve their ability to determine the root causes of identified problems, and do both of these in a more timely manner. The NRC should share the results of its industry-wide and generic analyses described in Section 2.4 with licensees so that both can enhance their programs.

Table 2.3 summarizes NRC and industry actions under way and actions proposed to be taken as well as the NRC/industry program for the assurance of quality in place when the major quality-related problems occurred (pre-1980).

TABLE 2.3. Comparison of Major Features of Former, Present and Proposed NRC and Industry Programs for Assurance of Quality in the Design and Construction of Nuclear Power Plants

Former Program (Pre 1980)	Present Program (1982-83)	Future Program	Application to Current or Future CP Holders
NRC ACTIVITY			
°Appendix B Rqmts.	°Appendix B Rqmts.	°Appendix B Rqmts.	Both
°Licensing Review	°Licensing Review	°Performance Objectives for QA Programs	Both
°Regional-Based Insp.	°Regional-Based Insp.	°Revised Regional-Based Inspection	Both
	°Resident Insp. Prog.	°Expanded Resident Insp. Program	Both
	°CAT Inspections-4/yr	°Interim Expanded CAT Inspection Program	Both
	°IDI Inspections-3/yr	°IDI Inspections-3/yr	Both
		°Enhanced Pre-CP Rev. (Mgmt & Adv. Board)	Future Only
		°Post-CP Demonstrations as Condition of License	Future Only
		°NRC Mgmt Assessments/ CAT Adjunct	Both
		°Periodic Third-Party Audits	Both
INDUSTRY ACTIVITY			
°Licensee QA Program	°Licensee QA Program	°Licensee QA Program	Both
°ASME Audits	°ASME Audits	°ASME Audits	Both
°NB Audits	°NB Audits	°NB Audits	Both
	°INPO Constr. Eval.	°INPO Audits	Both
	°IDVP Program	°Interim IDVP Program Pending Third-Party Audit Rule	Current

Note: The NRC actions that have been identified and recommended by the study are extremely comprehensive, and several of them could consume all of NRC's current budget and manpower allocated to development of the quality assurance program. It will be necessary to prioritize the quality assurance issues within the other issues faced by the NRC and to make resource allocations. As a result, some of the recommended actions may necessarily be deferred until the higher priority actions are completed.

2.6 POSSIBLE LEGISLATIVE INITIATIVES

Many knowledgeable people believe that any long-term solution to the problems of nuclear power in the U.S. involve major institutional changes to the structure of the nuclear industry itself. The institutional changes may require substantial legislative changes. This study confined itself only to the question of what changes, legislative or otherwise, should be made to improve quality and the assurance of quality in the commercial nuclear industry. Given this narrow scope, the study does not make any legislative recommendations at this time. However, further analysis of the impact of state Public Utility Commission decisions on construction quality and the issue of project ownership and management arrangements may require that legislation be proposed in the future. The relationship of state PUC actions to construction quality must be better understood before the need for a legislative proposal can be determined. Also, if further research indicates that public health and safety interests would be significantly better served if the owning, building, and operation of nuclear power plants were consolidated in the hands of fewer and stronger institutions, then legislation removing barriers to consolidating such interests might be proposed. Consolidation has long been widely discussed as a way of improving the quality of planning, financing, managing, designing, building and operating nuclear plants, but little concrete action has been taken in this area. Further analysis is clearly required and is proceeding.

3.0 QUALITY ASSURANCE CASE STUDIES AT CONSTRUCTION PROJECTS

To improve quality and quality assurance in the commercial nuclear industry, it is important to understand what caused the major quality-related problems of the past several years, and why some nuclear construction projects have apparently been successful in achieving quality and others have not without significant remedial action. In an August 1982 paper to the Commission (Secy-82-352, "Assurance of Quality"), the NRC staff proposed a long-term review and study of the quality problems in the nuclear industry. A key feature of this long-term review was a series of analyses of nuclear construction projects that have had varying degrees of success in achieving project quality in order to identify the underlying causal factors or root causes of quality success or failure. These analyses, which included site visits, were called case studies. They began in November 1982 and continued through August 1983. The case study activity was used by the NRC to satisfy a provision in the Ford Amendment requiring that successful quality assurance and quality control programs at representative sites be analyzed and that the reasons for their success be assessed. The case studies also provided the same analysis for projects that had had significant quality problems.

The utilities participating in the case study analysis and the projects analyzed were as follows:

Utility	Project
Arizona Public Service	Palo Verde
Florida Power and Light	St. Lucie 2
Georgia Power	Vogtle
Houston Lighting and Power	South Texas
Pacific Gas and Electric	Diablo Canyon
Public Service of Indiana	Marble Hill

A management analysis of a seventh project, Cincinnati Gas and Electric's Zimmer plant, was performed in 1983 by Torrey Pines Technology (TPT). Because the TPT findings on Zimmer are relevant to the questions addressed by the NRC case studies and the Ford Amendment alternatives, the results of TPT's evaluation of Zimmer are included as a part of this analysis.

This chapter describes the main findings from the case studies. Character- istics of projects that have had major quality problems and some that have not are highlighted, including root causes of apparent success or lack of it. Like all case study analyses, these findings are based on detailed analysis of a subset of a larger population, and the results may not be entirely general- izable to the population as a whole. In the case study analyses, four of the five projects identified in the legislative history of the Ford Amendment as having had major quality problems are examined, whereas the study examines only three of about sixty projects completed or under construction and not identified as having major quality problems in design or construction. There is always the possibility that as-yet-undiscovered problems would move projects from the "no significant problems" category to the "problem" category. Still, when similar characteristics are found consistently across disparate sites, confidence in them is increased. The case study conclusions have relied most heavily on these consistent findings. The case study approach, program, projects visited and results are described in more detail in Appendix A to this report.

3.1 PURPOSE

The primary purpose of the case studies was to determine the essential characteristics of both successful and less-than-successful commercial nuclear power plant construction projects, and to derive a set of lessons learned, good and bad, regarding the design and construction of commercial nuclear power plants. The studies are intended to provide a historical perspective on why certain licensees have had extensive quality problems while others have not. A by-product objective is to use the information to develop project organization and management criteria that may be applied to any future applicant for a construction permit (CP). The criteria, if properly applied, could result in applicants strengthening their programs and organizations before beginning the difficult job of constructing a nuclear power plant. When applied to projects currently under construction, the lessons learned from the case studies may also indicate projects that have a higher probability of incurring quality problems in design and construction and that should receive increased NRC scrutiny. Management appraisals, based on lessons learned from the case studies, are planned as an adjunct to future Construction Appraisal Team (CAT) inspections. See Section 2.3.1.

The purpose of the case studies was to answer "why", not "how". Accordingly, the case studies were not audits or inspections, so did not focus on such tangible items as records, manuals, and procedures. Rather, they focused more on other factors, some intangible, such as corporate attitude and commitment, management support for quality, utility management's understanding of the project and its responsibilities, project accountability, level of teamwork, appropriateness of staffing, and flow of project information horizontally and vertically. As a result of the intangibility of many of the aspects examined in the case studies, the results are also less tangible than inspection findings (e.g., poor project management vs. missing rebar).

By using actual examples, case study results tend to confirm the correctness of several widely held explanations for the major quality problems, e.g., shortcomings in utility and project management, lack of corporate commitment to quality, fossil approach to nuclear construction, and others. Case study results have also been useful in refuting some other widely held beliefs; e.g., the problem is craftsmanship. While poor craftsmanship was found to play a role in some of the quality problems studied, it was not the root cause. Craftsmanship problems observed were more the result of poor project management than lack of skill on the part of the craftsman. Craftsmanship is discussed in more detail in Section 2.1.1, Section 3.4 and in Chapter 8.

The case studies focused in particular on developing answers to the two underlying questions that were considered to be central to the study:

1. Why have certain nuclear construction projects experienced significant quality-related problems while others have not?

2. Why have the NRC and the utilities failed or been slow to detect and/or respond to these quality-related problems?

The first question is answered in two parts, in Sections 3.2 and 3.4. The second question is answered in Section 3.3.

3.2 WHY HAVE SEVERAL NUCLEAR CONSTRUCTION PROJECTS EXPERIENCED SIGNIFICANT QUALITY-RELATED PROBLEMS?

To determine the answers to this question, the NRC performed case study analysis on three of the five projects cited earlier as having experienced major quality problems in design or construction. These projects were Marble Hill, Diablo Canyon, and South Texas. Torrey Pines Technology (TPT) performed a management analysis of the Zimmer project, and the results of that review will also be used in this analysis. Of the five projects cited in the legislative history of the Ford Amendment as having experienced major quality problems, only Midland was not subjected to a complete case study analysis (by the NRC or others). This was due to time constraints. However, the study did include a review of inspection, licensing, and hearing records on Midland and interviews with cognizant NRC inspection personnel and management, past and present. The results of this partial analysis provided some insights into the quality problems experienced by the Midland project, but they are not as complete or in as much depth as were the results of the other four analyses.

Where appropriate, the results of this limited Midland analysis are factored into the following discussion. Information related to the Atomic Safety and Licensing Board (ASLB) decision not to issue an operating license to Commonwealth Edison for the Byron Station because of inadequacies in Commonwealth's quality assurance (QA) program is not included in the discussion. The ASLB decision in the Byron case is a licensing matter still to be considered by the Commission.

This section will focus on the results of the case study analysis of four projects (Marble Hill, Diablo Canyon, South Texas, and Zimmer) rather than on the background or history of these projects. Each project's history, the development of its quality-related problems, and the root causes of the problems as determined by the case studies or TPT are discussed in detail in Appendix A.

3.2.1 Lack of Prior Nuclear Experience

A common thread running through each of the four projects was a lack of prior nuclear experience of some key members of the project team (i.e., owner utility, architect-engineer (A/E), construction manager (CM), and constructor) in the role(s) they had assumed in the project. Moreover, in three of the four cases, lack of prior nuclear experience of the owner utility and/or other members of the project team in their assumed roles was a major contributor to the quality-related problems that developed.

While the study did conclude that assumption by project team members of project roles consistent with their prior nuclear design and construction experience seems necessary for project success in the future, it is not sufficient (see discussion at the end of this section and also Section 3.4.1).

Three of the four subject utilities were constructing their first nuclear plant. However, this by itself should not have precluded them from successfully completing their projects without developing major quality problems. Each owner utility of the approximately 80 nuclear plants now in operation in the U.S. was at some time a first-time owner. However, it is noteworthy that the first commercial nuclear reactor plant in the U.S. (Shippingport) was constructed under the management of people who had extensive prior nuclear design and construction experience in the Navy nuclear program. Moreover, a number of the early reactor plants constructed in the U.S. were "turnkey" plants, the construction of which was managed by a few large A/E and NSSS (nuclear steam supply system) firms. These firms, whose first reactor plants were far simpler than those of today, had developed a base of experience from which they could draw in constructing the increasingly more complex reactor plants that were ordered in the future.

In the early to mid-1970s when three of the four subject projects were conceived, there was a large block of orders for new reactor plants, and the demand for personnel and organizations with successful prior nuclear design and construction experience exceeded the supply. As a result, new or prospective owner utilities generally faced a choice of picking key project team members from either the "fourth or fifth team" of an experienced firm (i.e., personnel lacking depth and breadth of applicable experience) or the "first team" from a firm that was inexperienced in nuclear design and construction but that wanted to expand its business into the nuclear area.

This supply and demand problem for prior nuclear experience of non-owner members of the project team, coupled with the inexperience of the new owners themselves, led to situations in which some key members of the project team assumed project roles inappropriate with their past nuclear experience and exceeding their capabilities. The owner's inexperience is important because in at least three of the four cases the owner underestimated the complexity and difficulty of the nuclear project and treated it much as it would have another fossil project. As a result, the owner utilities followed management practices and project approaches that had been successful in non-nuclear projects but which, in retrospect, were not appropriate to successfully complete a nuclear project in the U.S. today.

In effect, these first-time owners were trying to construct a full-scale production facility of a new design without having overseen the construction of a prototype. Although such a task is possible in today's complex nuclear environment (see Section 3.4), it seems to require an owner utility who (1) fully appreciates that construction of nuclear plants is sufficiently "different" from construction of fossil plants, (2) is willing to change its corporate management approach to accommodate the project, and (3) requires strong nuclear experience of the other (non-owner) members of the project team.

Public Service of Indiana (Marble Hill) is a first-time nuclear utility that selected an A/E with nuclear experience, but selected as civil constructor a firm without prior nuclear experience in that role. In addition, Public Service of Indiana assumed the role of CM for the project, a role inconsistent with its lack of prior nuclear construction experience. Houston Lighting and Power (South Texas) is also a first-time nuclear utility. The utility assumed a project role consistent with its experience, that of project oversight, and delegated the A/E, CM and constructor functions to another firm. However, the firm selected as A/E, CM and constructor had prior nuclear experience only as a constructor, working under the management of another firm. Cincinnati Gas and Electric (Zimmer), also a first-time nuclear utility, assumed a project role consistent with its lack of expertise and experience, i.e., oversight only and selected an experienced A/E. However, it selected as CM and constructor a firm inexperienced in constructing commercial nuclear power plants.

Pacific Gas and Electric (PG&E) (Diablo Canyon) had a somewhat different situation. Its quality problem was in design (control of design documents), and it did not experience construction quality problems as did the other three projects. PG&E was not a first-time nuclear utility; it owned and operated a small turnkey reactor plant (Humboldt Bay) constructed by Bechtel in the early 1960s. PG&E had assumed an oversight role only on the Humboldt Bay project. For Diablo Canyon, PG&E assumed the roles of owner, CM and A/E. PG&E had extensive non-nuclear experience as CM and A/E, but no prior nuclear experience in these roles. As contractors, PG&E selected firms with prior nuclear construction experience.

For the other three plants, the case studies determined that assumption of a project role by one or more project team members who lacked appropriate prior nuclear experience was a causal factor in the development of the quality problem. For Diablo Canyon, it was a coincidental factor, but not a causal factor. Extensive reviews by NRC and independent auditors have shown that PG&E discharged its duties as A/E and CM well. The root of PG&E's quality problem was management oversight of the design process during a period of extensive design changes.

Table 3.1 summarizes the relationship of the project role to prior nuclear experience for each of the four project teams at the time the project's quality problem occurred. It should be noted that some inexperienced project team members at several of these projects have subsequently been replaced by more experienced organizations.

TABLE 3.1. Summary of Relationship of Project Role to Prior Nuclear Experience at the Time Quality Problems Occurred

Characteristics	Project			
	Marble Hill	South Texas	Zimmer	Diablo Canyon
Design quality problem(s)		X		X
Construction quality problem(s)	X	X	X	
First nuclear project	X	X	X	
Inexperienced nuclear A/E		X		X
Inexperienced nuclear CM	X	X	X	X
Inexperienced nuclear constructor	X		X	
Some member(s) of project team inexperienced in role assumed	X	X	X	X
Inexperience of project team member contributed to quality problem	X	X	X	

The issue of prior nuclear design and construction experience of key personnel of the project team is related to the issue of prior nuclear construction experience of corporate members of the project team. An inexperienced utility can compensate for its lack of prior corporate nuclear construction experience by hiring key personnel with appropriate prior experience, and by taking other management actions. For a more detailed discussion of this point, see the discussion of the Palo Verde project in Section 3.4. The key study finding on this issue is that while prior nuclear design and construction experience is important for all corporate members of the project team, it is essential for the key project individuals who work for them.

Given that lack of prior nuclear construction experience seems so important to the development of quality problems, it is reasonable to ask what additional insights the Midland project brings to the experience issue. Like PG&E, the owner utility for this project (Consumers Power) had prior nuclear experience. In addition, it selected an experienced A/E, CM and constructor.

Consumers Power has as operating plants Big Rock Point, a small (63 MW) GE-Bechtel turnkey plant that received its operating license in 1962, and Palisades, a medium-size (740 MW) plant designed and constructed for Consumers by Bechtel that went into commercial operation in 1971. In both cases, Bechtel was the A/E, CM and constructor; Consumers assumed an oversight role only and was not actively involved in managing the project. In effect, although Consumers had two operating plants, it had minimal nuclear construction experience, and Bechtel had been in firm control of the earlier projects. The respective roles of Consumers and Bechtel changed for the Midland project. Consumers took a more active management role in the project and Bechtel's management role was proportionately reduced. This was a major change in the roles of each from the prior projects, and it was a change to which neither adjusted quickly. NRC actions by the Midland ASLB hearing board and by the regional office thrust much more project and QA responsibility on Consumers for Midland than had been the case with the earlier plants. Consumers had limited experience within its staff to successfully discharge this responsibility.

A lesson of the Midland project is that while prior nuclear construction experience of each member of the project team may be necessary to avoid the development of quality-related problems and to successfully complete a commercial nuclear power plant in the U.S., experience alone is not sufficient. Many other factors, including management commitment to quality, effective oversight of contractors, qualifications of project staff, and a management attitude that does not view NRC requirements as the ultimate goals for performance, are important also. These and other factors will be discussed in subsequent sections.

3.2.2 Project Management Shortcomings

As suggested above, some utilities' lack of prior nuclear experience contributed to their failure to fully appreciate the complexity and difficulty of building or overseeing the construction of a large nuclear power plant. This inexperience contributed to but is not entirely the cause of several managerial mistakes or shortcomings that led to the quality problems at these four projects.

The principal finding of this study is that nuclear construction projects having significant problems in the quality of design or construction are characterized by the failure to effectively implement a management system that ensures adequate control over all aspects of a project.

To understand why utility management errors and shortcomings are such a dominant contributor to quality problems on construction projects, especially when coupled with lack of nuclear experience, it is useful to understand the underlying philosophy and character of a utility embarking on its first nuclear construction project. The following excerpt from one of the case studies explains one first-time owner's approach to nuclear power:

Utility Character and Background

Like many utilities, this utility had and has a conservative management philosophy and is adverse to taking unnecessary risks. As with many utilities, this one is quasi monopolistic, being protected from competition by public utility commission policies and practices. With this protection from competition, however, comes close scrutiny from the public utility commission regarding how the utility spends money and handles their finances. These factors contribute, in part, to a cost and schedule consciousness on the part of the utility. For many years the utility's hiring procedures provided for review and approval by several levels of management, including the chief executive officer for all new hires. All their contracts, including those for construction of generating plants, were fixed price contracts.

The utility's prior construction experience consisted of about twenty fossil-fired plants. In some cases the utility had served as construction manager. The utility had a construction department headed by a vice president, which was responsible for all construction utility wide. Over the years the utility developed a close working relationship with, and confidence in, several of the major construction contractors that worked on their fossil projects. The utility's fossil construction success was a source of pride: each plant had come on line on or before schedule and at or within budget. Each plant was of acceptable quality; after a few early bugs were worked out, each plant operated safely and reliably. This quality, incidentally, was something put into the plant by the builders - there was no formal program for quality or the assurance of quality. To the utility, quality was something that happened if you put good people on the project.

Reflecting the generally conservative management philosophy of the company was an adherence to tradition: if something seems to work, stick with it. The traditional way of building fossil plants seemed to be successful, and the company carried over many of its fossil construction practices to its nuclear project; e.g., the utility served as construction manager, and several of their key contractors on fossil plants were retained (although the utility had no nuclear experience and their contractors had limited nuclear experience); only fixed price contracts were let; the construction department was responsible for construction management except for a few people permanently assigned to the project; personnel from existing departments in the utility were matrixed in to work on the project as needed. They reported administratively and to some degree functionally to their department head, not to the project manager; the project was managed from corporate headquarters with a minimal utility presence at the site; and hiring and recruitment actions continued to be reviewed at the highest levels of the company.

This excerpt applies in varying degrees to the other utilities that had quality problems. In general, these utilities had managed or overseen the construction of several successful fossil projects. They approached their nuclear projects as extensions of the earlier fossil construction activity, i.e., to be managed, staffed, and contracted out in much the same way as fossil projects. The utilities did not fully appreciate or understand the differences in complexity, quality requirements, and regulations between fossil and nuclear projects and tended to treat the nuclear projects mentally and managerially as just another construction project.

One chief executive termed his utility's first planned nuclear plant as "just another tea kettle", i.e., just an alternative way to generate steam (this was before major quality problems arose at his project). Managerially, the utilities fit their nuclear projects into their corporations' traditional project management scheme, which, in retrospect, may not have been well suited for nuclear work. Generally, the utilities' lack of experience in and under-standing of nuclear construction manifested itself in some subset of the following characteristics (not all apply to each of the four utilities):

(1) inadequate staffing for the project, in numbers, in qualifications, and in applicable nuclear experience

(2) selection of contractors who may have been used successfully in building fossil plants but who had very limited applicable nuclear construction experience

(3) over-reliance on these same contractors in managing the project and evaluating its status and progress

(4) use of contracts that emphasized cost and schedule to the detriment of quality

(5) lack of management commitment to and understanding of how to achieve quality

(6) lack of management support for the quality program

(7) oversight of the project from corporate headquarters with only a minimal utility presence at the construction site

(8) lack of appreciation of ASME codes and other nuclear-related standards

(9) diffusion of project responsibility and diluted project accountability

(10) failure to delegate authority commensurate with responsibility

(11) misunderstanding of the NRC, its practices, its authority, and its role in nuclear safety

(12) tendency to view NRC requirements as performance goals, not lower thresholds of performance

(13) inability to recognize that recurring problems in the quality of construction were merely symptoms of much deeper, underlying programmatic deficiencies in the project, including project management.

Each of the four utilities had varying degrees of understanding of the project, its complexity, their role in it and how it should be managed. In several cases, utility management did not understand what was required for successful project completion and consequently could not provide effective oversight or leadership of their contractors. In some cases, no one was managing the project; the project had inertia but no guidance or direction. In several cases, the utility's project management approach failed to provide effective oversight of several aspects of the project, including planning, scheduling, procurement, cost control, degree of design completion, and quality. It is important to note that problems in quality and quality assurance were not the only management shortcomings at several of the projects; they fit into a larger pattern that evidences lack of effective overall project management. While some of the four projects studied had experienced extensive management problems, all had had problems implementing the quality assurance program, a key management control program for any complex project. Each nuclear construction project studied that had significant problems in the quality of design or construction was characterized by the failure to effectively implement a management system that provided effective oversight over all aspects of the project.

The pattern described above, which emerged from the four case studies (including the TPT study), fits the Midland project. A 1982 NRC staff report to the ACRS on Midland stated:

> The Region III inspection staff believes problems have kept recurring at Midland for the following reasons: (1) overreliance on the architect-engineer, (2) failure to recognize and correct root causes, (3) failure to recognize the significance of isolated events (4) failure to review isolated events for their generic application, and (5) lack of an aggressive quality assurance attitude.

Each of these five reasons was seen at one or more of the case study projects that experienced quality problems. The applicability of reasons (2), (3), and (4) to the case study projects is discussed in more detail in Section 3.3.

3.2.3 Shortcomings in NRC's Screening of Construction Permit Applications

Previous sections of this report have identified lack of prior nuclear experience and management shortcomings as two primary root causes of the major problems that led to this study. Given these findings, it is reasonable to ask what were the NRC/AEC screening practices for addressing experience and management capability when the construction permits (CPs) were issued for the plants that developed quality problems, and what were the results. Chapters 4 and 7 will address the former question. The latter question was addressed by the case studies.

As evidenced by the substantial remedial programs the NRC has required of several utilities after significant quality-related problems were discovered, it is clear in retrospect that some utilities that were granted CPs in the past would not, based on the same qualifications, be granted a CP today without substantial personnel and organizational improvements in experience level and management approach. In retrospect, it is apparent that NRC's screening process for these CP applicants failed to adequately address either the experience or management issue. This finding is relevant to at least three of the four projects in the case study population that experienced major quality problems.

The following excerpts from one of the case studies illustrate and provide background for this finding:

> For construction permits, NRC licensing review is limited largely to technical and engineering issues. NRC does not and did not in the case of the licensee, evaluate whether the applicant and his contractors had the experience, knowledge, staffing, or ability to effectively manage and consummate a project as complex as the construction of a nuclear power plant.

> NRC's licensing review for a construction permit is largely limited to technical issues and conformance with 10 CFR 50. NRC does not (and did not in the case of this utility) perform a formal review of the applicant's ability to manage, and carry through to completion, the construction of a nuclear reactor. The issues in this case are management capabilities and lack of experience, and NRC's formal licensing process failed to adequately address either.

> NRC contributed to the turnaround [after quality-related problems were uncovered], and its extent in a significant way by setting high standards for the resumption of the project. NRC's requirements for total restart of the project contained "hold points" corresponding to the different stages of recovery, each of which would be subject to intensive scrutiny by NRC inspectors.

> NRC's requirements for resumption of construction were more stringent than were NRC's initial requirements for CP issuance. For resumption of construction, NRC focused more on the issues of management and management capability, and required demonstrations of capability rather than statements of intent.

> NRC, in granting a CP, should look beyond the plant design, seismic criteria, and financial status to determine whether the utility is capable of managing a project having the scope and complexity of construction of a nuclear project.

> Opinions expressed by both regional and headquarters NRC personnel as well as licensee personnel suggest that the NRC

could have been more effective in some respects in avoiding the problems which occurred at this project. A recurrent theme was that the NRC licensing process does not do enough to address the ability and experience of project management as it relates to managing a nuclear construction project. The inspection process also tends to ignore management issues.

Although these excerpts are from one case study, they apply equally to three of the four case study projects that experienced major quality problems.

3.2.4 Other Factors Contributing to Major Quality Problems

Several other factors contributed to the development of major quality problems at the four projects studied. They include, but are not limited to the following: the changing regulatory, political, and economic environment surrounding nuclear power over the past several years and some licensees' inability to recognize and adjust to the changes as they were occurring; the failure of some licensees to treat quality assurance as a management tool, rather than as a paperwork exercise; and NRC's lack of effectiveness in convincing all licensees of the necessity to implement their quality assurance programs.

The major design quality problems that have arisen were related to shortcomings in management oversight of the design process, including failure to implement over the design process quality assurance controls that were adequate to prevent or detect mistakes in an environment of many design changes. Appendix A, the individual case study working papers, and the TPT report on Zimmer provide the basis for more information on these findings.

3.3 WHY HAVE THE NRC AND THE UTILITIES FAILED OR BEEN SLOW TO DETECT AND/OR RESPOND TO THESE QUALITY-RELATED PROBLEMS?

Determining answers to this question was part of the case study focus of the analysis of the four projects experiencing major quality problems. As with the first question (Section 3.2), several common threads emerged from the different case studies. Generally, these threads can be identified as shortcomings in utility programs and practices and shortcomings in NRC programs and practices.

3.3.1 Shortcomings in Utility Programs and Practices

The shortcomings in utility programs and practices that led to the utilities' failure or slowness to detect and/or respond to quality problems are largely outgrowths of the findings on lack of experience and management capability, discussed in the preceding section. As previously stated, the experience and management problems resulted in, among other problems, failure to adequately implement the quality assurance program. In 1969, the NRC established 18 criteria for an effective quality assurance program, and all subsequent license applications were required to describe a quality assurance program that met the 18 criteria. In some cases, these programs were simply not implemented. It is not surprising that those projects that failed to effectively implement a quality assurance program also did not detect or act on major quality problems in a timely fashion. The quality assurance program is the management system whose primary purpose is detecting and correcting such problems.

In several cases the poorly functioning quality assurance program had its roots in lack of management appreciation of or support for the quality function. This lack of support manifested itself in failure to adequately staff the quality assurance function in numbers, qualifications and nuclear experience. In each case senior management wanted a quality plant but generally did not see the quality function and quality assurance program as a vehicle to help achieve that end. Instead of seeing quality assurance as a management tool to help them exercise control over the project, some managers saw it as an extra government requirement that was not present in the construction of other (non-nuclear) projects. In one case, senior utility management had been warned that the quality assurance manager might try to establish a quality assurance "empire," and it consistently rejected his requests for additional quality control inspectors. Subsequent events proved the QA manager's requests to have been squarely on target. Cost and schedule considerations also contributed to weak management support for the quality function. Some senior managers saw quality assurance as an overhead expense that also had the potential for slowing the rate of construction.

The single most damaging manifestation of the lack of management support for quality assurance and the quality function is that in several cases management was not aware of vital information on the quality of construction which was known to the quality assurance staff. In some cases, management had pertinent information offered by the quality assurance organization (e.g., improper patching of concrete) but, seemingly, did not listen to it or believe it. In other cases the management chain, from the site quality assurance manager to the senior corporate official responsible for the project, contained so many layers (three to four) that vital information on inferior construction and design quality was severely attenuated when or if it reached top management.

The utilities studied did not take action on problems sooner because they generally had difficulty in aggregating seemingly isolated quality problems into a coherent picture that indicated the quality breakdown was pervasive and programmatic. The NRC suffered from this problem also (see Section 3.3.2).

3.3.2 Shortcomings in NRC's Programs and Practices

The case studies developed several findings on NRC's failure or slowness to detect and/or respond to quality problems in design and/or construction.

When the construction mistakes studied for this report were made, the then current Atomic Energy Commission (AEC)/NRC inspection program provided sporadic NRC inspection at construction sites. Each of the five major quality problems began or occurred before the resident inspector program for construction was implemented. The earlier sporadic NRC presence at construction sites made it unlikely that an NRC inspector would discover a quality problem on his own. It also meant that information on a project's performance was transmitted to NRC regional and headquarters offices in bits and pieces, making it difficult to aggregate and determine whether reported problems were isolated events or part of a larger problem pervading the project. Although individual inspectors may have sensed a pervasive quality problem at a site months or years before the NRC as an agency recognized it, isolated information from different inspectors in different disciplines inspecting at different times generally was not effectively aggregated and analyzed.

In most of the projects having major quality problems, neither the NRC nor the licensee adequately traced the more obvious quality problems to their root causes and devised a correction program. No project is without errors. These errors can be large or small, or there can be such an accumulation of small errors that the cumulative effect becomes large. The NRC treats small errors or "findings" as items that can be corrected within a licensee's normal quality assurance program. However, large errors question the adequacy of the licensee's entire quality assurance program. The point at which an inspection finding leaves the realm of "small" and becomes "large" is referred to as the inspection "threshold." Without a particularly glaring deficiency, it would take some time for the NRC to aggregate individual findings into a general conclusion that the overall construction effort was deficient. The inspection threshold has generally been higher for plants under construction than for operating plants; the rationale was that any major safety problems would be caught prior to operation through an intense pre-operational testing program. This approach was based upon upon the observation that a plant does not represent any potential hazard to public health and safety until it goes into operation.

For several of the projects having quality problems, the extent and magnitude of the quality problem was finally established by the NRC through a comprehensive NRC team inspection involving several inspectors in different disciplines and requiring several weeks of field work. In some cases, this kind of inspection effort was only applied after allegations of poor quality assurance were raised by parties independent of the NRC. Such comprehensive team inspections provide an opportunity for frequent interchange of information in a short period of time among inspectors looking at different areas. Team inspections facilitate the synthesis and integration of findings and the development of project-wide conclusions. These team-type inspections have now been made a regular part of the NRC inspection effort (see Chapters 2 and 7).

Historically, the NRC also did not perform inspections of any depth or frequency in the design area. Design was afforded less inspection attention than construction and construction less inspection attention than operating reactors. Reactors under construction were not afforded the degree of scrutiny given to operating reactors for the same reason the threshold for construction was set higher, as explained above. The lack of NRC inspection attention in the design area was due, in part, (1) to the need to prioritize the allocation of reactor inspection resources among operations, construction, and design, (2) to a shortage of inspectors technically qualified to review the design process, and (3) to a perception that design engineers did not need NRC inspection oversight as much as construction workers did.

In addition to NRC's slowness to recognize the extent of major quality problems, the NRC was slow to take strong enforcement action in some cases where such quality problems were identified. Historically, AEC/NRC has been slower to take enforcement action for construction problems than for operations problems since there is no immediate threat to the public health and safety posed by a plant that has no fuel or radioactive contamination. Problems identified by the NRC during construction were tracked and corrective action required before an operating license was issued. As explained above, it was believed that other quality-related problems that might affect plant safety would be detected during pre-operational testing of the plant. The NRC took strong action (shutdown of work, civil penalties, issuance of Show Cause

Orders) for significant construction quality deficiencies only after the quality problems were shown to be pervasive rather than isolated and to affect several aspects of the project. For the most part, such quality breakdowns were finally established through comprehensive NRC team inspections, not through the routine inspection program. The comprehensive team inspections in turn were often triggered by allegations of improper workmanship or poor quality of construction. In two cases, inspection findings by the National Board of Boiler and Pressure Vessel Inspectors on improper ASME code piping work were instrumental in the NRC eventually recognizing the extent and magnitude of the quality breakdown.

3.4 WHY HAVE SOME NUCLEAR CONSTRUCTION PROJECTS APPARENTLY BEEN SUCCESSFUL IN ACHIEVING QUALITY WHILE OTHERS HAVE NOT?

Determining answers to this question was a major part of the case study activity at each of the projects analyzed, both those having had major quality problems and those that had not. Note that the question uses the qualifier "apparently". The case studies did not demonstrate, nor were they intended to demonstrate, that the projects visited that had not experienced major quality problems were in some absolute sense "quality successes", while the other projects analyzed as case studies were not. The case study effort took as a given that the five projects specified in the legislative history of the Ford Amendment would form one category of projects for study and that all projects not in that set of five would form another category for study. Within the second category, one consideration was to select projects that had not experienced known design or construction problems to an extent greater than other projects under construction. No nuclear construction project is completed without some quality problems developing during construction, and identifying and correcting such problems can be a measure of success of the project and its quality program. It was assumed that all nuclear construction projects will experience some quality problems during their construction (which should be corrected before operation). Vogtle, St. Lucie 2 and Palo Verde were not expected to be exceptions. Thus, the the analysis focused on comparing their approaches to project management and quality assurance with those of Marble Hill, South Texas, Zimmer, and Diablo Canyon, and determining what lessons can be learned from the differences and similarities.

The case studies took as a given that Vogtle, St. Lucie 2 and Palo Verde were apparently successful projects from a quality perspective, even though each had experienced some minor quality problems. For these three projects, the case study findings tended to be almost a direct converse of the findings of the plants experiencing major quality problems. The main findings are contained in subsequent sections.

3.4.1 Prior Nuclear Experience

As discussed earlier, an essential characteristic of a successful nuclear construction project is the collective prior nuclear construction experience of the project team (utility owner, A/E, CM, and constructors). Within the project team, it is also essential that individual team members assume roles consistent with their prior level of nuclear experience and not overstep their capabilities. Prior nuclear construction experience of the utility owner is particularly helpful, although not mandatory if the rest of the project team is sufficiently experienced, and if the utility and the other members of the project team assume project roles consistent with their respective levels of nuclear experience. The following paragraphs discuss the experience levels for the three apparently successful projects.

Vogtle is the project of Georgia Power Company (GPC). GPC has two medium-sized operating plants, Hatch 1 and 2, which went into commercial operation in 1975 and 1979, respectively. GPC is part of the Southern Company, a consortium of four southern utilities that also own and operate the two Farley nuclear units (Alabama Power Company). The Southern Company has its own engineering arm, Southern Company Services, which supports the nuclear and non-nuclear engineering and construction activities of the four member utilities. The A/E for the Vogtle project and the other four Southern Company reactors is Bechtel. GPC started construction on Vogtle before the Hatch project was completed and has been able to maintain a core of personnel experienced in nuclear construction within the utility. The same is true of the Southern Company and Southern Company Services. GPC is the construction manager for the Vogtle project. All the major construction contractors (civil, mechanical and electrical) have had significant nuclear plant construction experience, as have many of the smaller contractors. In this project, each of the project team members has assumed a project role consistent with his level of nuclear experience and capability.

St. Lucie 2 is the fourth nuclear reactor constructed by Florida Power and Light (FP&L). The first two, Turkey Point 3 and 4, are medium-sized turnkey reactors constructed for FP&L by Bechtel Power Corporation. They were completed in 1972 and 1973, respectively. FP&L's role in their construction was oversight only, although they did participate in the startup activities. St. Lucie 1, which was completed in 1976, was designed and constructed for FP&L by Ebasco. FP&L was much more involved in the construction of St. Lucie 1 (although still in an oversight capacity) than in the construction of the Turkey Point plants. FP&L used all three projects as points on a learning curve, both as a corporation and for training utility personnel.

FP&L began construction of St. Lucie 2 shortly after St. Lucie 1 was finished. This was an advantage because the continuity of experienced FP&L and Ebasco project team personnel could be maintained from one project to the next. Another advantage was that the designs of St. Lucie 2 and St. Lucie 1 were very similar, so FP&L started the second project with a very advanced design. The nearly completed design and the construction experience gained from having completed an almost identical unit, together with a nine-month licensing delay, enabled FP&L to perform an unusually extensive amount of planning, scheduling, and procurement activity before actual construction of St. Lucie 2 began. This up-front planning was a significant contributor to the completion of St. Lucie 2 in a six-year period. During the licensing delay, FP&L decided to construct St. Lucie 2 with an integrated project team of experienced FP&L and Ebasco personnel, with FP&L assuming the role of CM. Ebasco was A/E and constructor. Again in this project, project team members assumed a project role consistent with their levels of experience and capability.

By the time five of the case studies had been completed, it was apparent that prior nuclear construction experience was a key factor in project success or lack of success. The Palo Verde project appeared to contradict the working hypothesis that prior nuclear construction experience of the owner was necessary in the present environment, so a case study was performed at the Palo Verde project to determine the reasons for this apparent anomaly. The Summer project was considered also for the same reason (apparently successful first-time owner/utility), but time did not permit case studies of both Palo Verde and Summer. Subsequent staff analysis of the Summer project indicates striking similarities to key aspects of the Palo Verde project.

Palo Verde is the first nuclear project of Arizona Public Service (APS). From the project's outset, senior APS management felt strongly that nuclear construction was sufficiently different from fossil construction that it would have to be managed differently. The utility did not have previous nuclear experience as a corporation, but it recruited a technically capable core group of project personnel with prior nuclear construction and A/E experience, reorganized the corporation to create a separate division dedicated to the nuclear construction project, and contracted for extensive applicable corporate and individual experience in each of the key project organizational roles of A/E, CM, and constructor. Bechtel occupies all three of these roles for the Palo Verde project. APS's role is one of oversight and active management involvement. Recognizing that the project oversight role requires managing the interfaces among the other project team members and recognizing its own inexperience, APS consolidated the roles of all the other project team members under one very experienced contractor to minimize problems across those interfaces.

In the construction portion of the Palo Verde project, each of the project team members assumed a project role consistent with his level of experience. However, this did not hold true as the operational phase approached. In the transition from construction to operations, APS appeared to commit managerial mistakes similar to those committed in the construction phase at some other plants studied.

At the time of the case study, APS was experiencing some difficulty in moving from the construction phase to the operation phase. These difficulties were not well known and were in addition to the highly publicized pump problem experienced by APS. Unlike construction, in which the owner-utility usually hires contractors to design and build the plant, the owner normally operates the plant itself. In this project, APS had assumed the responsibility for pre-operational checks and startup of the plant. However, APS did not apply all of the good management practices it had used in construction to startup and operations. Operational responsibility for the Palo Verde plant was not established in an organization separate from the rest of APS operations, and an existing APS vice president having only fossil experience was initially placed in charge of Palo Verde operations, before being replaced by someone with extensive nuclear operations experience. Both of these actions are in contrast to the APS construction project management decisions, and both contributed in part to the startup problems at Palo Verde.

The problems with startup were not anticipated and some delays ensued until APS recognized the nature of its problem. It separated Palo Verde operations from the remainder of APS operations and placed a senior-level APS management team with nuclear operations experience at the site. These startup problems were largely masked by technical problems with the reactor coolant pumps, but they served to support the study conclusion (see Section 3.4.2) that a separate nuclear organization staffed with personnel whose experience is consistent with the chosen project role is a key determinant for project success. The startup problems of this first-time utility underscored and corroborated the study findings on the importance of prior corporate nuclear experience and the necessity for personnel in key positions to have nuclear experience.

Subsequent to the case study, a regional CAT-type inspection was performed of the Palo Verde project. The CAT identified four major areas having deficiencies sufficient to warrant enforcement action, including civil penalties. Three of the four enforcement items dealt with start-up problems; the fourth was a collection of several individually minor construction quality program defic- iencies. No programmatic deficiencies or breakdowns were found in construction The proposed civil penalties arising for this special inspection were the first fines levied against APS in the life of the construction project.

After the case study and the CAT inspection, APS reorganized the management of the Palo Verde project to provide for more centralized control over construction, startup, and operations at a lower level in the organization. In effect, the Vice President who had been responsible for construction became responsible also for startup and operation.

3.4.2 Utility Management's Understanding of and Involvement in the Project

Another essential characteristic of a successful nuclear construction project is a project management approach that shows an understanding and appreciation of the complexities and difficulties of nuclear construction. Such an approach includes adequate financial and staffing support for the project, good planning and scheduling, and close management oversight of the project.

Management of two of the three apparently successful projects had nuclear construction experience and were able to develop an understanding and appreciation of the complexities and difficulties of nuclear construction. Senior management at the third project, Palo Verde, recognized from the outset that nuclear power plant construction was significantly different from fossil plant construction. As a result, APS changed project management practice to accommodate the nuclear project and its unique demands. APS management ensured that it had a full understanding of what the nuclear project entailed before committing to it. The following excerpt from one of the case studies illustrates how one licensee prepared itself for its first nuclear project:

Information provided by the Licensee showed that the project was started in the early 1970's with a small staff, all of whom were experienced in nuclear plant construction. This group analyzed what had gone wrong on the other nuclear projects and arrived at conclusions which played an important role in how the project was organized and carried out. First, it was important that there be a long-term commitment of qualified people to a project, both from the licensee as well as its contractors. Second, utilities typically tended to do the wrong things and get involved in the wrong places, such as wanting to approve everything. They often believed they knew more about all aspects of the projects than anyone else. Third, it was found that utilities were often very untimely in their actions and decisions, which caused costly delays. Fourth, they perceived that utilities have the wrong type of organization. For nuclear projects, the organization must be managed and detail oriented. Based on these general conclusions, the Licensee's staff came up with some recommendations which formed the basis for the project organization. First, there should be a strong project concept, both within the Licensee's and architect-engineer's (A-E's) organizations, but with a singleness of purpose. Second, the Licensee should manage the interfaces. Third, there should be single points of entry for all correspondence to each organization, and the communication channels should be monitored to ensure effectiveness. Fourth, clearly written design criteria should be established and maintained current as changes were made. Fifth, the Licensee should establish which documents produced by the A-E, and others, it would review. Sixth, the Licensee should be responsible for obtaining all project permits and licenses. Seventh, purchasing and construction work should be controlled through administrative procedures (such as having standard terms and conditions for contracts and purchase orders), a qualified bidders list, and work initiation procedures. Eighth, safety and quality must come ahead of schedule and cost, not only for the Licensee, but its contractors, also. These priorities must also be conveyed to the project regulators. Ninth, adequate systems and procedures must be established to monitor the project.

Of the projects studied that had not experienced major quality problems, the preferred project management approach was to set up a separate nuclear division responsible only for nuclear construction (and/or operations). This division had adequate financial and staffing resources to accomplish its mission and had administrative as well as functional control over project personnel (i.e., not a matrix arrangement). This approach contrasts that of several projects experiencing quality problems. The latter group generally tried to fit the nuclear project into an existing corporate framework for project management. In this case, the nuclear project did not have personnel or resources dedicated both functionally and administratively to the project and had to compete with other corporate activities for personnel and funding. After the discovery of significant quality problems and follow-on analysis of the causes of those problems, several of the projects with quality problems changed their project management approach to one similar to that preferred by the other group of utilities. In general, utilities that started their nuclear projects with other organizational forms eventually adopted the independent project form of organization.*

For the most part, the utilities that experienced major quality problems also experienced problems in other managerial aspects of the project, including planning and scheduling, procurement, oversight of vendors, material availability, etc. High-level attention to these management functions, including planning and scheduling, was a characteristic of the projects that did not experience quality problems.

---

*Electric Power Research Institute. 1983. "An Analysis of Power Plant Construction Lead Times." Vol. 1, Chapter 4, EPRI EA-2880, Palo Alto, California.

Another general characteristic of the projects not experiencing major quality problems was close management oversight of the project and the project's contractors. In general, this was not the case with projects that experienced major quality problems. In each of the three projects that have not experienced major quality problems, utility management was heavily involved in managing the project, was knowledgeable about the project, and had a strong appreciation for the differences between nuclear and fossil construction projects.

Licensing, design, engineering, construction management, construction, and startup are all much more difficult for nuclear plants than for conventional plants. More management attention and involvement is necessary (1) to understand the added complexities of nuclear construction, and (2) to take action to address small problems before they grow into big ones. Cost and schedule are project activities that compete with quality; they cannot be properly balanced without the licensee's strong management control and involvement. A licensee's contractors have neither the same overall responsibility that the licensee has nor the same authority and resources to deal with quality-related problems. When a licensee abdicates its role, some aspect of quality, cost and/or schedule is likely to be compromised.

In recent years, licensees have been forced to take more active roles in upgrading many aspects of the nuclear industry because of regulatory requirements—especially those aspects related to the quality of products or work from equipment suppliers and construction contractors. This has not been a role traditionally required of licensees for their fossil fuel plants. Where licensees have followed fossil fuel practices and have chosen not to be involved in supplier and contractor activities, quality-related problems were more prone to occur. The experience of several of the case study projects having quality problems strongly supports these findings.

3.4.3 Rising Standard of Performance/Commitment to Excellence

Of the projects studied there tended to be a direct correlation between the project's success and the utility's view of NRC requirements: more successful utilities tended to view NRC requirements as minimum levels of performance, not maximum, and they strove to establish and meet increasingly higher, self-imposed goals. This attitude covered all aspects of the project, including quality and quality assurance.

The following excerpts from one of the case study working papers illustrate this finding, as well as top management's commitment to quality, which filtered down to the worker level:

> The Licensee has an orientation toward, and an attitude supportive of quality in their nuclear project. The stated management philosophy of insisting on quality was not simply to satisfy the Nuclear Regulatory Commission (NRC), but to go beyond those requirements to have a reliable and safe operating plant. At higher levels in the management structure, the conviction appeared to prevail that public safety and company profitability demand quality in the construction (and operation) of nuclear plants, and that it is less expensive in the long run to "do the job right the first time." From the interviews conducted, both at the corporate offices and the site, it was evident that a sense of

commitment to quality pervades the Licensee's organization at all
levels. The Licensee volunteered to participate in the first INPO
construction pilot audit and has expanded on it with their own self-
initiated evaluation. The quality assurance staff has direct
access to an executive vice president. There was no indication
from the interviews of cost/schedule overriding QA/QC. At lower
levels, there was an expressed feeling that the company wants to
do the job right. Employees at all levels appeared to have a
constructive attitude toward the need for quality in general, and
quality assurance, in specific. A pro-company attitude and good
morale on the part of the employees appears to exist.

The Licensee is proactive in looking for improvements in its
assurance-of-quality practices. Key line managers were taken on
a retreat by the Executive Vice President for Power to consider new
approaches to the assurance-of-quality problem. This Licensee
volunteered to be the first to be evaluated under 10 CFR 50
Appendix B requirements in the early 1970s. Their own QA
organization was asked by senior utility management to study the QA
programs of other licensees for possible improvement as early as
1978.

While the Licensee's management seems very much aware of the importance
of complying with NRC requirements, the comment was made, "satisfy the
NRC and everything is okay is not true, you have to satisfy yourself."
There was recognition that a utility can be at considerable financial
risk with a nuclear plant, beginning at the highest levels of the
corporation and flowing downwards.

Other examples of how some utilities implemented their desire to improve their
standard of performance include improving programs by seeking information and
the benefit of other utilities' experience on a wide range of matters; creating
a work atmosphere that encourages looking for problems and solving them, rather
than ignoring them or putting them off; and expanding the quality assurance
program used for their nuclear plants to their non-nuclear plants.

3.4.4 Other Characteristics of Apparently Successful Projects

The case studies identified several other characteristics generally shared
by the projects that had not experienced major quality problems; these
characteristics were generally not evident when quality problems occurred at
the other projects. Some of these characteristics are summarized below.
Appendix A and the individual case study working papers provide additional
details on them.

Strong project management is required, with clearly defined responsibilities
and authorities. The personnel responsible for the project must have sufficient authority to accomplish their mission. Other characteristics
include management orientation toward quality and visible support of the
quality assurance program, including staffing and resources; an emphasis on
"doing it right the first time"; a philosophy that quality is everyone's
responsibility, especially the doer's, and that quality cannot be "inspected
in" by the QA/QC program; achievement of a minimal number of project interfaces; good public relations; constructive working relationships with the

NRC; appropriate contracting practices and labor relationships; careful selection of contractors; development of a project commitment and sense of team work on the part of the project staff, including contractors; and an ability to adjust to the changing political, economic, and regulatory environment surrounding nuclear power over the past decade.

Some individual members of senior management at utilities that had not experienced significant quality problems expressed the opinion that construction problems experienced by others in the nuclear industry could largely be attributed to management problems, not to regulatory requirements or to changes in requirements. A characteristic of the projects that had not experienced quality problems was a constructive working relationship with and understanding of the NRC. For example, Florida Power and Light established a special office in Bethesda staffed by engineers to facilitate exchange of information with the NRC during the St. Lucie 2 licensing process. Also, senior management of Arizona Public Service has established the following policies concerning the NRC:

> Don't treat NRC as an adversary; NRC is not here to bother us -- they see many more plants than the licensee sees; inform NRC of what we (APS) are doing and keep everything up front; and nuclear safety is more important than schedule.

3.4.5 Design Completion and Project Planning

The St. Lucie 2 experience results in several important lessons. The construction time for St. Lucie 2 was approximately half the industry average, and the cost to complete the plant will be less than half of that for some plants started before St. Lucie 2 and yet to be completed. St. Lucie 2 has been subjected to the identical regulatory process faced by plants yet to be completed. The case studies showed that the experience of the project team greatly aided the project, but this factor alone does not account for the atypical experience of St. Lucie 2.

The very complete design and the project planning and scheduling done during the nine-month delay in construction start were found to significantly contribute to the short construction time for St. Lucie 2. A 1979 study performed by the University of Texas for the Department of Energy* investigated declining work productivity and management of resources at ten single or multiple-unit power plants under construction and contained the following information:

---

*J. D. Borcherding and D. F. Gardner, University of Texas. 1979. "Work Force Motivation and Productivity on Large Jobs." Prepared for the U.S. Department of Energy, Washington, D.C.

	Average Time Losses in Hours Per Craftsman Per Week
Material Availability	6.27
Redoing Work	5.70
Overcrowded Work Areas	5.00
Total Availability	3.80
Crew Interfacing	3.29
Inspection Delays	2.66
TOTAL	26.72

Although other time losses were listed, the above listed losses are directly related to project planning and scheduling and were the kinds of losses that were minimized at St. Lucie 2 through the intensive project planning effort before construction started. It is important to note that the degree of project planning accomplished could not have been done if the design for St. Lucie 2 had not been at such an advanced stage.

Another lesson of St. Lucie 2 may be that it is not the regulatory process that causes the delays and poor quality of many commercial nuclear power plant construction projects. The results of St. Lucie 2 and the other case studies suggest that shortcomings in project management play a much larger role. Examples of project management shortcomings that can affect all three elements of cost, schedule, and quality include the following: starting construction before design is sufficiently complete; redoing work when there are interfaces between systems already built and systems whose designs are completed later; failure to supply construction materials and components to the job site when the workmen need them; failure to supply tools to workmen when they need them; scheduling two work crews to work in the same confined work spaces at the same time; and inability to get a QC inspector to a job in a timely manner when a task is finished.

The case study analysis concluded that pervasive quality problems were usually found in concert with other project management problems and that quality program performance was just one measure of the overall quality of the project management.

St. Lucie 2 demonstrates that even in today's regulatory environment, capable, experienced management with a very complete design and with adequate project planning can construct a quality nuclear plant, at a reasonably predictable cost, and in very little more actual construction time than is needed to construct a coal plant. FP&L management identified to the case study team what it thought to be the ten most important factors in completing the St. Lucie 2 plant essentially on schedule, within cost, and without major quality-related problems:

(1) management commitment

(2) a realistic and firm schedule

(3) clear decision-making authority

(4) flexible project control tools

(5) team work

(6) maintaining engineering ahead of construction

(7) early startup involvement

(8) organizational flexibility

(9) ongoing critique of the project

(10) close coordination with the NRC.

3.5 THE OVERLAP BETWEEN QA AND PROJECT MANAGEMENT

One consistent study finding was that shortcomings in quality assurance program implementation were linked to shortcomings in project management, and vice versa. This linkage is not surprising when one views QA in its simplest form: QA is a management tool for ensuring that a product is built as designed and that defects are corrected. Even if a formal QA program did not exist, prudent management of a complex project requires a management feedback system to know whether the product is being made correctly. Prudent managers would devise such a system because the information it provided would be essential to them in their role as managers. They would want such a management tool to contain features such as feedback on whether the design was being implemented correctly; whether design changes were reflected everywhere and when they should be; whether parts purchased from others were made properly and met specifications; whether appropriate corrective action was taken when mistakes or nonconformances were found; and whether the management feedback system itself was reliable and correct - all features that are required as part of a QA program for a nuclear plant.

Given that prudent management would create a system having many of the features of the required QA as part of their total project management system, why were there examples of management failure to listen to what their QA program was telling them, failure to adequately staff the QA program either in numbers or qualifications, and failure to support the QA program in general? Why were there repeated examples of lack of management commitment to QA?

There are several reasons. In most cases the answer is a combination of these reasons. The first reason is lack of prudence--not all the managers would have been sufficiently prudent to set up an effective management feedback system for the quality of the project if it were not required. These same managers would also fail to see the potential of the required QA program to fill this management need because they did not fully recognize the need. (The need is greater in nuclear than in fossil because the projects are more complex, the quality standards and requirements are more stringent, and the management challenges are greater.)

The second reason is that the QA program was a requirement. Some managers would treat the requirement as just a hurdle to be crossed. This perception leads management to focus not on the intent of the program, but on its details, e.g., a written manual, an independent QA manager, layers of procedures. Some

managers honestly felt they had met their responsibility when they had attended
to such details.

A third reason is that some viewed QA/QC not only as a requirement, but as an
adversary. A strong QC program can slow down construction and a rift sometimes
develops between construction workers and QC. FP&L addressed this by making QC
a part of construction and overchecking QC with QA. There was still a rift,
but it was at the QA-QC interface, and construction workers did not see QC as
the enemy. Some managers at other projects studied had viewed QA/QC as an
enemy: as previously noted, one utility executive had been warned by others to
watch the manager of the newly established QA/QC program to be sure he did not
create a QA/QC "empire".

The third point illustrates the fourth point: QA can be a management tool,
but to be so, it must be part of the team of engineering, construction management, and project management. To be effective as a management tool, QA must be
integrated into the project. A key lesson from the study of outside QA programs
(NASA, Gaseous Centrifuge Evolvement Plan, see Appendix D) is that not only
should QA be integrated into the project, it should be integrated early, at the
design phase.

The fifth reason is not so obvious as the others, but may be as important.
It is just the opposite of the first four findings: some managements have
recognized that QA is a management tool but have failed to execute some of the
project control that is appropriately their responsibility because they felt QA
would take care of it. That is, some managers have felt there were certain
aspects of the project they did not have to address because the QA system would
take care of them. In such a situation, attenuation of information flowing
from the QC program at the site to top management can be disasterous. Even if
such attentuation does not exist, reliance on the QA program to manage part of
the project can also be disasterous if top executives (1) do not fully understand the limitations and scope of the QA program; (2) are not personally
involved in oversight of the QA program at the detail level; (3) do not provide
for direct feedback from the program down to the QC inspector level; (4) do not
fully understand how the QA program relates to engineering, construction, and
the rest of project management; (5) do not integrate QA into the project,
making QA part of the team, (6) do not staff the QA function with qualified,
capable, motivated people; and (7) do not inspect the implementation of the
program personally.

3.6 IMPLICATIONS OF THE CASE STUDIES FOR FUTURE PLANTS

Having described the salient features and practices of those projects that
did and did not experience major quality problems in construction, it is
important to note that neither group did all things right or all things wrong.
The projects without major quality problems experienced quality failures and
project inefficiencies, and much of the work of the projects with major quality
failures appears to have been of good quality. The former did not have
experienced, dedicated personnel in every position, and their procedural
controls were not flawless. It cannot be said that their projects are exempt
from quality errors--only that the probability of the errors going uncorrected
and developing into a major quality breakdown was less because of appropriate
prior nuclear experience, management understanding of and involvement in the
project, dedication to quality, a problem-seeking and solving orientation, and

a view of a quality assurance program as a management tool rather than just a requirement.

The case studies have focused on what has happened in the past, or is happening now, to derive lessons to apply in the future. The increased industry and NRC experience and the lessons learned, if applied properly, should decrease the probability of major quality problems in future generations of nuclear plants. However, there are several conditions under which major quality problems might recur. These include the following:

(1) a first-time utility with a staff or A/E, CM, or constructor that have inadequate nuclear design and construction experience

(2) a very large growth in the number of nuclear plants being constructed that (again) overwhelms the industry's and NRC's capabilities

(3) a long delay before nuclear plant construction activities start agains, resulting in a dearth of experience in the industry

(4) regulatory actions at federal and state levels that undercut quality.

The NRC and the nuclear industry need to be aware of the implications for quality that these possibilities hold.

4.0 PILOT PROGRAMS: QUALITY ASSURANCE AUDITS PERFORMED BY INDEPENDENT INSPECTORS

Section 13(c) of the Ford Amendment directs the NRC to conduct a pilot program to better assess the feasibility and benefits of implementing alternatives 13(b)(1) - 13(b)(5). In particular, Section 13(c) directed that alternative b(5), which proposes the use of third-party audits, be tested through a pilot program. The text of the pilot program requirement is as follows:

Pilot Program

...the Commission shall undertake a pilot program to review and evaluate programs that include one or more of the alternative concepts identified in subsection (b) for the purposes of assessing the feasibility and benefits of their implementation. The pilot program shall include programs that use independent inspectors for auditing quality assurance responsibilities of the licensee for the construction of commercial nuclear power plants . . . .

The pilot program shall include at least three sites at which commercial nuclear powerplants are under construction. The Commission shall select at least one site at which quality assurance and quality control programs have operated satisfactorily, and at least two sites with remedial programs underway at which major construction, quality assurance or quality control deficiencies (or any combination thereof) have been identified in the past.

Before conducting the pilot program, the NRC staff reviewed the feasibility of testing each of the alternative concepts in a pilot program, with the following conclusions

Alternative b(1): More prescriptive architectural and engineering (A&E) criteria.

Because reactor plants under construction are in advanced stages of construction, a pilot program for testing the feasibility and benefits of more prescriptive A&E criteria could not be implemented. However, the NRC staff did analyze this alternative (Chapter 6).

Alternative b(2): Conditioning the construction permit (CP) on the applicant's demonstration of its ability to independently manage a quality assurance (QA) program.

No CP applications are currently pending, so this concept could not be tested on a current CP applicant, nor could a current CP application be conditioned on this requirement. This study considered two types of demonstrations of QA management capability. The first is a pre-CP issuance assessment, which evaluates potential management capability prospectively. The second is a post-CP demonstration, which assesses management capability and QA program effectiveness based on a review of the implementation of the QA program over some previous period of time. Because there are no new CP applicants currently, the pre-CP assessment could not be done as part of a pilot program. However, a post-CP test could be performed of this concept and was included as part of the pilot program.

Alternative b(3): Improved audits by associations of professionals.

The Institute of Nuclear Power Operations (INPO) has developed a new Construction Project Evaluation (CPE) Program, which represents a significant improvement in the capability of professional organizations to provide comprehensive evaluations of construction projects. To assess the new program's feasibility and benefits, senior NRC design and construction inspection staff monitored three INPO CPEs--Beaver Valley 2, Limerick, and Millstone 3. At these projects, INPO's methodology, and its depth and breadth were evaluated. Although NRC review of these INPO evaluations might be considered pilot programs, they are not treated as such in this report for two reasons: (1) the three plants covered do not meet the Ford Amendment pilot program criterion that at least two of the projects covered by the pilot have remedial programs under way, and (2) the CPE was past the pilot stage. INPO had tested an earlier version of their CPE program as a pilot in early 1982, and the industry had tested it later in 1982. The CPE program is now a routine INPO program, not a trial program. The role of INPO in the assurance of quality and NRC's analysis of the INPO CPE program are discussed in Chapter 5.

Alternative b(4): Improvements to NRC programs.

Several improvements to NRC's programs have been tested and implemented. Both the Construction Appraisal Team (CAT) and the Integrated Design Inspection (IDI) programs were fully implemented in June of 1983 after a pilot period that included several trial inspections. Chapter 7 discusses the CAT and IDI programs and several other improvements to the NRC program that were subject to trial periods before they were implemented, including the Resident Inspector Program and the Systematic Assessment of Licensee Performance (SALP) Program. Other future improvements to the NRC program suggested in Chapter 7, such as performance objectives for QA programs, will be subjected to a trial program before they are fully implemented. The case studies (see Chapter 3) also may be considered as a pilot for future NRC management assessments. However, for this study, the above activities are not treated as pilot programs in the sense of the Ford Amendment and are covered elsewhere in the report.

Alternative b(5): Conditioning the issuance of CPs for commercial nuclear power plants on the permittee entering into contracts or other arrangements with an independent auditor to audit the quality assurance program to verify quality assurance performance.

The Ford Amendment required that this alternative be tested as part of the pilot program. The Ford Amendment stipulated that at least two projects from the set consisting of Marble Hill, Midland, Zimmer, Diablo Canyon, and South Texas be selected for the pilot program, as well as at least one other project. These five projects were identified in the legislative history of the Ford Amendment as having had major quality-related problems.

In selecting sites for the pilot program, the NRC staff relied heavily on the legislative history of the Ford Amendment to try to be as fully responsive as possible to the intent of Congress. Statements made by sponsors of the Ford Amendment in introducing the amendment contributed heavily to developing the following general criteria for selecting sites for the pilot program:

(1) To the extent possible, sites will be selected that have qualifying programs already under way or that have in the past conducted such programs.

(2) To the extent possible, programs and sites will be selected to minimally disrupt ongoing construction activities.

(3) To the extent possible, sites will be selected whose owners will participate willingly in the pilot program. The legislative provision in Section 13 that allows the NRC to order participation would be used only if necessary.

(4) To the extent possible, sites will be selected with different architect/ engineer (A/E), constructor, and project management arrangements. Testing the pilot programs with a variety of participants should better indicate an alternative's potential.

Based on these criteria and the Congressional guidance that at least two sites must come from the list of five plants mentioned earlier, NRC staff contacted four utilities and obtained agreement from each to participate in the pilot program. The projects selected for the pilot program test of the third-party audit alternative, and the third-party auditor that each selected, are as follows:

Project	Auditor
Palo Verde	Torrey Pines Technology
Marble Hill	Torrey Pines Technology
South Texas	Gilbert Commonwealth Associates
Midland	TERA Incorporated

Each utility that participated in the pilot program did so willingly. Moreover, the four selection criteria were met in almost every case. The only exception was that Marble Hill did not meet criterion (1). The utility, Public Service of Indiana, did not have a qualifying program under way and contracted for this special review specifically in response to the NRC request that they participate in the pilot program. Two of the other three projects were conducting or had conducted a third-party review as part of the Independent Design Verification Program (IDVP). In these cases, the completed or ongoing IDVP was used as the third-party audit evaluated in the pilot program.

4.1 TECHNICAL APPROACH AND FINDINGS

As with NRC's evaluation of the INPO CPE methodology for this report (Chapter 5), the four third-party audits were monitored and/or reviewed by senior NRC inspectors having extensive construction, design, QA, and management backgrounds. For each NRC evaluation, the activities of the third-party auditor were monitored for several weeks at the plant site, at utility corporate headquarters, and/or at the offices of the A/E and the third party-auditor. The NRC evaluated the quality of the individual audits based on (1) the audit team's qualifications, (2) the audit team's competence and professionalism as demonstrated in the field, (3) the scope and depth of audit coverage in design, design control, construction procedures, completed construction work, quality assurance program implementation, and project management competence and capability, (4) the substance of audit findings, (5) the procedures used for reviewing and dispositioning audit findings, (6) the quality and content of the audit report, and (7) the independence of the inspector.

In conjunction with evaluating the quality of each audit, the NRC evaluated each audit considering the following questions:

(1) If this audit, or one like it, had taken place at an appropriate time in the project history of any of the five plants that experienced major quality-related problem(s), would the quality-related problem(s) at that plant have been detected earlier?

(2) Is this audit structured and conducted in such a way that it effectively verifies quality assurance program performance [i.e., alternative b(5)]?

(3) Could this audit, or some reasonable variation of it, be a way for a licensee to demonstrate that it is capable of independently managing the effective performance of all quality assurance and quality control responsibilities for the power plant [i.e., alternative b(2)]?

(4) Does this audit provide prevention, detection, and assurance capability beyond that provided by the NRC inspection program?

(5) If the answer to (4) is yes, are there more cost effective ways to bring about a comparable level of added detection and assurance capability?

(6) How often should such audits be conducted?

(7) Should such audits apply to future plants, to current CP holders, to both, or to neither?

The evaluation process led to the following conclusions and recommendations:

(1) Comprehensive audits of nuclear construction projects by qualified third-parties (independent inspectors) can significantly increase prevention and detection capability beyond that provided by the present NRC program. Such audits can also increase assurance that plants are built according to their design and licensing commitments.

(2) Alternative b(5) offers significant benefits over current and past practice. It should be adopted and applied to both future plants and current CP holders.

(3) Comprehensive third-party audits such as those examined in the pilot program, if modified to focus more on project management competence, present a viable mechanism for a new applicant to demonstrate in a post-CP

audit whether it can independently manage its QA program responsibilities and effectively manage the project. That is, an alternative b(5) audit could be used to satisfy the demonstration requirements of alternative b(2).

(4) Comprehensive audits of a construction project should be conducted about every two years.

(5) The present NRC CAT and IDI programs are limited in the extent of their coverage (4 CATs/yr, 3 IDIs/yr). Instituting a program of periodic third-party audits to supplement the present NRC program appears to be a more cost effective long-term approach than expanding NRC's program to a level that would provide the same degree of prevention, detection, and assurance coverage.

(6) The CAT and IDI programs should be used as overchecks of the third-party audit program.

(7) The NRC should develop criteria for independence of the third-party auditors and other criteria for the independent audit program, including qualifications of auditors, scope and depth of coverage, etc. Input from professionals having appropriate expertise and from other interested parties should be sought in developing these criteria.

(8) The NRC should monitor the actual performance of each audit and review its results.

(9) The depth and scope of each audit should be uniform and consistent to establish confidence in the third-party audit program. To achieve these goals and others, the third-party audit program should become a regulatory requirement.

4.2 PARAMETERS OF FUTURE THIRD-PARTY AUDITS

As a result of this study, the NRC staff has concluded that to provide sufficient preventive, detection, and assurance capability to feasibly supplement the NRC inspection program and affirmatively answer the first four questions in Section 4.1, the comprehensive independent audits recommended in the last section should, as a minimum, review the following areas in depth:

(1) experience, capability, and effectiveness of project management
(2) construction management
(3) management support of quality
(4) quality assurance program implementation
(5) qualifications of project personnel
(6) design process (A/E)
(7) design changes and control (A/E and site)
(8) quality of construction.

These categories are major areas relating to the ability of safety-related structures, systems, and components to function as required while in service. Other parts of this study also have identified other areas as being areas of weakness in the past (see Chapter 3, Case Studies). Design- and construction-related reviews in such an audit program should concentrate on whether the end product (design or system hardware) conforms with the technical requirements in the specifications and regulations, with licensee commitments made during the licensing process, and with the design basis. Such audits would measure the quality of the project team, project management, construction management, engineering, and the end results achieved by quality assurance programs. The design and construction quality reviews would be complemented by quality assurance program reviews that focused on implementing the procedural requirements of 10 CFR 50, Appendix B.

These reviews would be performed in conjunction with management reviews designed to assess the project team's effectiveness in managing all aspects of the project, including quality. The reviews should be both end-product oriented and process oriented. For example, designs would not only be audited to determine if they have been verified (a process required by Appendix B) but also reviewed for their technical adequacy (the end product). If the end product had deficiencies, then the process should be examined for generic implications. In the past, the NRC inspection program has concentrated too heavily on the quality program process and paper and not heavily enough on construction work in progress and the quality of the end product. Other measures being taken to address these shortcomings are discussed in Chapter 7.

Within the framework of the audit areas described above, a third-party audit should include sufficient review to satisfy the following performance objectives:

(1) assurance that the project team is capable of and is dedicated to constructing a nuclear power plant that, when operational, will not endanger public health and safety because of quality deficiencies that occurred during construction

(2) assurance that the project's programmatic controls for design and construction are adequate and have been adequately implemented

(3) assurance that the actual construction has been according to the design, and that design bases committed to by the applicant and approved by NRC have been translated correctly into the design

(4) assurance that the audit sample is broad enough to be reasonably representative of the plant as a whole.

Analysis of the results from the four independent audits revealed that while each has covered a part of the above proposed parameters and performance requirements for a third-party audit program, none has met all of them. Torrey Pines' construction assessment of Palo Verde did not include enough hardware verification, and Torrey Pines' assessment of the Marble Hill Project did not provide enough design or management review. The Gilbert review of South Texas was limited to programmatic controls, and the TERA review of Midland has not covered the areas of quality assurance or project management in enough detail.

A review of the four separate audit plans and their differences demonstrates that for future consideration NRC should develop audit criteria and should review in advance the audit plan of each auditor preparing for a third-party audit to determine adequacy of coverage. Also, in determining the audit's scope, the NRC should consider such factors as percent of design completion, significant types of work in progress, results of previous third-party audits at this and other plants, NRC inspection results, and the state of project completion. This approach is supported by the NRC's experience over the past two years both in its implementation of a program of independent design and construction reviews for those plants in the near-term operating license mode (IDVP program--see Chapter 7) and through the CAT and IDI programs (see Chapter 7).

Periodic audits by independent inspectors throughout the construction period are strongly preferred over a single audit occurring late in a project after design and construction are essentially complete, as is the case presently for most plants (for IDVP and generally for CAT and IDI programs).

4.2.1 Frequency of Future Third-Party Audits

For each of the plants at which serious construction quality-related problems developed, symptoms of the quality problem were evident early in the project. Based on the experience of those plants, the proposed third-party audit program should be conducted no later than two years into construction and preferably sooner to achieve maximum effectiveness. For example, Marble Hill was shut down for construction quality problems 16 months after the CP was issued. Viewed as a prevention measure, the third-party audit should be conducted as soon as construction work begins, before poor practices become ingrained in the project. Viewed as a detection measure and a way to satisfy alternative b(2)'s concept of a demonstration of management and QA effectiveness, the licensee must have enough time to make its program work before the first audit. Based on these considerations and the assumption that in the future, applicants would receive a much more searching pre-CP review by NRC (see Chapters 2 and 7), which should help prevent unqualified project teams from beginning construction, the study concludes that the first of the third-party audits should be conducted 12 to 20 months into the construction project. This timing is early enough that the audit would still have some prevention value but not so early that the project team's capability and its quality program effectiveness cannot be meaningfully evaluated.

In determining the frequency of subsequent audits, several factors were considered: changes in projects, project personnel, contractors, and level of project activity in different areas. (A project proceeds through a sequence in which the level of activity in the following areas is high at one project phase and low at others: civil/structural, mechanical, electrical, instrumentation and control, testing and startup, etc.) The study concluded that subsequent audits should be conducted about every two years, depending on those factors.

The last third-party audit should focus heavily on design implementation (hardware and process) as well as startup and testing activities. However, the final audit would focus less heavily on design issues than the present IDVP program because the present program provides, on a case-by-case basis, a single third-party audit near the end of construction to confirm the quality of design and/or construction from the project outset. Under the proposed program, a less retrospective look would be required by the final audit because a comprehensive audit would have been conducted about every two years over the project's life. Several of the areas covered by a current IDVP would have been covered under the new program in earlier audits. Those earlier audits would reduce the intensity of the final audit in some aspects from present practice. This reduced intensity would partly offset the increased effort the new program would require on the final audit in the areas of startup and testing, and management oversight of the transition from construction to operations. (For some background on transition problems, see Section 3.4.)

4.3. APPLICABILITY OF THE THIRD-PARTY AUDIT PROGRAM TO ALTERNATIVE b(2)

Section 13(b) of the Ford Amendment directs that the NRC analyze the following alternative approach to improving quality assurance and quality control in the construction of commercial nuclear power plants:

> Conditioning the issuance of construction permits for commercial nuclear power plants on a demonstration by the licensee that the licensee is capable of independently managing the effective performance of all quality assurance and quality control responsibilities for the powerplant.

The pilot program analysis included an evaluation of whether the third-party audit program proposed by alternative b(5) could also be used to satisfy the demonstration provision of this alternative. The study concluded that the first periodic audit conducted under the third-party audit program could be tailored to meet the demonstration requirement of alternative b(2) and that a third-party confirmation at this early point in construction was preferred to an NRC confirmation.

Including the b(2) demonstration as part of the third-party audit program is not the only way a licensee could achieve the performance objective implicit in alternative b(2). For example, the licensee could demonstrate this objective by an intensive NRC team inspection, such as a CAT modified to more directly address the issues of management capability and competence. However, the future program proposed by this study envisions a more rigorous screening by NRC before a CP is issued and an improved inspection program during construction. A third-party audit of the project 12 to 20 months into actual construction would not only provide assurance that the licensee's program is effective, but it would provide an independent test of the effectiveness of NRC's modified licensing and inspection programs. Such a third-party audit would provide Congress and the public increased assurance that not only the licensee but also the NRC met their responsibilities effectively.

When implemented, a post-CP demonstration of management capability and QA program effectiveness would significantly shift from present practice the method of determining whether the QA/QC program is being implemented as described in the Preliminary Safety Analysis Report (PSAR) and is producing an adequate level of quality. Including as a condition of the CP that such a demonstration occur 12 to 20 months after the CP is issued would place a "trip-wire" in front of the CP holder and the NRC. In effect, that "trip-wire" would specify that certain capabilities must have been demonstrated for plant construction to proceed beyond that point. Continuing construction activities would be contingent on the licensee successfully demonstrating its capability and program effectiveness in this post-CP audit. The licensee and the NRC would be fully aware at the onset of construction that such a demonstration was upcoming. This awareness could result in several significant and beneficial changes from current practices:

(1) The CP holder should better understand the necessity to provide trained and qualified personnel and commit sufficient resources to the project at the beginning of construction activity.

(2) The CP holder would have to act rather than to react. Not only would a management system and quality program have to be instituted, but the CP holder would also have to critically evaluate its performance and convince itself of its effectiveness in order to be prepared to convincingly demonstrate its adequacy to others.

(3) Under such a CP condition, especially if the alternative b(2) audit were to be conducted by an independent third party, the NRC would be motivated to more closely monitor the project's management effectiveness, the QA program's effectiveness, and overall construction quality before the first audit. Besides doing a better job than under current practice for achieving prevention, detection, and assurance objectives, the NRC would have current information and an understanding of management of quality program weaknesses and possible needed changes. Such information would help NRC evaluate the CP holder's demonstration of management and QA effectiveness, whoever performs the confirming audit.

4.4 DESCRIPTION OF THE FOUR THIRD-PARTY AUDITS

This section describes the independent inspection programs at each of the four sites selected and discusses improvements that could be made in future reviews by independent inspectors. Each program is summarized, and Table 4.1 at the end of the chapter provides a summary comparison of the characteristics of each. The title of the independent audit and the name of the auditing firm is listed in the title of each section. Copies of each audit are available from each licensee and should also be held in NRC's Public Document Room. Presently, on a case-by-case basis (see Chapter 7 discussion of IDVP programs), the NRC staff formally reviews and evaluates independent audits, including corrective actions for any identified deficiencies, as part of the process leading up to issuing an operating license.

Two of the four audits, Palo Verde and Midland, were conducted under the auspices of the IDVP program. The NRC review described in this section was separate from the routine NRC review of IDVPs for licensing purposes; it was for the broader purpose of assessing the utility of comprehensive third-party audits as a supplement to the regular NRC inspection program. In particular, the analysis focused on whether third-party audits represented a viable improvement over current practice and whether such audits by independent inspectors should be required by regulation for all plants under construction. The audits were intended as examples for which this evaluation was performed and while adequate for their intended purpose, some did not cover areas that a comprehensive audit would be required to cover. These areas have been identified in Table 4.1 under the heading "Comprehensiveness".

4.4.1. Independent Construction Review of Marble Hill Nuclear Generating Station Units 1 and 2, Torrey Pines Technology, San Diego, California, 1983

Early in its construction, the Marble Hill project experienced problems with work being performed by the concrete contractor. The problem was attributed to breakdowns in the utility's and the contractor's management of the quality assurance programs and eventually resulted in an NRC Stop Work Order. After an 18-month investigation and a remedial action program, which included instituting stronger management and quality assurance programs, safety-related construction work was permitted to restart. This project was particularly relevant for the pilot program because of the early stage in which the Stop Work Order was issued and the apparent success of the remedial action program. (For further discussion on the dramatic improvement in the Marble Hill program, see Appendix A.)

Torrey Pines Technology (TPT) was selected as the independent consultant to conduct the audit. TPT was experienced as a third-party auditor, having performed similar reviews for other plants, including San Onofre and Palo Verde. The objective of the TPT program was to conduct an independent audit of the quality of construction of the Marble Hill Nuclear Power Station and to evaluate compliance with approved design documents for systems, hardware, and structures. This construction audit program consisted of a detailed evaluation in five task areas:

(1) evaluation of QA organization and management policies toward QA

(2) construction design control and implementation

(3) physical verification of plant hardware and structures

(4) testing and inspection of ASME piping welds and concrete

(5) construction document review.

As a result of the review, several deficiencies were identified and referred to Public Service of Indiana for corrective action. The proposed corrective actions were reviewed and approved by TPT and further evaluated by the NRC pilot program review team. The corrective actions appeared to be satisfactory.

The NRC reviewers judged the TPT methodology, amount of hardware inspected, and detail of inspection to be satisfactory. The absence of significant electrical construction review is consistent with the plant construction status and is not viewed as a deficiency. This independent construction review was considered to be representative of a comprehensive third-party construction verification effort of a plant at this stage of construction. TPT conducted a limited, but beneficial, design review effort at Marble Hill; however, it would not constitute adequate coverage of the design process when compared to other plants in the pilot study. The NRC pilot program reviewers judged the TPT assessment of the Marble Hill project to be adequate in the five areas reviewed by TPT. Design was not reviewed by TPT as a part of this audit because a similar plant of essentially the same design and having the same A/E had undergone an extensive design review by the NRC IDI team in June 1983.

The management assessment was confined mainly to the quality assurance organization and functions. Management issues would have to be more broadly evaluated to meet the evaluation parameters for future third-party audits described in Section 4.2.

4.4.2 Independent Design and Construction Verification Program - Midland Units 1 and 2, Monthly Status Reports Numbers 1 through 6, TERA Corporation, Bethesda, Maryland, 1983

The Midland Plant has experienced several quality-related problems during its construction, including excessive settlement of the diesel generator building and other safety-related structures. The licensee is currently conducting an extensive correction program to correct all deficiencies.

The TERA Corporation was selected to perform this review, which is still ongoing. TERA Corporation is a professional services and systems engineering organization that provides engineering and environmental consulting, project management, and software to industry and government.

The objective of the TERA review is to conduct an independent assessment of the quality of design and construction of the Midland Plant. The utility, TERA, and NRC staff defined the scope of review. The approach selected by TERA is to review and evaluate a detailed "vertical slice" (indepth review of many aspects of a selected system from design assumptions through completed construction, in contrast to a "horizontal slice," which looks at a few similar aspects of several systems) of three safety-related systems, and extrapolate from this review an overall assessment of the adequacy of the plant's design and construction.

Three areas were examined in the design assessment: the design criteria and commitments, their accuracy and consistency, and the implementing documents for design. Original calculations were checked, alternative calculations performed, and completed designs, including drawings and specifications, verified. Independent calculations performed by TERA incorporated both similar and different methods from the original design calculations.

The construction program review looked at supplier documentation, storage and maintenance documentation, and construction and installation documentation, and physically verified configuration and installation of selected systems and components.

As of January 1984, about 50% of the work scope of the TERA review had been completed, covering mainly Auxiliary Feedwater System design verification. Several deviations and deficiencies have been identified and some will require corrective action by the licensee. The disposition of these will be reviewed by the staff before the license is issued.

The TERA methodology, extent of design review, and the amount of hardware inspected were found to be satisfactory. TERA's review of the Consumers Power Company's (the utility) quality assurance program and management was limited, however. Coverage in these areas would have to be expanded to meet the parameters of future third-party audits described in Section 4.2. The use of checklists, periodic quality assurance audits of the independent inspectors, and critiques of the TERA audit by senior level TERA management should result in a satisfactory review for the scope it covers. A final assessment of the adequacy of this audit will be made when it is completed.

4.4.3 Evaluation of South Texas Project - Units 1 and 2 Construction Project, Gilbert Commonwealth Associates, et al., 1983

The South Texas Project experienced several design and construction deficiencies in the late 1970s. These problems and allegations, some of which were later substantiated, and decisions by Houston Lighting and Power led to the replacement of the project's original A/E and construction manager (CM), Brown & Root. The engineering effort was transferred to the Bechtel Power Corporation, which was also designated as the CM, and Ebasco was assigned the constructor responsibilities.

Gilbert-Commonwealth Associates was selected as the independent audit team manager. Nineteen persons from Gilbert-Commonwealth Associates, Management Analysis Company, Nutech, and Energy Incorporated were selected to conduct the evaluation. The objective of the evaluation was to conduct an independent quality assurance evaluation of the South Texas Plant to ensure the adequacy of the design and construction. This audit was unique among the four in that the INPO evaluation criteria were used.

Two methods were used in the detailed design examination. First, INPO criteria were used to analyze the control of each step of the design process to determine whether it was sound and if it met the established requirements. Second, the evaluators reviewed a "vertical slice" of design activity. The system reviewed, the Component Cooling Water System, was examined in detail. The design team, in cooperation with the construction team, conducted a walkdown of the Component Cooling Water System to verify that it was constructed as the design specified. In addition, various in-process work activities were observed. The independent audit revealed weaknesses in design controls in interfaces with other contractors, engineering responses to Field Change Requests, construction drawings that were incomplete, and the utility's limited control of design changes. Several construction weaknesses were also identified.

The audit of the South Texas Project used the INPO performance objectives and criteria, which are mainly programmatic. The audit preparation, competence of evaluators, and review techniques were judged to be satisfactory. In the construction evaluation, only a limited number of weld radiographs were reviewed by the team. In the design evaluation, the scope of the review devoted to design was judged to be limited. Because of the known engineering problems of this site, a more substantial effort could have been performed in this area. In that regard, the staff understands that the licensee has a separate, continuing audit process for design. This audit would have to expand its coverage in these areas, as well as in management, to meet the parameters of future third-party audits (see Section 4.2).

4.4.4 Independent Quality Assurance Evaluation of Palo Verde Nuclear Generating Station, Units 1, 2, and 3, Torrey Pines Technology, San Diego, California, 1983

The Palo Verde Nuclear Generating Station is considered to be an example of a site at which quality assurance and quality control programs have operated satisfactorily.

TPT was also selected to perform this review. The overall objective of this effort was to independently evaluate project organization, management, quality assurance, design, and construction activities. The scope of TPT's review included activities of Arizona Public Service (APS) Company, Bechtel Power Corporation, and Combustion Engineering Corporation (the owner, A-E/CM, and nuclear steam supply system vendor, respectively). In the overall audit plan, which incorporated NRC comments and was approved by the NRC, five task areas were to be evaluated in detail:

(1) evaluation of project management organization

(2) evaluation of management's policies toward quality assurance

(3) evaluation of quality assurance activities

(4) design verification

(5) construction verification.

The objective of the first task was to evaluate APS's project management organization to determine the adequacy of its structure and organization and whether it could assure that the high standards required for nuclear power plant design, procurement, and construction had been met. The objectives of the second and third tasks were to review APS management policies that affect quality assurance and to assess the degree to which the policies ensure an effective quality assurance program. Also, specific elements of the APS quality assurance program were evaluated to determine if those elements were adequately defined and implemented.

The goal of the design verification, the fourth task, was to verify that the design bases contained in the Final Safety Analysis Report (FSAR) had been adequately converted into design documents for the constructor and fabricator. This task was divided into three subtasks consisting of design procedure review, design procedure implementation review, and a detailed technical review.

The final task, the construction verification review, was to verify the compliance of construction-related quality assurance procedures and controls with NRC requirements. Compliance was verified to evaluate the implementation of these procedures and controls and to determine whether selected safety-related systems and components were constructed according to design documents.

Valid deficiencies were referred to APS and their proposed corrective action was reviewed and approved by the TPT. The NRC review team further reviewed the corrective action, which appeared to be satisfactory.

The scope of review could have been broader. Specific areas not covered in this review are listed in Table 4.1, which summarizes the comparison of the independent audits of the four pilot programs. For example, more coverage of management issues, including the management of transition from construction to operations, would be required for this audit to meet the parameters for future third-party audits (see Section 4.2).

TABLE 4.1. Summary Comparison of Pilot Program Independent Audits

Project and Utility	A/E	Construction Manager	Constructor
Marble Hill, Public Service of Indiana	Sargent & Lundy	Utility	Various Contractors
Midland, Consumers Power Company	Bechtel (Ann Arbor)	Bechtel (Ann Arbor)	Bechtel (Ann Arbor)
South Texas Project, Houston Lighting & Power	Bechtel (San Francisco) (was Brown & Root)	Bechtel (San Francisco) (was Brown & Root)	Ebasco (was Brown & Root)
Palo Verde, Arizona Public Service	Bechtel (Los Angeles)	Bechtel (Los Angeles)	Bechtel (Los Angeles)

Project and Utility	Evaluation Consultant	Evaluation Schedule and Level of Effort
Marble Hill, Public Service of Indiana	Torrey Pines Technology (TPT) Average nuclear experience per team member was 10 years and each had partici-pated in one or more similar evaluations.	6/3 - 7/23/83 8,000 person-hours total effort
Midland, Consumers Power Company	TERA Average nuclear experience per team member is 10 years with most of team having an average of 15 years.	6/83 - Mid 84 Total effort as of 9/83 estimated to be 20,000 person-hours
South Texas Project, Houston Lighting and Power	Gilbert-Commonwealth, Management Analysis Company NuTech, and Energy, Inc. Average nuclear experience per team member was 17 years. Members had on average participated in two similar evaluations.	8/22 - 9/2/83 4,000 person-hours total effort
Palo Verde, Arizona Public Service	Torrey Pines Technology (TPT) Average years of nuclear experience not identified - however, the Project Team Leader and key inspection team members were inter-viewed by NRC and found to be qualified and sufficiently experienced.	6/82 - 11/82 16,000 person-hours total effort

Project and Utility	Evaluation Scope

Marble Hill, Public Service of Indiana	o QA organization & management policies
	o Construction design control & implementation
	o Physical verification of plant hardware
	  – Reactor coolant
	  – Auxiliary feedwater
	  – Component cooling
	  – RHR
	  – Fuel handling & auxiliary building
	  – Ultimate heat sink
	o Testing & inspection of ASME piping welds
	o Construction document review

Midland, Consumers Power Company	o Design verification & construction verification o Auxiliary feedwater, standby electric power, control room HVAC systems examined

South Texas Project, Houston Lighting and Power	o Design & construction evaluation
	  – Component Cooling Water System
	  – Used INPO methodology

Palo Verde, Arizona Public Service	o Project management organization o Management's policies towards QA o QA activities o Design verification o Construction verification

Project and Utility	Physical Verification Statistics
Marble Hill, Public Service of Indiana	° 21,000+ documents reviewed ° 13,000+ checks performed ° 56 welds visually inspected ° 49 weld radiographs reviewed ° 11 welders & welding inspector qualifications reviewed ° 67 hangers - installation features inspected ° 70 valves inspected ° 34 structural members inspected (beams, columns, guides, bracings, etc.) ° 34 areas of rebar inspected for proper location ° 50 areas of concrete tested for strength ° 22 hangers - detail verification ° 16 pieces of equipment inspected ° 25 cable tray hangers inspected ° 1800 feet of piping runs inspected
Midland, Consumers Power Company	° 50% of work scope conducted at time of preparation of this report; therefore, physical verification statistics not available
South Texas Project, Houston Lighting and Power	° 165 welds visually inspected ° 25 radiographs reviewed ° 15 welder qualifications reviewed ° 3000 feet of piping runs inspected ° 850 feet of cable trays inspected ° 140 pipe supports and cable tray hangers inspected ° 160 valves inspected ° 45 pumps inspected
Palo Verde, Arizona Public Service	° 15,000+ documents reviewed ° 15,000+ checks performed ° 55 welds visually inspected ° 48 welder or inspector qualifications reviewed ° 900 feet of piping runs inspected ° 68 hangers inspected ° 7 pieces of equipment inspected ° 50 feet of cable tray inspected ° 132 valves inspected ° 15 instrument wiring terminations inspected ° 55 instrument sensing elements, indicators and transmitters inspected

Project and Utility	Deficiencies Identified By Consultant
Marble Hill, Public Service of Indiana	° 19 Potential Deficiencies - 2 Valid - 8 Invalid - 9 Observations
Midland, Consumers Power Company	° 50% of work scope conducted at time of this report. Number of deficiencies identified to date is 10.
South Texas Project, Houston Lighting and Power	° 43 Potential Deficiencies - 13 safety-related - 30 nonsafety-related
Palo Verde, Arizona Public Service	° 89 Potential Deficiencies - 17 Valid - 31 Invalid - 41 Observations

Note: The four independent audits differed in scope, depth, and number of manhours (range of 4,000 to 20,000). Moreover, the evaluation criteria and the definitions of deficiencies varied from audit to audit. The reader should be aware of these nonuniformities in audits in evaluating the statistics on this page. The proposed third-party audit program would establish uniform audit criteria that would reduce the variations among audits and permit a more valid comparison among projects.

Project and Utility		Evaluation Program		
		Strengths		Comprehensiveness
Marble Hill, Public Service of Indiana	°	Methodology, amount of hardware inspected and detail of inspections were judged to be satisfactory (absence of significant electrical construction review consistent with project status), and representative of a comprehensive third-party construction verification effort.	° °	Limited, but beneficial design review effort. However, the coverage afforded was not comparable to other programs evaluated under the pilot program. Management assessment was limited.
Midland, Consumers Power Company	°	Program plan, methodology, extent of design review, amount of hardware inspected, use of checklists, use of periodic program plan QA audits and critiques by senior level management were judged to be satisfactory. A final assessment of the evaluation's adequacy will be conducted when the evaluation program is completed.	°	Quality assurance and project management could have been reviewed in greater detail.

Evaluation Program continued on next page.

Project and Utility	Evaluation Program	
	Strengths	Comprehensiveness
South Texas Project, Houston Lighting and Power	° Preparation, competence of evaluators and inspection techniques were judged to be satisfactory.	° In construction evaluation a limited number of radiographs were reviewed. ° Limited level of effort devoted to design evaluation. ° Limited coverage of design controls and their implementation by the NSSS vendor. ° Review was limited to programmatic controls.
Palo Verde, Arizona Public Service	° Methodology, competence of evaluators, conduct of review under a QA program, which included periodic audits and reviews by a senior technical review committee and use of checklists for design review and physical verifications, were judged as satisfactory.	° The following areas would be expanded in the contemplated independent audit program: cross-section of welder qualifications, sample of weld radiographs, HVAC contractor's QA program, fire protection design, and broader look at critical equipment supplied by the NSSS vendor.

5.0 AUDITS BY ASSOCIATIONS OF PROFESSIONALS

Section 13(b) of the Ford Amendment directs the NRC to analyze the following alternative approach to improving quality assurance and quality control in the construction of commercial nuclear power plants:

Alternative b(3)

Evaluations, inspections or audits of commercial nuclear power plant construction by organizations comprised of professionals having expertise in appropriate fields which evaluations, inspections, or audits are more effective than those under current practice.

The major associations of professionals currently conducting evaluations, inspections or audits of commercial nuclear power plants are the Institute of Nuclear Power Operations (INPO), the American Society of Mechanical Engineers (ASME), and the National Board of Boiler & Pressure Vessel Inspectors (NB). The analysis of alternative b(3) included an evaluation of the audits conducted by these organizations.

Many U.S. associations of professionals also participate in developing national consensus standards for different aspects of quality assurance. Applicable national standards are endorsed by the NRC and represent the core of many inspections and audits. However, no changes to this process are contemplated, and these standard-making activities are not covered in the analysis of alternative b(3) because they do not constitute audits, inspections or evaluations.

The evaluation, inspection, and audit activities of the three organizations identified above supplement NRC inspection activities and provide detection and assurance capability beyond that provided by NRC's inspection program. For example, in the early phases of construction at Marble Hill, the NB confirmed ASME code compliance problems with piping installation and brought this quality problem to NRC's attention. At Zimmer, the ASME identified and brought to NRC's attention problems in the quality of safety-related piping welds.

During the past two years, INPO has tested and implemented an extensive evaluation program of plants under construction. Because of NRC's familiarity with the long-established ASME and NB programs, the relative newness of the INPO program, and the broader spectrum of construction activities examined by the INPO program, field work to support the analysis of Congressional Alternative b(3) concentrated on the INPO evaluation activity. The analysis of all three organizations sought to determine how these efforts can best be used to enhance the overall level of assurance provided the public. Some consideration was given to whether any of these programs could act as a surrogate for the NRC program, rather than as a complement to the program, but this was a secondary consideration. Section 5.1 presents the conclusions and recommendations resulting from this analysis, and Section 5.2 describes the separate analyses.

5.1 CONCLUSIONS AND RECOMMENDATIONS

In this section, the conclusions and recommendations of an analysis of ASME's and NB's audits and inspections are discussed first, followed by a more detailed discussion of the analysis of INPO's Construction Project Evaluation program.

5.1.1 ASME/NB Audits and Inspections

The ASME and NB audit and inspection programs cover a limited number of areas in more depth than the routine NRC inspection program, thereby providing a valuable supplement to the NRC inspection program. The ASME and NB audit and inspection programs have a proven record of providing detection and assurance capability beyond that provided by the routine NRC program. The NRC should continue to use this narrower but deeper oversight capability in the limited areas in which they work, thus permitting better focus of NRC resources in other areas.

The NRC, ASME and NB should continue earlier efforts to coordinate selected inspection activity to avoid unnecessary duplication. However, the ASME and NB effort provides a valuable additional independent measure of assurance beyond the NRC inspection program, and any coordination initiatives should not compromise the independence of the ASME and NB nuclear inspection program.

5.1.2 INPO Construction Project Evaluation Program

The new INPO Construction Project Evaluation (CPE) program fits the alternative b(3) criteria of "evaluations...by organizations comprised of professionals having expertise in appropriate fields which evaluations... are more effective than those under current practice." INPO implemented its CPE program after enactment of Public Law 97-415, and this program represents a significant enhancement of efforts by the nuclear industry to improve quality assurance and quality control in design and construction. The CPE program is consistent with INPO's stated mission of promoting the highest levels of safety and reliability and encouraging excellence in all phases of construction, design control, and operation.

Consideration was given to suggesting alterations in the CPE program to make it more like NRC construction audits and thereby to allow the INPO program to directly substitute for portions of NRC's inspection program. However, this idea was rejected on the basis that INPO's current mission of improving industry performance and raising the industry's standards better serves the public interest. The NRC can and does set minimum standards that meet the requirements of law, but a regulatory agency is not equipped to adopt the counseling and advisory role required to move industry practice above those minimums. INPO was established for just such an advisory and counseling role. The study concluded that any attempt to use INPO as a surrogate for NRC construction inspections would limit the ability of INPO's CPE program to provide candid assessments to licensees and would damage this industry-initiated mechanism for improving overall performance of the nuclear industry for establishing industry-wide standards of excellence.

Some consideration was also given to INPO's ability to qualify as an independent auditor for performance of independent audits similar to those tested in the pilot projects. The apparent weakness of this proposal--INPO's "independence" from the licensee--becomes INPO's strength in the counseling and advisory role.

The study concluded that public health and safety interests seem best served presently by INPO continuing in its role of "inside" independent auditor for the nuclear utilities--which is useful and necessary in assuring excellence and upgrading of industry's programs for achieving and assuring safety and quality. INPO is seen as a very important contributor to this result, rather than as a substitute for NRC regulation and inspection of the utilities' safety and QA programs and results thereof. However, NRC's and INPO's respective roles, which presently are fixed and separate, are not immutable and over time they may change.

This study has confirmed a widely held impression that INPO is developing into an effective industry instrument with significant potential for raising the quality of design and construction of nuclear power plants. Because INPO's potential is not yet fully realized, the NRC should remain alert to future changes in INPO's program that would justify NRC's placing greater reliance on it and which would lessen the combined impact on the industry of NRC Construction Appraisal Team (CAT) inspections, INPO CPE evaluations, and the proposed program of periodic third-party audits. Such action is not without precedent. Past successes in the INPO program for operating reactors have allowed NRC to reduce some inspection activity because industry improvements attributable to INPO resulted in a less intensive inspection presence needed by the NRC. Improved industry performance resulting from INPO activities at operating reactors led to a reduction in NRC Performance Appraisal Team (PAT) inspections from 14 to 4 per year.

5.2 TECHNICAL APPROACH IN EVALUATING ORGANIZATIONS OF PROFESSIONALS

Letters were sent to 15 organizations having various nuclear-related interests to draw their attention to the NRC study required by the Ford Amendment. Each letter provided a copy of the Federal Register Notice requesting public comments and information about the alternative programs in the NRC study. The letter requested their review and comments on methods to improve quality in the construction of nuclear power plants. Among those organizations receiving letters were the ASME, the Institute of Electrical and Electronics Engineers (IEEE), the National Board, the American Welding Society (AWS), INPO, and the American Society for Quality Control (ASQC).

The programs of the ASME, NB, and INPO were selected for evaluation because they were in place and either currently do supplement or have the potential to supplement the NRC inspection program. The IEEE, which was suggested for consideration as a possible candidate professional organization for conducting audits when the Ford Amendment was debated in Congress, recommended instead that ASME and INPO perform the evaluations by organizations of professionals. The IEEE stated that alternative b(3) was already in effect:

> The evaluations performed by INPO and the ASME 'N' Stamp Program
> in addition to independent verifications for near-term license
> plants have been quite effective in identifying and correcting
> areas requiring attention. There is evidence in the reports
> generated by each of these that the programs provide an adequate
> and effective means of monitoring and evaluating licensee's quality

assurance program in addition to the Commission's evaluations.
We recommend the use of these programs to satisfy this alternative.

This section describes NRC's process of evaluating the potential of each of
these three organizations of professionals (1) for supplmenting NRC's
inspection program for nuclear power plant construction, and/or (2) for acting
as a third party, and (3) for performing comprehensive construction audits
similar to those recommended for the future in Chapters 2 and 4.

5.2.1 ASME/NB

ASME's and the NB's current audit and inspection programs provide valuable
supplements to NRC's inspection program. In areas such as ASME code work and
pressure vessel and primary coolant boundary welding, these programs inspect in
more depth than the NRC inspection program, except for CAT or other special
inspections. However, the ASME/NB programs are narrower in focus than the
overall NRC inspection program and do not cover many of the areas covered by
the NRC. Because there is some overlap between the ASME/NB, and NRC inspection
programs, each can use the results of the other's audits and inspections to
check the effectiveness of its own program.

Because of the current narrower focus of the ASME and NB programs, they are not
considered to be viable substitutes for the comprehensive third-party audits
described in Ford Amendment Alternative b(5) and the pilot program analysis in
Chapter 4. The ASME/NB programs would have to be considerably expanded in scope
to reach the level of comprehensiveness of the recommended third-party audit
program. Such expansion is not considered to be as feasible as adoption of
alternative b(5) with private companies performing the audits because of the
start-up time and additional ASME/NB resources that would be required. In
either case, the NRC has no control over the ASME/NB inspection programs. In
contrast, a third-party program such as that recommended from the pilot program
has already been partially implemented (the Independent Design Verification
Program). Moreover, expanding the ASME/NB program rather than implementing
the recommended comprehensive third-party audit program is considered to
have less overall benefit because the total level of detection capability
and assurance provided by an expanded ASME/NB program and the NRC program would
be less than that provided collectively by the present ASME/NB programs, the NRC
program, and the recommended third-party audit program. The NRC has the necessary
authority to require third-party audits.

5.2.2 Institute of Nuclear Power Operations (INPO)

INPO, a utility-sponsored and funded organization, was established in 1979
to promote improved safety and reliability in operating nuclear power plants.
INPO's Institutional Plan (May 1983) states that INPO's mission "is to promote
the highest level of safety and reliability in the operation of electric
generating plants. In carrying out its mission, the Institute strives to
encourage excellence in all phases of construction, design control, and
operation..."

In 1982, INPO developed performance objectives and criteria to evaluate design
control, construction activities and other related areas in the construction of
nuclear plants. INPO initiated and conducted a pilot program consisting of
several evaluations. Following training sessions with utilities on the new

evaluation methodology, about 20 self-initiated evaluations were conducted by utilities to evaluate their construction performance using INPO criteria. Subsequently, in early 1983, INPO began a formal program of INPO construction evaluations. This program was named the Construction Projects Evaluation (CPE) Program, and evaluations of 22 plants in an 18-month period are planned under this program. INPO further established guidelines that plants under construction would be evaluated every 18 months thereafter, except those in the near-term operating license phase. The CPE evaluations are conducted by INPO evaluation teams, which may be supplemented by utility-appointed personnel or by third-party evaluation teams contracted by the utility and monitored by INPO.

The NRC's evaluation of the INPO effort for this Congressional study is based on NRC staff observation and review of the Beaver Valley 2, Limerick and Millstone 3 evaluation efforts. These efforts were conducted in the following time frames:

Beaver Valley 2	-	May 16 through May 27, 1983
Limerick	-	July 11 through July 22, 1983
Millstone 3	-	August 22 through September 2, 1983

This new INPO program and NRC's evaluation of it was in a sense a pilot program as defined in the Ford Amendment. However, the three plants reviewed did not include projects identified as having had major quality-related problems. Therefore, the INPO CPE program is discussed here rather than in the discussion of pilot programs in Chapter 4.

The INPO performance objectives and criteria require review of the following areas: Licensee Organization and Administration, Design Control, Construction Control, Project Support, Training, Quality Programs and Test Control. INPO's design review is essentially an effort to identify in the management control systems deficiencies and weaknesses that could permit design or construction deficiencies to occur. This approach is different from the NRC integrated design inspections (IDI) methodology, which includes detailed examination of equipment and system design, including the checking of design calculations. INPO's position is that programmatic review is superior and more productive than a verification approach, which consists of examining a limited sample of design details.

INPO's construction review emphasizes observation of work "in-process" as well as detailed review of programmatic controls to determine the effectiveness of management control of the construction process. INPO limits its review of actual construction to work in progress during the course of the two weeks the INPO team is on site. There is a limited retrospective look at completed work to assure that installed hardware conforms to design and specifications, which is a characteristic of the new NRC construction and design inspection programs (CAT and IDI). INPO's findings concentrate on ways to improve the construction process and are not, in many cases, directly applicable to assessing that completed work conforms to NRC requirements. Therefore, NRC's ability to rely on these evaluations in support of the licensing process is limited.

The INPO teams used for an evaluation usually consist of a team leader plus 4 or 5 evaluators for the design review at the A/E's office and a team leader and

10 to 12 evaluators at the site for the construction evaluation. INPO prepares detailed work schedules for each evaluator so that each of the INPO performance objectives and criteria are reviewed. The licensee provides any pre-licensing documentation needed. Approximately two weeks after the evaluation is completed, an exit meeting is held with the utility to discuss in detail the evaluation team's findings and to permit utility management to respond to those findings. The utility further responds in writing to each finding and prepares a corrective action plan that is reviewed by INPO. INPO then prepares a final report and sends it to the utility. INPO encourages licensees to make the report available to the public, but the member utility may withhold the report from the NRC and the public. To date, the NRC has received a copy of all final reports that have been prepared.

To be an acceptable alternative to the third-party audits recommended under alternative b(5), INPO's CPE methodology would have to be modified and expanded. The current program focuses on identifying deficiencies and weaknesses in the management control system. While management control is a key factor in the design and construction of nuclear power plants, an acceptable comprehensive audit must also examine the end product in depth to be assured that it meets the design intent and is of acceptable quality. The design review program would need to be more comprehensive and include checks of the calculation of selected design features. Where there are subcontractors to the A/E, the INPO evaluation would also need to review their activities. In the construction area, the programmatic and "in-process" observations would have to be supplemented by an increased retrospective detailed examination of representative plant hardware. For example, various sample sizes of welds, radiographs, structural steel, concrete, pipe runs, hangers, mechanical equipment, cables, terminations, cable trays, tray supports and other representative hardware would have to selected and inspected. The final INPO report would have to be comprehensive enough to include not only the current information provided, but the amount and condition of hardware and equipment inspected and the detailed findings. The reports would also have to be made available to the public, without exception.

This analysis has been presumptive in that it hypothesized that INPO's Board of Directors may find it in their organization's best interests to act as a third-party auditor, part of whose mission is to confirm compliance with NRC regulations. Such action was not envisioned by INPO's founders, nor does it necessarily seem to be in the public interest to have INPO act as such a third party or as a substitute for NRC. This study concludes that there is great value in having a separate industry-sponsored body that performs, in effect, management reviews and project diagnoses for the nuclear industry and then provides advice and support in a cooperative atmosphere for improvement. Assumption of a quasi-regulatory role would significantly hamper self-improvement activities. The great value of INPO is its acceptance by utilities as a peer that they believe is there to help. The study concludes that NRC should not attempt to burden INPO at this time with roles that are inconsistent with this very valuable aspect of its mission.

A thoughtful analysis of the relationship between NRC and INPO was offered by Robert V. Laney, a member of the special review group established to provide advice to the study staff on this project. Excerpts from his comments on the analyses leading to this report appear below. The full text of Mr. Laney's comments may be found in Section 10.4.

Fostering an effective relationship between the NRC and INPO, one which allows each to do that which it can do best, should continue to be a constant goal of both organizations. This consideration is most compelling during a period of changing roles and expanding activities, such as that described in the NRC study. It is desirable for the NRC to allow ample scope to the industry's move to improve construction quality represented by INPO's Construction Project Evaluations (CPE).

INPO is the central feature of industry's determined commitment to self-improvement and self-regulation. Simultaneously, INPO is the industry's chosen instrument for achieving rising standards of performance in all phases of nuclear power, including, most recently, design and construction. Thus it is particularly important that, when setting a new agenda for strengthening the quality of nuclear construction, all concerned should recognize that INPO is similarly engaged. In deciding what inspections, audits, or evaluations it will do, the NRC should encourage INPO to do those which INPO might do as well or better. If this requires modifying the scope or methods INPO now uses, as the CPE's, NRC should discuss this possibility with INPO, as an alternative to continuing both CAT's and CPE's.

The present study includes...excellent descriptions and discussions of the respective NRC and INPO roles in achieving construction quality. The study concludes that the present role differentiation should continue, with INPO in a "counseling and advisory role" and the NRC in its statutory role of setting standards and inspecting to assure that those standards are met. This may be the appropriate conclusion at the present time. However, in my opinion, this section of the report would be improved if it were amplified to recognize that there are circumstances which, in the future, might argue for adjusting the NRC/INPO interface and their respective inspection activites.

...INPO is exploring ways by which it might exert pressure on member utilities to respond constructively to correct faults revealed by INPO's evaluations. In addition, INPO appears to be moving towards a performance "ranking" system which will provide a utility management with a specific measure of relative success in achieving rising standards. These and related INPO initiatives, as they mature, will benefit from NRC recognition and a willingness to consider role adjustment as appropriate."

> RECOMMENDATION. This report is the appropriate place for the NRC to acknowledge that (1) INPO is developing into an effective industry instrument for raising the quality of operations and construction, and (2) since INPO's potential is not yet fully realized, the NRC should remain alert to future improvement in INPO's program which would justify the NRC's placing greater reliance on it.

The study concurs in this recommendation and carries it forward to the study findings, conclusions, and recommendations appearing in Chapter 2.

6.0 MORE PRESCRIPTIVE ARCHITECTURAL AND ENGINEERING CRITERIA

Section 13(b)(1) of the Ford Amendment directs the NRC to analyze the following alternative approach to improving quality assurance and quality control in the construction of commercial nuclear power plants:

Alternative (b)(1)

Providing a basis for quality assurance and quality control, inspection, and enforcement actions through the adoption of an approach which is more prescriptive than that currently in practice for defining principal architectural and engineering criteria for the construction of commercial nuclear power plants.

The discussion of the amendment contained in the Congressional Record indicates that some of the amendment's sponsors had in mind an approach similar to NRC's technical specifications for operating plants. The NRC provides guidance for developing technical specifications as part of the process leading up to issuing an operating license; the applicant/licensee develops them for the specific plant; and the NRC reviews and, subject to further review and revision, approves them. Any licensee desiring to continue operating under a condition that does not comply with its technical specifications must receive prior NRC approval. For the construction process, the NRC does not have similar requirements for controlling licensee performance. In particular, design and construction commitments made in the Preliminary Safety Analysis Report (PSAR) are not equivalent to technical specifications in terms of requiring licensee compliance with them.

Under the current regulatory process, the applicant for a construction permit (CP) generally submits required information in the application and makes whatever commitments are necessary to have the CP application accepted and issued. The design description contained in the application (which includes a PSAR) includes the principal architectural and engineering (A&E) criteria. Although not defined in the regulations, these A&E criteria may be thought of as the performance specifications for the safety systems and major components, and for commitments to consensus codes and standards, NRC branch technical positions, and NRC regulatory guides. The applicant then commits to implementing the design and to constructing the plant as described in the application. Under current regulatory procedures, the CP holder can unilaterally modify those portions of the PSAR that are not explicitly stated to be conditions of the CP without notifying the NRC. All changes to the PSAR must be included in the Final Safety Analysis Report (FSAR), which must be submitted as part of the process of applying for an operating license, but the FSAR is submitted after much of the plant construction has already been completed. Under current practice, detailed information and prescriptive commitments, in general, and A&E criteria, in particular, usually are not conditions of the CP.

6.1 SUMMARY, CONCLUSIONS, AND FINDINGS

The NRC's case study analyses of quality assurance and quality control programs at selected sites having had satisfactory programs and at sites that have not did not show a direct connection between prescriptive criteria (A&E or other-wise) and the achieved level of quality in the design and construction of nuclear power plants. Rather, the analyses have shown that changes in the design basis or the design, for whatever reason, increase the likelihood of reduced quality in construction. Several NRC initiatives (see Section 6.2.2) are under way to systematically reduce opportunities for either the NRC or the licensee to change a design, once completed and approved. These initiatives require a stringent review of the impact of required design changes and encourage a much greater degree of design completion at the time of CP application.

This study concluded that requiring a substantially completed design, including design changes made because of initial procurement activity, before construction begins would significantly enhance the achievement of quality during construction. Such a requirement would significantly reduce the amount of change associated with completing the design after construction has begun, thus enhancing design/construction interfaces, reducing rework, improving the basis for planning and scheduling, and generally making it much easier for the project to cope with and manage change. However, more prescriptive criteria, short of a requirement for a completed design before construction, would have proportionally less effect on controlling the level of change and hence on improving the environment for achieving quality.

Current practice does not provide a strong basis for NRC inspection of PSAR commitments or any resulting enforcement activities during the construction phase. A much improved basis for NRC inspection activity in this regard can be established by adopting an approach that makes a licensee's significant commitments in its PSAR conditions of the CP. This study recommends that staff review practice be changed to provide that, during NRC's review of the licensee's quality assurance program, the licensee's commitments to certain codes, standards and regulatory guide positions in the PSAR would be reviewed for inspectability and enforceability. Selected commitments would be designated as mandatory and made conditions of the CP. The designated commitments would then be binding and readily inspectable and enforceable. It should be noted that any changes to such commitments would require a license amendment and a concomitant notice procedure under Section 189a of the Atomic Energy Act. This would result in a reopened CP hearing under Section 189a if one were requested by an interested party.

This study also recommends that the NRC further evaluate the impact of changes in general (regulatory, technical, procedural, etc.) on the NRC, industry and project management structure to develop further guidelines for controlling unnecessary changes and for better managing necessary changes. The study also recommends that NRC should further examine the feasibility and benefits of requiring a substantially completed design at the time of CP application. See Chapters 3 and 9 for more discussion of the enhancement to quality available with an advanced design early in the construction process.

6.2 ANALYSIS

The issue of more prescriptive A&E criteria has been approached from two aspects: (1) should the requirements to which licensees are committed during design and construction be more prescriptive? and (2) should the NRC be more prescriptive in its procedures dealing with changes to those commitments?

This section discusses earlier attempts to define "principal A&E criteria", current initiatives concerning prescriptiveness, the relationship of A&E criteria to major quality-related problems, and the industry's management of change. Also discussed are NRC's inspectability and enforceability of changes to design criteria and other licensee commitments, and the amount of prescriptiveness that is appropriate in regulations.

6.2.1 Earlier Attempts to Define "Principal A&E Criteria"

For some time the NRC has been aware of the need for better controls on the licensee's type of design commitment and the extent of changes to design commitments and of NRC's changes to the design basis. The first attempt at improving the situation took place in 1969 as part of an effort to stabilize the licensing process. As part of the proposed rule on backfitting, 10 CFR 50.109, the staff included a more prescriptive definition of principal A&E criteria. However, when the final rule was issued, the more prescriptive definition was not included because the Commission decided that the definition needed further study. As a result of this determination, two studies were conducted to define principal A&E criteria. The results of the first report were published in December 1975 and the results of the second in March 1977. No formal staff action on these studies was taken because of difficulties with implementing the recommended definition and because of other priorities.

While the earlier action did deal with more prescriptive A&E criteria, a December 1979 action addressed the issue of control of design changes. As part of the Commission's decision on the need for a hearing and/or a CP amendment on the Bailey Nuclear Station short pile issue (SECY-A-79-24 and 24A), the staff was requested to prepare a proposal on precisely what design and other changes a CP holder could make without (a) notifying the NRC, (b) securing prior NRC staff approval, and/or (c) obtaining a CP amendment. In response to the Commission's request, the staff developed Commission Paper SECY-80-90, which detailed the historical background (the 1969 proposed rule, the 1975 and the 1977 studies) and proposed five alternatives for addressing the problem:

(1) Maintain the status quo.

(2) Borrowing from 10 CFR Part 50.55(e) (dealing with notifications of significant deficiencies having safety significance) and 50.59 (dealing with changes to previously approved designs having safety significance), adopt a rule that establishes general criteria for determining circumstances requiring a CP amendment.

(3) Adopt a rule defining "principal architectural and engineering criteria" (in effect reviving the 1969 rulemaking on this subject) using information learned to date, including the 1975 and 1977 staff studies.

(4) Adopt a rule stipulating that all details of the application, including the PSAR, be made conditions of the CP and may not be changed without prior NRC approval.

(5) Restructure the licensing process to require that complete plant design details be provided in the PSAR (i.e., essentially a final design), which,

upon review and approval, would be made conditions of the CP and could not be changed without prior NRC approval.

The staff then presented the five alternatives to the Commission for publication for public comment as an Advance Notice of Proposed Rulemaking. In approving the publication of the Advance Notice of Proposed Rulemaking, the Commission added the following statement: "The Commission tentatively prefers Alternative 3 now, with a shift to Alternative 5 in three years."

6.2.2 Current Initiatives

The specific rulemaking described above has been subsumed into a series of new initiatives. The initiatives include, in order of occurrence, establishing the Committee to Review Generic Requirements, submitting legislation on one-step licensing, issuing a proposed policy statement on severe accidents, which includes standardization of design, and issuing an Advance Notice of Proposed Rulemaking on the backfitting of new requirements to operating plants and plants under construction. All requirements proposed by the staff for imposition on one or more classes of power reactors is reviewed by the Committee to Review Generic Requirements, which compares the improvement in operational safety to the cost of the change and recommends their approval or disapproval to the Executive Director for Operations.

Both the legislation on one-step licensing submitted to Congress and the policy statement on standardization contained in the proposed severe accident policy statement would require that a much more complete design be submitted for approval at the CP application stage. However, both would stop short of requiring that the design be complete to the point that it incorporate changes made to the initial design as a result of feedback from the procurement process. (To accommodate available equipment that may not satisfy initial design assumptions and to provide an acceptable level of safety, the design may have to be changed.) The most prescriptive A&E criterion, of course, would be requiring a complete design including the characteristics of specific components to be submitted as part of the CP application. The design approval granted under the one-step licensing proposal would be for 10 years, and the design could not be changed in that time frame by either the licensee or the NRC without going through the hearing process again. The Advance Notice of Proposed Rulemaking on backfitting would require the NRC staff to justify any change in requirements they wish to impose on operating plants. The incremental improvement to operating safety would have to be weighed against the cost of the change in terms of dollars and exposure.

6.2.3 Relationship of A&E Criteria to Major Quality-Related Problems

Previous NRC efforts in the area of more prescriptive A&E criteria have been directed towards stabilizing the licensing process rather than towards improving the basis for quality control, quality assurance, inspection and enforcement actions. While more prescriptive A&E criteria may be the answer to the licensing issue, this study did not show them to be an answer to quality problems. NRC's case studies and regional inspections have shown that the welding and masonry construction problems at Zimmer, the soil compaction problems at Midland, and the voids in the concrete at Marble Hill were not related to either the prescriptiveness or the enforceability of the principal A&E criteria. In these three cases, the problems were caused by inadequate management of the construction process to assure that the design criteria were met. See Chapter 3 and Appendix A for more discussion of this point.

South Texas had both design and construction problems. The design problems resulted from failure to effectively manage the design process and to keep the design sufficiently ahead of construction to avoid redesign and rework due to physical interferences. The problem was not so much that the design did not meet the NRC's criteria, but that design changes coupled with an improperly managed design/construction interface made construction problems almost a certainty.

The errors identified in the seismic analysis at Diablo Canyon are generally considered to be an example of design errors. However, those errors occurred mainly in areas that had to be redesigned after a previously unknown geologic fault was discovered. The errors occurred because of needed design changes, coupled with deficiencies in management oversight of the design process, rather than from a lack of prescriptiveness in A&E criteria.

6.2.4 Management of Change

As the South Texas and Diablo Canyon cases illustrate, the difficulties inherent in managing complex projects are exacerbated by having to deal also with a rapidly changing project environment. Besides design changes, management of a nuclear power plant construction project must also cope with technical, regulatory, and procedural changes. The following excerpt from a letter written by a member of the study's special review group, Dr. George Coulbourn, expresses the author's viewpoint on the analyses leading to this report (see Section 10.4 for the text of entire letter):

> There is a level of change action (technical, regulatory, and procedural) beyond which any program management structure can no longer prosecute its program. Utility management has consistently been faulted for quality assurance breakdowns. In some instances, the charge is well founded. However, in most instances, I believe the root cause is found in the circumstances which produced rampant, uncontrolled change. I submit that most of the utility management structures assembled to build the nuclear power plants of the past decade could have performed adequately in a more stable design and construction environment.

While not endorsing Dr. Coulbourn's position in total, the study has concluded that historically neither the industry nor the Atomic Energy Commission (AEC)/NRC have done a good job in managing change. The nuclear industry grew rapidly and was subject to rapid changes in technology and sizing of reactors. Also having to make these changes were several established industries comfortable with their routine methods of operation and not always amenable to the changes to their way of doing business required by the new and developing technology. These industries include the utilities, A/E firms, construction firms, and power plant equipment suppliers (see Section 9.2 for more discussion of this point).

The quality problems of several licensees can be directly attributed to their inability to foresee and adapt to changes to their traditional methods of power plant construction and project management required for nuclear construction (see Chapter 3 for more discussion of this point). The AEC's understanding of safety issues grew along with the industry, and new requirements were provided at an increasing rate as the nuclear power "state of the art" developed, increasing the difficulty for either licensees or regulators to stay current. During study interviews, licensee management and staff most frequently mentioned stabilizing the process that had produced the constantly changing (and increasing) level of requirements as being an area in which NRC programs and policies could be improved. Issuance of new requirements reached a peak after the 1979 accident at Three Mile Island. To control the flow of require- ments and to examine them for benefits and feasibility, the Commission established the Committee to Review Generic Requirements, discussed earlier. That committee is generally credited with providing a rigorous analysis of new requirements over the past two years and with bringing a much greater degree of stability to the regulatory process. In so doing, it has reduced the impact of change on both the industry and regulators, making it somewhat easier for both to manage and to cope with the level of remaining change.

Dr. Coulbourn's thoughts on the management of change conclude with the following recommendation, which the study endorses:

> Accordingly, I recommend that NRC commission an examination of the change management process itself, both within NRC and in the other parts of the industry. This examination should focus on both the management of change as a discipline (elsewhere called configuration management) and upon the reduction of the volume of change. The latter can have numerous constituents; for example, higher percent design completion prior to start of construction, more restraint regarding in-process change, standardization, etc. All of these constituents require disciplined and consistent management.

6.2.5 Inspectability and Enforceability

When considering the use of criteria that are more prescriptive to improve inspection and enforcement, it becomes apparent that existing procedures for handling changes to design criteria and other licensee commitments do not provide a strong basis for inspections and enforcement against PSAR commitments during the construction phase. The NRC's enforcement policy contained in 10 CFR 2, Appendix C -"General Policy and Procedures for NRC enforcement actions," paragraph IV.E(3) states:

> Notices of Deviation are written notices describing a licensee's or vendor's failure to satisfy a commitment. The commitment involved has not been made a legally binding requirement. The notice of deviation requests the licensee or vendor to provide a written explanation or statement describing corrective steps taken (or planned), the results achieved, and the date when corrective action will be completed.

Because the licensee can unilaterally modify the PSAR commitments that are not conditions of the CP and therefore not legally binding, a licensee's answer to a Notice of Deviation may be nothing but a change in the commitment. Changes in commitments should be based on factors other than a desire to legitimize nonconforming work. The basis for NRC inspection and resulting enforcement action can be improved by adopting an approach that makes significant licensee PSAR commitments conditions of the CP.

6.2.6 Prescriptiveness

One of the difficulties faced by regulators is determining the amount of prescriptiveness appropriate in regulations. In recent years, the NRC has favored performance-oriented regulations that state the level of performance to be achieved but not the way to achieve that level. To provide supplementary guidance, NRC regulatory guides are also issued to describe acceptable ways the performance objective may be met, but those guides do not require any one method to be used. Maintaining this separation between saying what must be achieved and how it is to be done is fundamental to preserving the existing statutory roles of NRC and the industry. The industry is primarily responsible for the safety of nuclear power (e.g., reactor operators are employees of a utility, not the NRC), and the NRC is responsible for regulating the use of nuclear power in a manner consistent with maintaining public health and safety.

In this context, the NRC has two strong reasons to be careful about being more prescriptive in its regulations for design criteria. First, there is usually more than one satisfactory way to perform most design activities, and prescription would unreasonably limit the designer's choices. Second, too much prescription tends to put the NRC into an industry management role, where it does not belong, and tends to shift some of the licensee's responsibility for safety to the NRC.

These arguments against prescriptiveness apply to areas other than design also, e.g., quality assurance. Although the study concluded that the NRC could provide better guidance to licensees on acceptable ways to meet NRC QA requirements, the study did not find that the QA requirements should be made more prescriptive. The study recommends the opposite: rather than more prescriptive requirements that say "how" something should be done, the study concluded that present QA programs should be reoriented to meet performance objectives based on Appendix B, which say what is to be achieved but do not specify how it should be done. See Section 2.3.1.

6.2.7 Summary

The study has concluded that increased quality in the construction of nuclear power plants will result from a more careful coordination of changes in design criteria and design during construction rather than from more prescriptive criteria. Several initiatives are under way to systematically reduce opportunities for either the NRC or the licensee to change a design, once completed and approved.

The study has also concluded that the basis for inspection and enforcement during construction would be improved by including certain licensee commitments contained in the PSAR as conditions of the CP and that staff review practice should be revised to provide such conditioning. Such procedures should only apply to new CP applicants. The study concluded that this condition does not need to be applied to plants currently under construction because they will have passed the point where changes to principal A&E criteria are likely to occur by the time implementing regulations could be made effective.

The study also concluded that the NRC should examine the change management process itself, both within the NRC and the nuclear industry, to evaluate the impact of changes on the collective NRC-industry regulatory and project management structure. The goal of this examination would be to develop further guidelines for controlling excessive change and for better management of necessary change. The aerospace industry's apparently successful approach to configuration management should be a principal focus of study in this area (see Chapter 9 and Appendix D). Moreover, the study concluded that NRC should further analyze the feasibility and benefits resulting from requiring that plant design of future CP applicants be substantially complete before construction activities begin.

7.0 REVIEW OF NRC'S PROGRAM FOR ASSURANCE OF QUALITY

Section 13(b) of the Ford Amendment directs the NRC to analyze the following alternative approach to improving quality assurance (QA) and quality control (QC) in the construction of commercial nuclear power plants:

Alternative b(4)

Improvement of the Commission's organization, methods, and programs for quality assurance development, review, and inspection.

This chapter presents the analysis and findings of the study for this alternative. In Sections 13(b)(1), (b)(2), (b)(3), and (b)(5) of the Ford Amendment, Congress was specific in identifying the alternative concepts for NRC to analyze. Specific improvements to NRC's organization, methods, and programs were not specified in 13(b)(4), although several improvements are suggested by the other alternatives and by the debate during Congress' consideration of the Ford Amendment. However, a review of the legislative history of the Ford Amendment did not indicate that the sponsors had any specific NRC program improvements in mind other than those already described in Section 13(b).

Because there is no specific direction of possible improvements to pursue in analyzing this alternative and because events have shown that NRC's approach to the assurance of quality in the design and construction of nuclear power plants needs improvement, this study interpreted alternative b(4) as a broad mandate to determine shortcomings in NRC's approach to QA and to recommend improvements. While the charter of alternative b(4) was interpreted as being limited to assurance of quality in design and construction, some of the results have implications for more than just the NRC's QA program. In devising a study approach to address alternative b(4), the NRC used the following question introduced in Chapter 1 as a study focus:

What changes should be made to the current policies, practices, and procedures governing commercial nuclear power plant design, construction, and regulation to prevent major quality problems in the future or to provide more timely detection and correction of problems that have occurred?

This question directly addresses the issues of prevention and detection and, as a corollary, assurance. These objectives of the NRC QA program were introduced in Section 2.2.

7.1 TECHNICAL APPROACH

To determine how to prevent major quality-related problems in the future and to provide more timely detection and correction of developing problems, the study first tried to determine why these problems occurred and why they were not discovered and corrected earlier. A series of case studies, which are described in Chapter 3 and Appendix A, was the primary means for answering "why." (See in particular Sections 3.2, 3.3 and 3.4.)

This analysis (see Chapters 2 and 3) showed that in prevention the NRC's underlying shortcoming was in its pre- and post-construction permit (CP) licensing reviews and inspections. The NRC had not performed searching analyses of (1) the applicant's capability to manage or provide effective management oversight over a nuclear construction project, or (2) whether project team members have the requisite nuclear construction experience to properly execute their various project roles. Several improvements to the NRC program were identified to address this prevention problem: enhancing pre-CP review by NRC staff; establishing a special advisory committee to help screen new applicants; conditioning the CP on a licensee's satisfactory post-CP demonstration that it can effectively manage all quality-related aspects of the project; and directing more NRC attention in general to the issues of management capability and prior applicable experience of members of the project team and their project staffs. These improvements are addressed in more detail in Chapters 2 and 4 and in the remainder of this chapter.

The NRC was also slow to detect and/or take strong action for significant quality problems that developed at each of the five projects cited as experiencing major quality problems. Reasons for this slowness included the following: (1) sporadic, NRC inspection presence at construction sites (before the NRC resident inspector program was implemented), (2) inability of the NRC inspection program to coalesce scattered quality program-related inspection findings coming in over a period of time into a comprehensive picture of a project-wide breakdown, (3) a prioritization of limited NRC inspection resources to address operations first, construction second, and design last, which resulted in an almost total neglect of design and the design process, (4) setting the threshold for reacting to construction-related problems higher than for operational problems because of the lack of an immediate threat to health and safety, because of an attitude that construction problems would be found during an intensive period of startup testing before an operating license was issued, and because of an attitude that required the demonstration of a project-wide breakdown before enforcement action would be taken for construction quality problems, (5) an orientation of the inspection program to focus heavily on programmatic matters and paperwork at the expense of examining actual work in progress and program implementation, and (6) the NRC's reluctance to address the issue of capability of utility management until problems grew so large that a remedial program became necessary.

Several improvements to the NRC program were identified to address these detection problems: expanding the resident inspector program; increasing team inspections; training inspectors and supervisors to better relate individual inspection findings to programmatic weaknesses; increasing inspection attention to construction and design; reorienting the inspection program to emphasize paper less and hardware quality more; and increasing inspection attention to management issues. These improvements are discussed in more detail in Chapter 2 and later in this chapter.

Although the case studies were useful in identifying why the prevention and detection problems occurred and in suggesting possible fixes, the overall study plan called for a broader analysis by an outside organization of the NRC's organization, methods, and programs for QA. This outside analysis was purposely lagged behind the first several case studies so that information from the case studies would be available as input to the outside analysis. The desirability of such an analysis was emphasized by the comments of the individual review group members at the June 1983 review group meeting. The next section discusses the results of that analysis.

The firm selected to perform the management analysis of the NRC's QA program was N. C. Kist and Associates, a management consulting firm experienced in performing QA audits and program reviews for industry but which had not done work for the NRC prior to the Ford Amendment Study. Senior members of Kist Associates participated as team members in each case study. This experience enhanced their understanding of the problem under study and helped them to focus on weaknesses in NRC's approach to QA. Although the NRC staff provided logistical support to Kist in their analysis of NRC's QA activities and participated in some of the interviews, the Kist Report is entirely the product of N. C. Kist and Associates. The Kist Report further confirms and supports many of this study's findings and identifies several areas for improvements not identified in the case studies or other project activities. The major recommendations of the Kist Report are summarized in the next section, along with planned NRC actions or responses. The Kist Report is included in its entirety as Appendix B to this report. The findings upon which the Kist Report recommendations are based are found on pages 5-11 of Appendix B.

7.2 ABSTRACT OF APPENDIX B, THE KIST REPORT

Appendix B reports the results of Kist's review of the NRC's QA organization, methods, policies and programs. Kist's management analysis of NRC's QA program was based on (1) review of literature pertaining to past and present Atomic Energy Commission (AEC)/NRC programs for assurance of quality in design and construction of commercial nuclear power plants, including previous studies of those programs, (2) participation in the NRC case studies, and (3) interviews with the staff of the Office of Inspection and Enforcement (IE) in Bethesda, Maryland; Region II offices in Atlanta, Georgia; Region III offices in Glen Ellyn, Illinois; Region IV offices in Arlington, Texas; and Region V offices in Walnut Creek, California. The management analysis was limited to NRC programs for assurance of quality in design and construction of commercial nuclear power plants and did not include other NRC programs. The analysis included the perceptions of licensees, contractors, and NRC inspection staff and management regarding problems with the NRC and QA program. It also included suggestions for improvements obtained during the NRC case studies described in Chapter 3 and Appendix A.

Based on this review, several items were identified as candidate areas for revision, deletion and/or development to improve the NRC's policies and programs for the assurance of quality in the design and construction of nuclear power plants. These areas are summarized in the following section.

7.2.1 Recommendations of the Kist Report for Improvements in NRC's Organization, Methods, and Programs for Quality Assurance Development, Review, and Inspection

N. C. Kist and Associates' analysis of (1) NRC's implementation of management programs and practices for QA, past and present, and (2) the root causes of the NRC's inability to prevent problems and slowness to identify and act on problems resulted in the following Kist recommendations:

(1) Stabilize the regulatory process through more preventive action and planning.

(2) Streamline regulations and guidance documents and make them more prescriptive and definitive in terms of required elements of control without specifying how the elements of control must be implemented. Regulations that can stand on their own would eliminate the need for many guidance documents. Clearly define the applicability of quality program requirements, safety-related items and items important to safety.

(3) Make the quality assurance program and licensee commitments a condition of authorizations and permits.

(4) Replace the licensing review of the quality assurance program described in the Preliminary Safety Analysis Report (PSAR) with a licensing or IE review of the licensee's quality assurance manual and require the manual to detail how the quality assurance program shall be implemented. Require licensing or IE approval of quality assurance manual changes. Establish definitive acceptance criteria for manual reviews, specifying required elements of control but not methods for accomplishing them. Do not permit work to be performed until the quality assurance manual is approved.

(5) Evaluate licensee and contractor experience, attitude and management capability before authorizations and permits are issued. Establish parameters of acceptance criteria.

(6) Require demonstration of the licensee's capability to implement the quality assurance program before authorizations or permits are issued.

(7) Devote greater attention to design activities.

(8) Develop programs based upon what must be done and then obtain the necessary resources to implement the programs.

(9) Establish mandatory requirements in inspection programs and reduce dependency upon individual engineering judgment.

(10) Require an Inspection Plan of licensees and contractors and establish NRC hold points.

(11) Re-evaluate NRC personnel practices, including salaries.

(12) Change regulations to permit industry organizations rather than individual licensees to evaluate vendors and monitor their activities or establish licensing or certification programs for vendors. Extend the program to include material manufacturers and suppliers.

(13) Take stronger enforcement action. Require expeditious handling of corrective action, including determining the magnitude of problems and correcting their root causes.

(14) Perform detailed annual audits of the licensee's implementation of the quality assurance program.

(15) Review functions to be performed by the Quality Assurance Branch and Construction Programs Branches of IE to assure that efforts are not duplicated.

(16) Eliminate differences in basic regional office structures and job titles to assure uniformity of functional responsibilities.

(17) Increase the training of inspectors in quality assurance, auditing, and implementation of inspection modules. Broaden the inspectors' capabilities to encompass all disciplines or provide additional support.

(18) Establish an audit program of NRC activities, using qualified personnel not having responsibility in the areas audited.

(19) Establish a quality assurance program within the NRC.

These areas for improvement of NRC's QA policies and programs were extracted from pages 11 to 13 of Appendix B. The findings that form the bases for these recommendations are discussed in detail in Appendix B and are summarized on pages 5 to 11. The findings cover the following areas: organization; management practices; the QA standards program; the QA licensing program; the QA inspection program; the licensee, contractor and vendor inspection program; the QA enforcement program; and NRC's inability to prevent problems and slowness to identify and act on problems.

Many of Kist's recommendations are consistent with results from the NRC case study reviews (Chapter 3 and Appendix A) and the review of the quality and quality assurance programs of other government agencies and industries (Chapter 9 and Appendix D). For example, recommendations 1, 5, 6, 7, 13 and 14 corroborate case study findings and have been carried forward into Chapter 2 as major recommendations of the report. Recommendations 1, 2, 3, 4, 10, 12, and 17 are consistent with results of the study of outside programs (Chapter 9), and further action and/or analysis is planned in each area.

Recent NRC actions also address several of Kist's findings. For example, as discussed earlier, the Committee to Review Generic Requirements was established in 1981 to, among other purposes, stabilize the flow of new and/or revised NRC regulatory requirements and to ensure that the impact and resultant benefits of regulatory changes are fully assessed (recommendation 1). Also, in recent years, the NRC enforcement program has been bolstered by Congressional legislation that permits stronger enforcement and penalties for licensees' failure to comply with NRC requirements (recommendation 13). Another example of recent improvements is two new training courses developed in 1983 in the area of QA for operations, construction, and modification (recommendation 17).

Not all of Kist's findings were considered of sufficient importance to be carried forward into Chapter 2. In some cases, the recommendations and their feasibility need to be further evaluated. Each of the above findings will be evaluated and pursued, collectively, with the findings of other QA study reviews (the pilot program, the case studies, analyses of Alternatives b(1) - b(5) and review of outside programs), to identify the most effective areas for improving NRC's policies and programs for assurance of quality.

Section 7.3 identifies actions that the study recommends to improve NRC's programs for assurance of quality and Sections 7.4 and 7.5 identify additional improvements to NRC's QA policies and programs that have recently been implemented or are under development, respectively. Several of the actions discussed in those sections address Kist's recommendations; those that are not addressed will be analyzed by the NRC staff and may result in subsequent action.

7.3 ACTIONS RECOMMENDED TO IMPROVE NRC PROGRAMS

This section discusses two groups of actions recommended to improve NRC programs. The first group discusses the recommendations resulting from the NRC case studies, the review of NRC QA policies and programs, and a review of outside programs. The second group discusses additional areas identified in the study and needing further consideration.

7.3.1 Recommendations of NRC Case Studies, Review of NRC QA Policies and Programs, and Review of Outside Programs

The findings from the NRC case studies (Chapter 3), review of NRC QA policies and programs (the Kist Report), and the review of outside programs (Chapter 9) form the basis for the following recommended changes to NRC's program for the assurance of quality. Recommended changes (1) to (6) address the prevention issue, changes (7) to (9) address the detection issue, and change (10) addresses the assurance issue (see Chapter 2 for a discussion of prevention, detection, and assurance). Because much of the rest of this report addresses improvements to NRC's program, this section will reference other parts of the report in which certain improvements are more fully discussed.

(1) Enhanced Pre-Construction Permit Reviews

The study recommends that NRC improve its pre-CP review of an applicant's capability for managing or overseeing the management of a commercial nuclear reactor construction project. In particular, future NRC reviews of CP applicants should focus much more heavily on the project team's prior nuclear construction experience and on management capability. The pre-CP review should also cover planning, design, design control and planned construction control processes. This recommendation is described in more detail in Chapter 4 (Pilot Programs) and in Section 2.4.1.

(2) Post-CP Demonstrations of Ability to Manage an Effective Program.

As a condition of their CP, new applicants should be required to successfully demonstrate their ability to manage the implementation of an effective quality assurance and quality control program. This capability should be demonstrated and verified in the first periodic independent audit, approximately 12 to 20 months after the CP is issued. This recommended action is also described in more detail in Chapter 4 (Pilot Programs) and in Section 2.4.1.

(3) Performance Objectives for QA Programs

NRC currently establishes prescriptive review requirements for a "QA program" in Chapter 17 of the Standard Review Plan (SRP). Once NRC has approved a licensee's QA program description of how 10 CFR 50 Appendix B will be met, the licensee develops a set of detailed implementation procedures that the licensee's employees use in performing their jobs.

A licensee is inspected against the requirements of Appendix B to 10 CFR 50 and against the commitments made by that licensee in its approved QA program description. The QA program must address each of the elements described in the SRP. If licensees elect to describe a QA program that has elements going beyond the SRP requirements, the NRC regards those additional elements as commitments that are also subject to enforcement. Because of this, licensees have tended to maintain their QA programs at a level designed to satisfy NRC requirements only, i.e., the minimum required to protect public health and safety. It is inevitable that human endeavor will sometimes fall short of targeted performance. If the target is NRC's requirements, licensees will inevitably fail to meet these requirements on occasion. NRC's current QA licensing practices can thus be counter-productive to 100% attainment of NRC objectives.

The NRC should consider revising current practices by developing a set of inspectable performance objectives and criteria that would meet NRC's requirements for a QA program. These inspectable performance objectives would describe what NRC wants the licensee's QA activities to actually accomplish. The licensee would then develop detailed procedures designed to meet or exceed NRC's performance objectives. NRC's intermediate step of reviewing and accepting an applicant's QA program description would therefore be eliminated. The performance objectives would replace the current Chapter 17 of the SRP. A licensee could elect to establish procedures that exceed NRC's performance objectives. However, inspection and enforcement of a licensee's actual performance would be against NRC's performance criteria rather than the procedures, which could exceed NRC's performance objectives.

If the NRC evaluates a licensee's actions against a nationally uniform set of inspectable performance criteria rather than against the licensee's commitments (which are different for each licensee and sometimes for each plant), there is a greater likelihood that licensees will set their targets (i.e., the detailed procedures) higher than NRC's minimums. There would then be a greater likelihood of licensees consistently exceeding NRC's minimums, even when their actual performance sometimes falls short of their targets. This practice would also indicate to licensees that the NRC is more concerned with what a QA program accomplishes rather than with how it is described, as some believe.

A reform of NRC's current practice for quality assurance becomes even more important if current legislative initiatives are enacted to revise the licensing process by limiting the operating license hearing essentially to operator qualifications and quality assurance matters. The effectiveness of the licensee's quality assurance activities will be vitally important to that kind of process. This recommendation is also discussed in Section 2.4.1.

(4) Management Appraisals by NRC

The study recommends that NRC address the issue of management competence
more directly. The NRC should incorporate management lessons learned from
the case studies, remedial program experience and other sources into the
NRC inspection program to improve NRC's capability to assess the capability and effectiveness of utility and project management. In particular,
NRC should (1) develop an inspection module to evaluate the capability,
effectiveness, understanding and qualifications of utility management, and
(2) implement this management inspection approach by applying it to plants
currently under construction.

This recommendation would address a shortcoming in the NRC inspection
program. Although this study and years of NRC inspection experience
suggest that a primary cause of problems in construction and operation
is shortcomings in some utility management, the NRC inspection programs'
focus on compliance with requirements addresses the management issue, at
best, indirectly and generally after the fact. Developing an inspection
approach that looks primarily at the sources of problems rather than the
effects should lead to earlier detection and possibly prevention. This
recommendation is discussed in more detail in Section 2.4.1.

(5) Retrospective Look/Inspection Prioritization of Plants Currently Under
Construction

Besides applying management lessons, the NRC should apply the Ford study
lessons to analyze plants currently under construction to improve NRC's
and licensee's diagnostic capability and to better prioritize the NRC
inspection effort. In particular, the NRC should examine the current
population of plants under construction to determine which seem to most
exhibit the characteristics of plants that had major quality problems
in the past and use this information to help prioritize its inspection
program for those plants. Although at the beginning this prioritization
would be based upon Ford study lessons, it should be sharpened over time
by feedback from the inspection program and the development of a trend
analysis capability (discussed below). This recommendation is discussed
in Section 2.4.1.

(6) Perform Trend Analysis of Construction Indicators

The NRC has been slow to detect major quality breakdowns in the past. One
cause of this slowness has been its inability to synthesize scattered bits
of information into a comprehensive picture of the health of a construction project. To synthesize information and to develop a closer picture
of management effectiveness, the NRC should develop a set of construction
performance indicators that could be monitored, trended, and evaluated by
the licensee and the NRC. Such indicators should be oriented toward
measuring the effectiveness of activities that contribute to, control, or
verify construction quality.

Efforts in this area are presently under way (1) to analyze inspection
program data, including manhours per site per activity vs. inspection
findings, and (2) to develop a computerized NRC capability to analyze
licensee construction events and vendor events reported to the NRC under

10 CFR 50.55(e) and 10 CFR 21, respectively. This recommended action would combine these efforts with analyses of other indicators, some arising from the case studies and some yet to be determined, into a comprehensive NRC management information capability.

Besides using the system for observing trends, NRC inspection groups will be able to use data in the system as followup for determining whether plants acceptably resolve outstanding reports and whether deficiencies reported by one plant may potentially apply to other plants. The quality of licensee management of safety deficiency reporting in design and construction may be used as one measure of its commitment to quality and the effectiveness of its QA program.

Some NRC resources need to be redirected to this area, including training, to ensure close attention to detecting problems. QA problems at any site should be clearly and accurately identified, including root causes, and that information should be provided to all sites immediately. Competent and prompt followup to ensure that proper actions are taken is mandatory. Knowledge of the problems by NRC managers is vital and should be stressed. Success of this program will be enhanced by selecting results-oriented NRC managers to lead this activity. See Section 2.4.1 for more discussion of this recommendation.

(7) Independent Audits

Periodic independent audits should be required of all commercial nuclear power plants under construction. This requirement should be imposed on both all current construction permittees and all future applicants by conditioning the CP on the applicant's agreement to employ periodic independent audits. See Chapter 4 and Section 2.4.2 for a complete discussion of the third-party audit recommendations.

(8) Regional Inspections

The regional inspection program should be supplemented with additional contractor support for its regular inspection program. Such support would allow more NRC staff time for reactive inspections such as allegation followup, remedial program inspections, and special regional construction team inspections. Increased use of regional team inspections is being tested in one NRC regional office. Pending results of this trial program, the NRC inspection program in all regions may be reoriented to greater emphasize team inspections. This recommendation is discussed in more detail in Section 2.4.2.

(9) Resident Inspectors

The study found that for new applicants or for the restart of construction at projects presently delayed, resident inspectors should be assigned to the site as early as possible, preferably before the CP is issued and before safety-related construction activities are started. This study finding will be considered for NRC's future policy on placing residents at construction sites. The NRC is also in the process of establishing a pilot program in one of its regional offices. That program would place more resident inspectors at plant sites and correspondingly reduce the size of the regional inspection staff. Pending the outcome of this trial program, the NRC inspection program may be reoriented to more heavily emphasize resident inspectors. See Section 2.4.2 for more discussion of this recommendation.

(10) Audits of Implementation of NRC Programs

The NRC should strengthen its programs for conducting audits of NRC Program and Regional Offices to assure that NRC programs are being implemented consistently, adequately, and uniformly. Besides providing information to NRC management on the status of that implementation, the audits could be an evaluation tool for feedback on appropriate areas for program revision and as an aid for prioritizing programs. NRC should also arrange for periodic independent management audits of the NRC program relating to QA. See Section 2.4.3 for a discussion of this recommendation.

7.3.2 Additional Areas Requiring Further Evaluation

In the review of NRC programs, some additional areas were identified which the NRC should further consider and evaluate as potential methods for improving NRC's program for the assurance of quality in the design and construction of nuclear power plants. These areas include the following:

(1) Inspection Planning

Better methods of planning quality assurance inspections should be pursued to plan and use the limited inspection resources in these most important areas. Possible methods include applicability of probabilistic risk analysis and qualitative and deterministic risk assessments and development of an overall "inspection plan" that would bring coherence to NRC headquarter's inspections, regional inspections, resident inspections, independent audits and the licensee's regular inspection program.

(2) Readiness Reviews

The NRC should consider requiring formal "readiness reviews" during nuclear power plant construction. Plant designers, construction managers, owner/operators, and possibly the NRC could participate in the reviews, which would be required at key points in the project, beginning with "design ready for construction". The reviews' purpose would be to ensure the coordination of all parties involved and the readiness of the project team to proceed with each new construction phase. This recommendation is also discussed in Chapter 9 and Section 2.4.5.

(3) Training

The NRC should consider additional training for the NRC staff in quality assurance, auditing, conduct of inspections, and analysis of inspection findings to determine programmatic weaknesses. These training programs would help the staff to implement the inspection program more effectively and to develop the ability to detect more readily causes of problems that go beyond surface symptoms. This recommendation is also discussed in Chapter 9, Appendixes B and D, and Section 2.4.1 (item 6).

(4) Control Over Vendors

The NRC holds the licensee responsible for all aspects of the nuclear
power plant, including all parts and equipment furnished from vendors and
suppliers. The NRC's current vendor program and near-term focus are
discussed in Chapter 2. The longer-term implications of the changing
supplier-vendor-contractor-utility infrastructure is changing with unknown
implications for the future. The NRC should be aware of these changes and
their implications so that it can take prudent action to prevent future
problems rather than react to them. Assurance of the quality of vendor
and supplier activities could be improved by the NRC's stricter enforcement against deficiencies in the licensee's required vendor control and
inspection programs and by more NRC inspection of the licensee's control
of vendors and suppliers. The NRC should explore different institutional
arrangements for oversight of component suppliers, such as changing
regulations to permit industry organizations to be responsible for evaluating component suppliers (see the Kist Report). The NRC should support
continued development of a data bank on performance of and problems with
vendor-supplied components, as suggested by the Battelle report on outside
QA programs (Chapter 9 and Appendix D).

(5) Design Completion

NRC should consider requiring that plant designs be well advanced before
construction activities begin. Besides permitting better construction
planning and scheduling, the more completed design should result in fewer
design changes and better design interfaces. See Chapter 6 and Sections
2.2 and 2.4.5 for more discussion of this recommendation.

7.4 RECENT IMPROVEMENTS TO NRC'S QA PROGRAM

After a series of quality-related problems were identified in the design or
construction of several nuclear power plants, the NRC staff initiated a series
of QA improvements to the NRC QA program designed to improve the assurance of
quality in the design and construction of nuclear power plants. The following
paragraphs discuss recent improvements to NRC's QA programs stemming from these
initiatives as well as some improvements that were already in place, such as
the resident inspector program and the Systematic Assessment of Licensee
Performance (SALP) Program. As noted earlier, many of these improvements
specifically address some of Kist's findings.

(1) Resident Inspector Program

In the 1960s and early to mid-1970s, the reactor inspection program was
carried out by inspectors assigned to NRC Regional Offices. In 1974, a
two-year trial resident inspection program was initiated to test the
concept of placing NRC inspectors at a nuclear power plant site. The
program's purpose was to derive benefits accruing from increased onsite
inspection time, to improve NRC's awareness of site activities and status,
and to increase inspector efficiency. The program demonstrated that the
resident inspector concept was viable, and in 1977 the NRC adopted the
program as a central feature of the inspection program. At first,
resident inspectors were placed at operating reactors, and in 1979 they
began to be stationed at nuclear power plants under construction.

The resident inspector program currently includes one inspector for each reactor site at which plant construction is more than 15% complete and one for each operating reactor. The resident inspector performs a significant part of the total inspection effort. As a "generalist" (as opposed to a "specialist"), the resident inspector monitors day-to-day activities and performs the parts of the inspection program in which he is knowledgeable. Specialists from the regional office conduct inspections in specific technical areas to complement the resident inspector's activities.

This study found the resident inspector program to be the backbone of the current NRC inspection program. The resident's constant presence at a site enables him to more comprehensively understand the project's health and status and better enables NRC to analyze individual inspection findings to determine if they represent only isolated deficiencies, a programmatic problem, or a quality assurance breakdown.

The resident program is one aspect of NRC's approach to improving its detection (and prevention) capabilities. The study recommends that for future CP applicants, experienced NRC residents should be assigned to the site before the CP is issued, as soon as preliminary site work begins. The resident inspector program and recommendations above are discussed further in Section 2.4.2.

(2) Construction Appraisal Teams

In 1980, on a trial basis the NRC initiated Construction Appraisal Team (CAT) inspections to provide in-depth inspections of the quality of the implementation of management and quality controls at a nuclear construction project. In a CAT inspection, a multi-disciplinary team of specialists assess program implementation by examining safety-related hardware after it is installed and after the licensee's QA/QC inspection is completed. The principal objective of the CAT program is to evaluate the effectiveness of design controls, construction practices, and other management controls used to ensure that as-built conditions are according to the plant's design.

During 1980-1981, eight trial CAT inspections were performed by 5-man teams from Regional Offices. Each inspection included about 2 weeks of onsite inspection time. In 1982-1983, the CAT program was revised and CAT inspections are now performed by NRC headquarters using teams of NRC personnel and consultants. A team generally consists of a team leader and 10 engineers and spends approximately 4 weeks at the site. Each inspection entails approximately 1,600 to 2,000 manhours of direct inspection time onsite. In 1982-83 NRC performed about 4 CAT inspections per year.

The CAT inspection program is another aspect of NRC's effort to improve its detection capabilities and to address the "threshold" problem for taking action for quality problems in construction. The headquarters-based CAT inspection partially, but not completely, addresses Kist recommendations 18 and 19, serving as both an audit of the performance of the licensee inspected and as an overcheck of the implementation of the NRC resident and regional-based inspection program. The CAT program is further discussed in Section 2.4.2.

(3) Systematic Assessment of Licensee Performance

Following the Three Mile Island accident, the NRC initiated a program for the Systematic Assessment of Licensee Performance (SALP). The SALP program consists of periodic reviews of regulatory performance of nuclear power plants (both under construction and in operation) by a team of inspectors, licensing staff and regional supervisors and management. The SALP assessment is intended to be sufficiently diagnostic to provide a rational basis for assessing licensee performance, for allocating NRC inspection resources, and for providing meaningful guidance to licensee management. The SALP assessment is based on a review of inspection data, licensing staff input, licensee performance in areas such as deficiency reports (Licensee Event Reports and reports submitted pursuant to 10 CFR 21 and 10 CFR 50.55e reporting requirements), and licensee responsiveness to Inspection and Enforcement Bulletins and other suggestions for improvement. Each of nine or ten functional areas is evaluated and is assigned to one of three categories to indicate whether more, less, or about the same level of NRC inspection attention and licensee attention is appropriate for the coming period. The SALP program represents an effort by the NRC inspection program to better address management capability and competence. The SALP program is also discussed in Section 2.4.1.

(4) Integrated Design Inspection (IDI)

NRC has recently developed a special design inspection program to assess the quality of design activities. The design area has received little inspection attention in the past, and recent experience has suggested that it should receive greater attention. This design inspection program also uses the team approach and encompasses the total design process on a selected system, from formulating design and A&E criteria through developing and translating the design to actually performing site construction. While the NRC staff evaluates a great deal of basic design information in the licensing reviews, it has not previously verified that this basic information has been properly incorporated in the actual design drawings. This new design inspection program examines the adequacy and consistency of the integration of all the design details within a selected sample area. The focus of the inspection is on the completed drawings and includes such things as independent calculations to verify piping and tank sizes, seismic support strengths and failure modes. Where errors are found in designs, the design process is examined to determine if there are generic problems. It is believed that conclusions about the adequacy of the overall design process can be drawn from this very detailed audit of a selected sample. Each IDI requires about twelve persons and four months to complete. Current plans are to conduct three IDIs per year.

The IDI program is the main NRC initiative aimed at addressing the problem of insufficient past NRC inspection attention to design. The IDI program is another aspect of NRC's effort to improve its detection capability. The IDI program is also discussed in Section 2.4.2.

(5) Revised Construction Inspection Program

The construction inspection program was recently revised for two reasons:
(1) a recognition that inspection requirements exceeded inspection
resources; and (2) programmatic review was being emphasized at the expense
of observing work and inspecting hardware. In 1982 the NRC staff began
revising the individual inspection procedures in the construction inspection program to better match the budgeted resources. The main goals of
the revision program, which is to be an ongoing program of review with
the first cycle of review to be completed in the spring of 1984, are as
follows: (1) to shift emphasis of inspection from reviewing records to
observing work; (2) to facilitate performance of certain procedures by
resident inspectors; (3) to re-examine the scope and frequency of some
inspections based on limitations of inspector resources; and (4) to
eliminate redundancies in the procedures. With current plans, the first
review cycle will consolidate 115 inspection procedures to 61 procedures.
The revised inspection program is also discussed in Section 2.4.2.

(6) Quality Assurance Staff Consolidation

In the fall of 1982, the quality assurance responsibility and functions of
the NRC Office of Research were assigned to the Office of Inspection and
Enforcement (IE). These responsibilities included regulatory development,
standards development, liaison with code and standards making organizations, and research. In January of 1983, the quality assurance licensing
functions for power reactors were also assigned from NRC's Office of
Nuclear Reactor Regulation to IE. These re-assignments of personnel and
functions are intended to consolidate responsibility for all NRC quality
assurance matters in one NRC line office. Consolidating NRC QA functions
and responsibilities has been a long-standing issue within the AEC and the
NRC. Programmatic weaknesses in the AEC's QA program resulting from
diffusion of QA responsibilities among several AEC program offices was
first identified as an issue in a 1973 assessment of QA regulatory programs.*

(7) Independent Design Verification Program (IDVP)

On a case-by-case basis, the NRC staff has requested that an applicant for
an operating license provide additional assurance that the design process
used in constructing the plant has fully complied with NRC regulations and
licensing commitments. Many licensees have responded by initiating a
design review through an independent third-party contractor. This review
program has been termed the Independent Design Verification Program
(IDVP). The independent review evaluates the quality of design based on a
detailed examination of a small sample. The independent review has also
addressed programmatic areas, for example, classification of systems and
components, design and verification records, interface control and interdisciplinary review, consistency with the Final Safety Analysis Report
(FSAR), nonconformances and corrective actions, and audit findings and

---

*Davis, J. G. and H. H. Brown. 1973. "Quality Assurance and the Utilities:
  Is Regulatory Doing Enough?" Prepared for the Director of Regulation.

resolutions. The review includes verifying specific design features by independent calculations and by comparing installations against as-built drawings. The NRC staff reviews the selection of the independent review organization and the audit plan before they are implemented, reviews the completed report, and assesses the applicant's response to the audit findings. In all cases to date, the NRC staff has concluded that the applicant has complied with NRC regulations and licensing commitments.

Some licensees have expanded their IDVP to cover construction quality as well as design, and these are referred to as Independent Design and Construction Verification Programs (IDCVP). THE IDCVP conducted at Palo Verde and the one in process at Midland were selected for special review by the NRC staff in conjunction with the Ford Amendment Pilot Program (see Chapter 4). The scope of the IDVPs (IDCVPs) has varied from plant to plant. THE IDCVP at Palo Verde was of greater scope than the average and involved about 120 manmonths of review.

The third parties selected to perform the IDVPs or IDCVPs must meet strict NRC-established criteria to ensure they are independent of the licensee. In particular, the organization selected and each individual participating in the review must not have had any responsibility for or involvement in the project's design or construction, and safeguards are established around the review of draft inspection reports. Plants that have received an IDI or that are replicates of plants that have already been subjected to an independent design review have generally been able to provide sufficient assurance that the design process has complied with NRC requirements without performing a second design review.

The usefulness of these audits has varied from site to site because of the variability among each audit's scope and methodology. With the transfer of IDVP responsibility to the same NRC program office (IE) responsible for the IDI program, future IDVPs will be patterned more like IDIs and the variability should decrease.

This study concluded that a series of comprehensive third-party audits, using a clearly established set of audit criteria, will better enable the NRC to meet its responsibilities than the current IDVP practice. Until this regulation has been established, however, the NRC should continue to encourage licensees to perform voluntary independent design reviews. This recommendation is discussed in more detail in Chapter 4 and Sections 2.2 and 2.4.2. The IDVP program is also discussed in Sections 2.4.2 and 2.4.3.

(8) Quality Assurance Surveys on Computer Code Development and Use

Since 1978, the NRC has been developing and implementing a program to assure that vendors, national laboratories and utilities that develop or use thermal-hydraulic computer codes apply quality assurance programs that provide traceability and independent review of calculations used for the design of plant systems.

The licensing staff, with the assistance of Region IV, has conducted inspections at vendor facilities, national laboratories, and selected utilities. These inspections have not revealed any major deficiencies in the quality of the work performed with various codes. However, QA practices applied in developing and using codes varied significantly among national laboratories, while the practices of vendors and utilities were consistent with staff and industry guidelines. As a result of work done to date, the staff is in the process of proposing a uniform QA program for the national laboratories and will continue the inspection of vendors and utilities with an expanded scope that will include other types of codes (e.g., seismic, radiological).

7.5 PROGRAMS UNDER DEVELOPMENT

The previous section identified NRC initiatives that the staff has implemented as methods to improve NRC's assurance of the quality in the design and construction of nuclear power plants. The initiatives presented in this section are additional efforts that the staff has under preparation, in varying stages of development and implementation. These efforts are in addition to the areas identified in Section 7.3.2.

(1) Regional Administrator's Evaluation

To provide additional confidence in the quality of design and construction to the regions, the NRC staff has taken steps to improve its guidance in the NRC program of pre-operating license review. In this program the NRC Regional Administrator comprehensively evaluates the licensee's performance and plant construction status shortly before an operating license is issued. Based on inspection and enforcement history and other licensee performance information, the new evaluation guidance helps identify areas requiring additional inspections. A report of this evaluation is forwarded from the cognizant Regional Administrator to the Director of the Office of Nuclear Reactor Regulation (NRR) to provide information relevant to NRR's considerations in plant licensing. This procedure is currently being revised to incorporate the results of the periodic SALP evaluations.

(2) Qualification and Certification of QA/QC Personnel

Inadequate qualifications of some personnel working in quality assurance areas have been noted as a contributing factor to quality-related problems in NRC investigations or inspections of quality problems at Marble Hill, South Texas, Zimmer, and Midland. To better understand and characterize the significance of this issue, the NRC is conducting a study to determine the extent and magnitude of the problem, the underlying causes for it, and the extent and quality of existing standards for QA/QC personnel qualifications to develop recommended actions for NRC program improvement. The staff also has efforts under way to direct more NRC attention to enforcing the existing standards for qualifications of quality assurance personnel, to work with the industry in developing improved qualification standards, and to further consider the benefits and feasibility of requiring formal qualification and certification of QA/QC personnel.

(3) Craftsmanship and the Importance of Feeling Personally Responsible for Quality

The NRC recognizes the important role that craftsmanship plays in putting quality into a product. Improving craftsmanship in nuclear construction is a high priority. The study concluded that improving management will improve craftsmanship more than any other single factor. The University of Texas study of craft productivity in power plant construction cited in Chapter 3 strongly supports this conclusion.

Clearly, ultimate responsibility for performing high-quality work rests with the actual doer. However, management must provide the directions and supporting conditions that allow and encourage the individual to attain quality. The individual must feel personally responsible for attaining quality. If management does not carry out its responsibilities such as, for construction, giving a qualified craftsman a complete and accurate set of drawings, the proper tools and materials, valid acceptance criteria and confidence that enough time is available to do the job correctly, the craftsman is unlikely to feel the degree of personal responsibility that has the greatest probability of yielding quality work. The primary role of the quality control inspector then shifts from providing assurance that the work has been done properly to screening out improperly performed work. While it has been established that many nuclear power plant construction projects suffer from poor craftsmanship, this report concludes that improving management in nuclear construction is a necessary precursor to significantly improving the job done by the craftsman.

The importance of feeling responsible for quality extends from the craftsman upward to all levels of management, including first-line supervisors. First-line and higher supervisors should be held accountable for the quality of work under their direction. These supervisors should be appropriately trained to provide instruction on how to achieve quality work and to recognize project activities or practices that may degrade quality.

The feeling of personal responsibility for the successful outcome of a project, whether it is large or small, applies equally to the NRC. NRC management is also required to establish a framework for its inspectors in which those inspectors feel a sense of personal responsibility for determining the effectiveness of the QA programs of their assigned plants.

During this study, some labor unions involved in nuclear construction were contacted to explore potential methods and incentives to enhance the crafts role in assuring the quality of construction activities. Meetings with union officials and discussion with union training officials highlighted the following points:

(1) Craftsmen are generally not well informed of their role in the QA/QC process.

(2) Continuous rework because of changes has a demoralizing effect on craftsmen and affects the quality of the final work.

(3) Utilities and contractors have not provided adequate training on quality for craftsmen.

(4) Utilities are not convinced that quality assurance is a cost-effective approach to construction. Labor perceived that utilities think QA/QC is a "high-cost" item rather than a "cost-saving" tool.

(5) Improved front-end engineering and procurement would reduce the amount of change and rework.

The staff has used this input from the unions and crafts in preparing the changes to NRC programs discussed in this report. The NRC will further study improving the management of crafts.

(4) Improved NRC Management Reviews

The case studies identified management experience, competence, and commitment to quality as fundamental for assuring an effective quality assurance program on a nuclear project. CPs have been issued to licensees who, in retrospect, experienced difficulty in managing their projects, including the quality program, because of inexperienced personnel in major project organizations and lack of understanding of the complexity of designing, constructing, and licensing a nuclear plant. Moreover, the NRC has been slow to determine the extent and magnitude of the results of inadequate management.

The SALP program discussed above performs periodic appraisals of the quality of licensee and licensee management performance, based on inspection findings and other indicators. CAT inspections and Performance Appraisal Team (PAT) inspections for operating plants also measure management effectiveness. The NRC staff is currently examining how to incorporate lessons learned from the case studies into the inspection program to improve NRC's capability to assess the quality and effectiveness of utility and project management. See Section 2.4.1.

Chapters 2 and 4 discuss some of the improvements being considered (enhanced pre-CP reviews, post-CP demonstrations, and third-party audits), to improve (1) the focus of the NRC review of management capabilities before a CP is issued, (2) confirmation of management capabilities shortly after site construction is begun, and (3) management effectiveness throughout the project.

(5) Prioritization of QA Efforts and Integration of QA

The NRC has three QA research projects planned or under way to address the applicability of QA requirements to various structures, systems, and components in a nuclear power plant. One project is attempting to develop a methodology to prioritize QA coverage commensurate with the relative importance of equipment and components to prevent or mitigate postulated accidents. The second project is a test application to a nuclear power plant of the National Aeronautics and Space Administration's (NASA) approach to analyzing system safety and reliability. The NASA approach requires establishing safety and reliability goals and objectives, analyzing the system's capability to meet those goals and objectives, and developing a quality plan to specify the QA requirements necessary to obtain the safety and reliability goals and objectives. A third project planned for this area is an NRC survey of existing utility practices for applying QA to nonsafety-related items. The goals of the project are (1) to increase NRC staff understanding of current industry practice, (2) to identify strengths of existing programs, and (3) to establish a practical basis for considering any generic actions in this area.

It is hoped that the three projects will help NRC identify the optimum areas for applying QA requirements, the extent to which QA should be applied, and a more quantified basis for applying QA. The end objective is for the nuclear industry to have definitive guidance on practical ways to prioritize QA measures. Prioritization of QA efforts is discussed also in Section 2.4.5.

(6) Designated Representatives

The Federal Aviation Administration (FAA) uses a system of designated representatives (DR) to achieve extensive oversight of the design and manufacture of commercial aircraft. These representatives, who are employees of the manufacturer but are certified by the FAA, perform examinations, inspections, and tests on behalf of the FAA and report results of such activities to the manufacturer and the FAA. The NRC is considering variations of a DR program to increase NRC inspection capabilities. Several legal, technical, and programmatic issues remain to be addressed before NRC decides whether an FAA-like DR program or some variant of it is feasible.

8.0 CONTRACTUAL, ORGANIZATIONAL, AND INSTITUTIONAL ISSUES

In the course of conducting the quality assurance study mandated by the Ford Amendment, it became clear that a study of some of the indirect factors that shape the environment in which utility management must operate during the design and construction of nuclear power plants would be desirable. Such a study would contribute to a better understanding of the management capability issue and would provide a broader base of information from which to develop approaches to improve the achievement of quality. Battelle Memorial Institute's Human Affairs Research Center (HARC) was selected as the lead contractor to conduct this study, and their interim report constitutes Appendix C to this report. This chapter summarizes the study approach of this special review and its preliminary findings and conclusions. Where appropriate, these findings and conclusions have been incorporated into the findings and conclusions of Chapter 2. From this special review, some issues that merit futher study were also identified in Chapter 2.

8.1 ABSTRACT OF APPENDIX C

Appendix C presents preliminary findings, analyses, and conclusions of a study of the contracting and procurement process used in constructing nuclear power plants and selected organizational and institutional issues associated with nuclear construction. The objectives of the study were as follows:

(1) to characterize the aspects of contracts and procurement that appear to affect the quality during construction of a nuclear power plant

(2) to determine the types of contract and procurement provisions and arrangements that could contribute most to enhanced quality

(3) to develop guidelines for construction contracts and procurement that could assist in achieving overall quality objectives

(4) to examine the contributions of selected organizational and institutional arrangements to nuclear construction projects.

To accomplish these objectives, a series of site visits to utilities constructing nuclear power plants, architectural-engineering (A/E) firms, constructors, and subtier contractors was planned and partially implemented. (The study is still in process.) Specific contractual, organizational, and institutional factors were investigated at each site. The findings and conclusions contained in Appendix C and summarized here are based upon four such visits (three to nuclear construction projects and one to an A/E firm). Also, much information used in the analyses was obtained from secondary source materials and from telephone and personal contacts with informed sources, including 16 state Public Utility Commissions (PUCs).

8.2 PRELIMINARY FINDINGS, CONCLUSIONS AND RECOMMENDATIONS

From the Appendix C study by HARC, the following preliminary findings and conclusions were reached:

(1) Previous nuclear experience appears to provide a significant advantage in a nuclear construction effort. Utilities that do not possess such experience internally should consider hiring either a project staff or contractors who can provide such expertise.

(2) A nuclear construction project appears to benefit when its procurement entity is large enough and experienced enough to exert "marketplace presence". A large procurement entity offers the advantages of market familiarity and commercial power (based upon frequency and continuity of purchasing) as well as the expertise needed to secure satisfactory performance on procurements.

(3) Bid evaluation and selection processes should be based upon functional criteria related to the work to be performed.

(4) To achieve quality objectives in contracting and procurement, clearly defined requirements, program implementation and oversight are important.

   ° The level of detail in QA/QC requirements in procurement documents is extremely important.

   ° A contractors' ability to perform to these requirements must be evaluated before issuing a contract.

   ° Followup is essential to evaluate contractors' and subcontractors' performance against these requirements.

(5) Because designs are usually not complete before construction is begun and nuclear construction projects are subject to unanticipated changes due to changes in the state of the art and regulatory requirements, fixed-price contracting for most aspects of nuclear power plant construction projects is not appropriate. Instead, cost-reimbursable contracts with fixed fees are recommended most frequently by those involved in nuclear construction projects, particularly for assuring quality performance. Except in special cases where the work scope can be clearly specified in advance and will not be impacted by change, fixed-price contracting for nuclear construction work tends to be a disincentive to achieving high quality because under a fixed-price contract, the contractor has to pay for rework out of his profits.

(6) Along with the NRC, state PUCs provide a major source of regulatory oversight for nuclear construction projects. Regulatory influence in this case is exercised through the rate base treatment of such projects. Historically, state PUCs do not appear to have been active in disallowing construction costs that may have resulted from lapses in quality assurance or project management. This position results in shifting the risks of quality lapses from the utility to its ratepayers. Recent actions by several PUCs suggest that this position is changing with unknown implications for the course of nuclear projects under construction.

Possible recommendations resulting from these preliminary findings and the Appendix C study by HARC are given below. This study has adopted several of these recommendations and the more important ones appear, in the same or in a similar form, in Chapter 2.

(1) As part of its management review, the NRC should consider requiring applicants for construction permits to explain their proposed contracting methods, their bid evaluation and selection procedures, and their reasons for choosing them.

Given the overwhelming consensus about contractor selection processes and cost-reimbursement contracting, this item clearly seems to warrant NRC attention. The contracts study found that utilities would be well advised to require bidders to demonstrate their approach and commitment to a project, and that NRC should require the same of licensees. This would force the potential licensee to think through the contracting process with all its implications for risk sharing, cost control, and quality performance requirements.

(2) The NRC should examine methods to focus more attention on the way a licensee proposes to ensure that quality work is being performed rather than on the documents that describe general QA and QC programs.

An overemphasis on what is written about quality assurance and quality control appears to contribute little to the actual assurance of quality and may be detrimental. This is particularly true if such an emphasis diverts attention from how the elements of QA and QC programs will be implemented. The issue here is the difference between examining a utility's written QA program description and examining the number and qualifications of the staff it assigns to QA functions. The former audits writing ability; the latter contributes to an assessment of the capacity to carry out a QA objective.

(3) The NRC should examine the implications for its own mission of state PUC scrutiny of and policies toward nuclear construction project costs and management.

State PUCs appear to be taking more action in examining and disallowing what they view as unnecessary and unwarranted expenses. How this new posture affects execution of the NRC's safety mission, PUCs expectations of the NRC, and the assurance of quality in nuclear construction projects is not yet clear. This shift represents what may be a major change in the institutional environment of nuclear power plant construction; thus, the NRC should carefully examine its implications.

(4) Nuclear construction projects appear to benefit significantly when the owners and members of the project team possess strong management capabilities, seasoned by prior nuclear construction experience. The advantages to a project under these circumstances appear great enough to warrant NRC's examination of how such beneficial ownership and management arrangements can be stimulated and fostered.

One suggestion frequently made is to encourage greater consolidation within the nuclear industry (along the lines of the more centralized nuclear industries in foreign countries, for example). However, before any course is adopted, the specific advantages/disadvantages of various ownership and management arrangements for assuring safe and successful nuclear projects need careful study.

9.0 REVIEW OF OTHER EXISTING AND ALTERNATIVE PROGRAMS FOR THE ASSURANCE OF QUALITY

In conducting the quality assurance study mandated by the Ford Amendment to the NRC Authorization Act, it became clear that a review of the programs for assurance of quality of other government agencies, other industries, and foreign countries would provide a broader base of information from which to develop approaches to improving NRC's program for assurance of quality. Pacific Northwest Laboratory (PNL) was selected as the lead contractor for this review. A PNL-prepared report on this review constitutes Appendix D to this report. This chapter summarizes the study approach of the outside program review and its findings and conclusions. Where appropriate, these findings and conclusions have been incorporated into the findings and conclusions of this report in Chapter 2. Some issues that merit further study from this special review are also identified in Chapter 2.

9.1 ABSTRACT OF APPENDIX D

Appendix D reports the results of a study of the assurance of quality programs of five other U.S. government agencies and of NRC counterparts in six foreign countries. Based on features found in these outside programs, several items were identified as deserving of further consideration to potentially enhance the program to assure quality in the design and construction of nuclear power plants in the United States.

An important element in the study of outside QA programs is selecting the industries and programs to be examined. One organizational category of interest is nuclear endeavors that are not under NRC jurisdiction. This category includes the Department of Energy (DOE) and the nuclear programs in foreign countries. A second organizational category is non-nuclear endeavors that involve highly complex technology requiring high-quality standards in design and manufacture and that strive for low failure probability because the consequences of failure may be substantial. This category includes aircraft manufacturing regulated by the Federal Aviation Administration (FAA); non-nuclear shipbuilding under both the U.S. Navy (USN) and the Maritime Administration (MarAd); and spacecraft under the National Aeronautics and Space Administration (NASA).

The DOE, NASA and the USN parts of the shipbuilding industry represent examples in which a government agency is the owner and/or operator of products or facilities generally produced by the private sector under government contract. The FAA and the MarAd are examples of private sector endeavors regulated by a government agency. The foreign nuclear programs reviewed include both government and private ownership and operation of nuclear power plants. The foreign nuclear programs examined were those in Canada, the Federal Republic of Germany, France, Japan, Sweden, and the United Kingdom.

The Appendix D study by PNL was conducted by reviewing published information on each of the programs selected for study and supplementing this review with information obtained from interviews with FAA and DOE representatives. Limited interviews were also conducted with the NASA staff in Washington, D.C. Published information and interviews with those in the private sector organizations corresponding with these government agencies were also used.

The reviews of the foreign nuclear programs were based almost entirely on publicly available information. Subcontractors with experience in the countries of interest conducted these reviews. There were also limited contacts with foreign nationals in developing the necessary information. Studies of the shipbuilding programs in the United States, both USN and commercial, were conducted entirely through reviews of publicly available documents.

The Appendix D study was not intended to, nor did it attempt to, evaluate the effectiveness of the other programs studied. Rather, it focused on identifying features in those programs that had the potential to improve and translate to the NRC program. In general, these were features that program administrators viewed as positive factors in their respective programs.

9.2 RESULTS AND RECOMMENDATIONS

There are several significant differences among the programs investigated in Appendix D:

(1) The nature and extent of the interfaces differ between the government sector and the private sector.

(2) The incentive systems for achieving quality vary.

(3) In some cases, the major thrust for quality needs arises from safety considerations; in others, it arises from a need for reliable performance. However, safety and reliability are frequently closely intermixed.

Each of the programs reviewed in Appendix D operates within its own cultural environment and such differences profoundly affect the resulting program for assuring quality. This is particularly evident in the foreign nuclear programs. In spite of such differences, there are also identifiable areas of commonality. For example, all of the programs studied are quite dynamic. Although each program has experienced its own evolutionary process and some are much older than others, changes aimed at improving the effectiveness of the quality assurance programs are ongoing.

One observation from Appendix D is that the FAA, NASA, USN and MarAd ship-building regulatory programs are directed towards industries that have evolved as specific entities. These industries are, respectively, the aircraft manufacturing industry, the aerospace industry, and the shipbuilding industry. Design and fabrication are normally performed by industrial sectors that have generally evolved in parallel with the corresponding regulatory programs. In contrast, the NRC program is directed towards regulating the "nuclear industry"--a construct that has never evolved as a specific industrial entity in the traditional sense. Nuclear power plants are designed and constructed as an offshoot activity from several traditionally established industries, i.e., the electrical utilities, the architect-engineers (A/Es), the major power plant equipment suppliers, and the construction industry. Each has its own historical methods of doing business. Implementing the NRC program in these industries has required major changes in traditional practices for what might be a limited segment of total activities. Furthermore, NRC's regulations are directly applied only to the utility that chooses to build a nuclear power plant with the stipulation that it will be responsible for all other participants' compliance with NRC's regulations.

One result of the complex institutional arrangement for building nuclear power plants has been that major changes in long-established ways of doing business have been imposed across many business-management interfaces. Pursuing such a complex issue to the point of developing recommendations was beyond the scope of the PNL study; however, PNL reported it as an issue deserving further study.

Although significant differences exist between the NRC's assurance of quality program and the other programs reviewed, some elements of the other programs may be applicable to the NRC program. The major results in Appendix D were derived from studies of the various individual programs. It must be emphasized that the scope of these studies was limited to general concepts. Therefore, these findings should be viewed as features deserving NRC consideration for its assurance of quality program, rather than as features that should be immediately adopted.

In formulating these results, consideration was given to the institutional differences that exist between the NRC and the outside programs reviewed. For example, the relationship between the government and the private sector is regulatory in some cases (FAA, NRC, MarAd) and contractual in others (DOE, NASA, USN). Other intrinsic aspects of the various programs studied include cultural differences, as observed in the foreign nuclear programs, and national commitment to developing the product, as observed in the USN shipbuilding, NASA, and foreign nuclear programs.

Results and recommendations for further study arising from Appendix D are categorized below by design and quality engineering, quality programs, program reviews, vendors, inspection programs and making management more responsible for quality. The NRC agreed with many of these recommendations, and the most important appear, in the same or shortened form, in Chapter 2.

9.2.1 Design and Quality Engineering

The NRC should consider requiring that plant design be well advanced before initiating construction activities. Design requirements should include the completion of safety, reliability, and availability analyses including failure mode and effect analyses, fault tree and hazards analyses, and safety analyses. The analyses should be integrated with QA and should be completed before construction begins. This recommendation is based upon findings from the DOE, NASA, FAA, foreign nuclear, and shipbuilding programs.

9.2.2 Quality Programs

The NRC should consider requiring the establishment of a QA system that prioritizes quality efforts, quality measures and QA coverage commensurate with the relative importance of equipment, components and systems. This importance would be determined by the safety, reliability and availability analyses discussed under "Design and Quality Engineering" above. This recommendation derives from findings of the DOE, NASA, and shipbuilding programs.

9.2.3 Program Reviews

The NRC should consider adopting the following recommendations relating to program reviews:

(1) The NRC program should require "readiness reviews" during nuclear power plant construction. In some industries, readiness reviews are conducted before embarking on a major new phase of a project to ensure that appropriate planning, coordination and necessary previous work has been completed and that the project team is "ready" to proceed to the new phase. These reviews might involve plant designers, construction managers, owner-operators, and (possibly) NRC staff and should be required at key points in the project, beginning with "design ready for construction". Additional reviews at selected key milestone points may be useful. This recommendation is based upon findings from the DOE, NASA, and shipbuilding programs.

(2) The NRC should study ways to better integrate NRC inspection functions with system design reviews, test program reviews, and test program evaluations. This recommendation is based upon findings from the USN, FAA, DOE, and NASA programs.

9.2.4 Vendors

Consideration should be given to enhancing the NRC's vendor inspection program. The licensee should continue to be held fully responsible for vendor-supplied items, with necessary enforcement actions relevant to vendors applied to the licensee. The NRC should continue supporting the development of a data bank on performance of and problems with vendor-supplied components. These data should be analyzed and the results published periodically. This recommendation is based on findings from the FAA, the USN, and the foreign nuclear programs.

9.2.5 Inspection Programs

The NRC should consider adopting the following inspection-related suggestions:

(1) The NRC should expand its inspector training program to increase emphasis on "how to inspect". The training program should concentrate on such areas as conducting inspections and use of time, and should include specific guidance on identifying possible indicators of developing problems. This recommendation is based upon findings from the USN program.

(2) The NRC should consider requiring inspections of nuclear power plants by independent inspecting agencies. This recommendation is based on findings from the foreign nuclear programs.

9.2.6 Making Management More Responsible For Quality

The NRC should re-examine its posture on quality assurance to emphasize to the licensee that quality and assurance of quality are responsibilities of overall management rather than responsibilities that can be delegated to the QA/QC organization. This recommendation is based on findings from the DOE and NASA programs.

10.0 PUBLIC COMMENTS

Section 13(b) requires the NRC to obtain comments on the Ford Amendment from the public, licensees of nuclear power plants, the Advisory Committee on Reactor Safeguards (ACRS), and organizations comprised of professionals having expertise in appropriate fields. In response to that requirement, the NRC took the following actions:

(1) on March 3, 1983, published a Federal Register Notice (FRN) that detailed Sections 13(b), (c), and (d) of the Ford Amendment and requested comments by May 1, 1983

(2) issued a public announcement detailing the Ford Amendment requirements and requesting comments by May 1, 1983

(3) sent copies of the FRN to fifteen organizations of professionals as an enclosure to a letter requesting comments by May 1, 1983

(4) on July 18, 1983, briefed an ACRS Subcommittee in a public meeting about the Ford Amendment study plan. On December 6, 1983, briefed an ACRS Subcommittee in a public meeting on progress made toward completing the study. On February 24, 1984, briefed an ACRS subcommittee on the results of the study. On March 15, 1984, briefed the ACRS on the results of the study.

Thirty-four sets of comments were received as a result of these actions--nine from private citizens, five from citizen organizations, seven from licensees, three from professional organizations, nine from other industry groups, and comments from both members of the ACRS and the ACRS. The NRC also established a review group of distinguished professionals having a broad range of expertise in related fields to provide an ongoing peer review of the study while it was in progress. Comments from the public, licensees and associations appear in Section 10.2. ACRS comments appear in Section 10.3, and written comments from the review group appear in Section 10.4. For convenience, the public comments are consolidated and grouped according to each of the alternatives contained in Sections 13(b)(1) through (b)(5) of the Ford Amendment. In addition to comments on Sections 13(b)(1) through 13(b)(5) and 13(c), comments were received on a variety of related and unrelated subjects. These comments are included in Section 10.2.7, General Comments. The NRC response appears below each comment. The source(s) of each comment appears to the right of the comment. The commenters and abbreviations used in the discussions are listed and categorized just before the comments in Section 10.2.

One public comment concerned NRC's failure to provide both a program plan in the Federal Register Notice for the conduct of the Ford Amendment study and an opportunity for comment on that program plan. Such a study plan had not been completed at the time of request for comments. Moreover, the NRC staff wished to use the comments received on the Ford Amendment as part of the study plan development process. Once the plan was developed, it was presented to the ACRS at a public meeting. To provide an ongoing review of the study by persons outside the NRC and the government, the NRC staff established the review group of professionals mentioned earlier to review the study's plans and progress. This review was performed by nine distinguished professionals having expertise in nuclear power plant quality assurance, project management, engineering, and other relevant areas. The review group represented a broad spectrum of expertise, experience and viewpoints. Section 10.4 provides the names and positions of the reviewers and a summary of their comments on major issues.

Public and other comments were sought early in the study effort. Several of those comments were used in conducting the study and many of the study conclusions supported comments received. As a result, many of the comments received have been adopted within NRC's planned actions or included in issues slated for further study.

10.1 TEXT OF THE AMENDMENT

The Federal Register Notice contained Sections 13(b), (c), and (d) of the Ford Amendment. The accompanying text of the Federal Register Notice invited the public to provide to the NRC any comments on the quality assurance study by May 1, 1983. For convenience, the text of Sections 13(b), (c), and (d) is reproduced below:

Sec. 13(b) The Commission shall conduct a study of existing and alternative programs for improving quality assurance and quality control in the construction of commercial nuclear powerplants. In conducting the study, the Commission shall obtain the comments of the public, licensees of nuclear powerplants, the Advisory Committee on Reactor Safeguards, and organizations comprised of professionals having expertise in appropriate fields. The study shall include an analysis of the following:

(1) providing a basis for quality assurance and quality control, inspection, and enforcement actions through the adoption of an approach which is more prescriptive than that currently in practice for defining principal architectural and engineering criteria for the construction of commercial nuclear powerplants;

(2) conditioning the issuance of construction permits for commercial nuclear powerplants on a demonstration by the licensee that the licensee is capable of independently managing the effective performance of all quality assurance and quality control responsibilities for the powerplant;

(3) evaluations, inspections, or audits of commercial nuclear powerplant construction by organizations comprised of professionals having expertise in appropriate fields which evaluations, inspections, or audits are more effective than those under current practice;

(4) improvement of the Commission's organization, methods, and programs for quality assurance development, review, and inspection; and

(5) conditioning the issuance of construction permits for commercial nuclear powerplants on the permittee entering into contracts or other arrangements with an independent inspector to audit the quality assurance program to verify quality assurance performance.

10-2

For purposes of paragraph (5), the term "independent inspector" means a person or other entity having no responsibility for the design or construction of the plant involved. The study shall also include an analysis of quality assurance and quality control programs at representative sites at which such programs are operating satisfactorily and an assessment of the reasons therefor.

(c) For purposes of --

(1) determining the best means of assuring that commercial nuclear powerplants are constructed in accordance with the applicable safety requirements in effect pursuant to the Atomic Energy Act of 1954; and

(2) assessing the feasibility and benefits of the various means listed in subsection (b);

the Commission shall undertake a pilot program to review and evaluate programs that include one or more of the alternative concepts identified in subsection (b) for the purposes of assessing the feasibility and benefits of their implementation. The pilot program shall include programs that use independent inspectors for auditing quality assurance responsibilities of the licensee for the construction of commercial nuclear power-plants, as described in paragraph (5) of subsection (b). The pilot program shall include at least three sites at which commercial nuclear powerplants are under construction. The Commission shall select at least one site at which quality assurance and quality control programs have operated satis-factorily, and at least two sites with remedial programs under-way at which major construction, quality assurance, or quality control deficiencies (or any combination thereof) have been identified in the past. The Commission may require any changes in existing quality assurance and quality control organizations and relationships that may be necessary at the selected sites to implement the pilot program.

(d) Not later than fifteen months after the date of the enact-ment of this Act, the Commission shall complete the study required under subsection (b) and submit to the United States Senate and House of Representatives a report setting forth the results of the study. The report shall include a brief summary of the information received from the public and from other persons referred to in subsection (b) and a statement of the Commission's response to the significant comments received. The report shall also set forth an analysis of the results of the pilot program required under subsection (c). The report shall be accompanied by the recommendations of the Commission, including any legislative recommendations, and a description of any administrative actions that the Commission has undertaken or intends to undertake, for improving quality assurance and quality control programs that are applicable during the con-struction of nuclear powerplants.

10.2 COMMENTS OF THE PUBLIC, LICENSEES OF NUCLEAR POWER PLANTS, ORGANIZATIONS OF PROFESSIONALS AND OTHERS, AND NRC'S RESPONSE TO THESE COMMENTS

The following is a listing of the persons and organizations providing comments to the March 3, 1983, Federal Register Notice:

Private Citizens (P.C.)
Christine Simmers, MD
Marvin Lewis, PA
L. H. Wilkie, Jr., AZ
Daniel Garland, WA
Nancy Compton, TN
L. D. Gustafson, WA
John O'Neill, MI
Wells Eddleman, NC
Scott Bullock, NY

Citizen Organizations (C.O.)
Suffolk Nuclear Study Group (SNSG), NY
Union of Concerned Scientists (UCS), DC
Audubon Society, The Indiana Sassafras (ISAS), IN
Sinnissippi Alliance for the Environment (SAFE), IL
Ohio Citizens for Responsible Energy (OCFRE), OH

Utilities/Licensees
San Diego Gas and Electric (SDG&E), CA
Public Service Electric and Gas (PSG&E), NJ
Houston Lighting and Power (HL&P), TX
Florida Power and Light (FP&L), FL
Duke Power Company, NC
Baltimore Gas and Electric (BG&E), MD
Cleveland Electric Illuminating Co. (CEI), OH

Professional Organizations (P.O.)
The National Board of Boiler and PV Inspectors (NB), OH
American Society of Mechanical Engineers (ASME), NY
Institute of Electrical and Electronics Engineers (IEEE), NY
Nuclear Power Engineering Committee (NPEC)

Other Industry Groups (Other)
Townsend & Bottum, MI
Automatic Switch Co (ASC), NY
Stone and Webster (S&W), MA
Atomic Industrial Forum (AIF), D.C.
Management Analysis Corp. (MAC), CA
Institute of Nuclear Power Operations (INPO), GA
American Nuclear Insurers (ANI), CN
Edison Electric Institute (EEI), D.C.
Commonwealth-Lord J.V.C. (CWL), IN

10.2.1 Section 13(b)(1): More Prescriptive Architectural and Engineering Criteria

Comment	Commenter *

(1) The NRC should establish more prescriptive requirements for QC inspections, process control, records, etc.

1 utility
1 other

Response: The NRC is in the process of endorsing an American National Standard (ANSI), ANSI/ASME, NQA-1, "Quality Assurance Program Requirements for Nuclear Power Plants," which includes requirements and guidance for establishing and carrying out quality assurance programs during the design, construction, operation, and decommissioning of nuclear facilities. The NRC intends to continue to work with ANSI to produce needed standards. The NRC also is initiating research efforts to analyze alternative approaches to inspection planning and prioritization. The appropriate level of prescriptiveness in required inspection activities will be one of the issues considered in the research. See the response to Comment 2 on standardizing QA/QC paperwork.

(2) The NRC should standardize all QA/QC paperwork.   1 other

Response: The NRC has certain minimum reporting and recordkeeping requirements and attempts to standardize its own requirements. Each project is sufficiently different that it would be difficult, if not impossible, to devise a system of standardized QA/QC paperwork that would be appropriate for all projects. Prescribing paperwork at the QA/QC record level would have the effect of putting the NRC into a quasi-management rather than an oversight role. The NRC should set standards and performance objectives for QA/QC systems and their paperwork but should not prescribe how to achieve those objectives. If the NRC were to standardize all QA/QC paperwork, the flexibility of licensees would be limited in recording any additional data for their own needs. This could lead to the licensees developing a second set of records for this additional data. Duplicate records markedly increase the chances for errors and would also increase costs without an increase in safety.

(3) The NRC should require scale models and computer-   2 utilities
assisted drawings.

Response: With the advancement of computer modeling, scale models may find less use. Computer-assisted drawings are currently used at some plants and should be considered for all future plants primarily because of the technique's economics. This study also strongly endorses the use of models and found them to be extremely useful in helping some licensees manage the construction of their projects. The use of models would be considered to be a good project management practice. The NRC's follow-on research to this study will further evaluate this suggestion.

---

*For abbreviations and codes used to identify commenters, see Section 10.2.

(4)   NRC should institute more prescriptive architectural   1 P.C.
and engineering (A&E) criteria.   1 C.O.

Response: The topic of more prescriptive A&E criteria is discussed at length in the report. In particular, see Chapters 2 and 6. The study found that more prescriptive A&E criteria would not have prevented or led to earlier detection of the five major quality-related problems that stimulated this study. The QA study did find that a more complete plant design would help facilitate better planning and scheduling, would minimize design changes and potentially would reduce rework caused by the design not being far enough ahead of construction. This study suggests further efforts to improve the management of change in nuclear design and construction and to examine whether future applicants should be required to submit an essentially complete design at the construction permit (CP) application stage.

(5)   One-step licensing and more prescriptive criteria   1 utility
are a logical combination.

Response: The study agrees, see Chapter 6.

(6)   The QA study should investigate the amount of   1 C.O.
engineering and design review the NRC could perform.

Response: The case studies, pilot programs, and NRC inspections have shown that most often design interfaces and implementation are the problems, not the design itself. To address this problem, the NRC initiated a new program for integrated design inspections (IDI) in 1983. The IDIs examine the design interfaces and implementation as well as provide selected overchecks for the design itself. The IDI program is described in greater detail in Chapters 2 and 7. As with construction and operation, the licensee is primarily responsible for ensuring that the design and engineering work is adequate. The IDI program is an NRC overcheck of the effectiveness of the design and engineering review and process of the licensee and its architect-engineer (A/E), and it in no way relieves the licensee and contractors of their design responsibilities, nor does it replace any licensee activity.

(7)   More prescriptive A&E criteria will not solve the   1 P.C.
quality problem; do not add more requirements.   2 other
                                                4 utilities

Response: The study agrees. As noted in the report (see Chapters 2 and 6), the design quality problem does not appear to be with the criteria for the plant design as much as with design changes and with design and engineering work not staying sufficiently ahead of construction work.

(8)   More prescriptive criteria would negate designers'   1 other
flexibility and creativity and may also lead to less   1 utility
rigorous design and more reliance on the NRC.   1 P.O.

• Response: The study agrees. A significant level of design completion at the outset of construction, e.g., through the procurement phase, was found to be more important in avoiding quality problems than rigorous adherence to specific criteria. See Chapter 6.

10.2.2 Section 13(b)(2): Demonstration of Capability to Independently and Effectively Manage the QA/QC Function

Comment	Commenter *
(1) Licensees should be required to demonstrate their capability to independently and effectively manage the QA/QC function.	2 utilities 1 P.C. 1 P.O. 1 other 1 C.O.

Response: The study agrees. The study recommends increasing NRC's efforts in reviewing the applicant's management capability before CP issuance, which presumably would include an audit of pre-CP design activities. The study also recommends requiring an independent review in the first 12 to 20 months of construction in which the licensee and the project team must demonstrate their capability to independently and effectively manage the project, including the quality function. Subsequent NRC or third-party audits would be conducted about every two years thereafter, and management capability would be one of the areas of appraisal. See Chapters 2 and 4.

(2) The NRC should perform a cost/benefit analysis on requiring a demonstration of QA/QC management capability.	1 utility

Response: A cost/benefit analysis is a routine administrative requirement for the NRC when it performs a regulatory analysis of proposed changes to NRC requirements. A cost/benefit analysis will be performed as a part of the regulatory analysis for any rulemaking activity in this area.

(3) Adequate provisions are already in place to evaluate management's capability to manage a QA/QC program.	1 utility 1 P.O.

Response: The requirement for the NRC to evaluate management capability is already in place. However, the method, criteria, and approach for implementing the requirement to evaluate management's capability should be revised to reflect the results of this study. The study showed that the NRC has historically performed little pre-CP evaluation of management capability and has not focused on management issues in inspection until after major problems have occurred. The study recommends that NRC focus more on the management capability issue both before and after the CP is issued. See Chapters 2 and 3.

---

*For abbreviations and codes used to identify commenters, see Section 10.2.

(4) Evaluation criteria for management capability   1 other
    should be flexible enough to permit the Institute
    of Nuclear Power Operations (INPO) to perform
    this function.

Response: The study has concluded that using INPO to perform a quasi-regulatory function is not consistent with INPO's current mission or in the best interests of improving quality and the assurance of quality in the nuclear industry. INPO is in the process of seeking to raise the overall standard of performance and to achieve excellence in the nuclear industry through evaluation, education, and counseling. Placing INPO in a position where it would be a determinant in the licensing process could significantly damage its primary function. See discussion of INPO in Chapters 2 and 5.

(5) Requiring the licensee to demonstrate its   1 utility
    management of the quality function would unnecessarily
    restrict the owner/licensee to managing the project
    itself.

Response: The study has not interpreted this provision as a requirement for the licensee to solely manage the project. There are varying arrangements under which the licensee can choose to manage the project. The case studies (Chapter 3.0 and Appendix A) provide examples of different organizational arrangements that have worked, including the owner/licensees in an oversight role. However, this study has indicated that certain functions must be retained by the licensee to properly discharge its responsibilities. The ability to independently confirm that an effective quality assurance and quality control program is in place and is being properly implemented is one of those functions. Criterion I to 10 CFR 50, Appendix B, specifically states that although a licensee may delegate establishment and execution of its QA program to contractors, the licensee is still responsible for it. Both this Ford Amendment alternative and the study's recommendations are consistent with this criterion. See discussion of this alternative in Chapter 2.

10.2.3 Section 13(b)(3): Evaluations by Organizations of Professionals Which are More Effective Than Those Currently Performed

Comment	Commenter *

(1) The NRC should establish a requirement similar to   1 other
    ASME "N" Stamp for electrical equipment and eliminate
    utility audits.

Response: The American Society of Mechanical Engineers (ASME) "N" stamp program was initiated and implemented by the nuclear industry and was not an NRC requirement. If any other professional association wants to implement such a program, the NRC could participate in its development

*For abbreviations and codes used to identify commenters, see Section 10.2.

and consider endorsing such a program through a regulatory guide or its regulations, as the NRC has done for the Boiler and Pressure Vessel Code. It is NRC policy to adopt national consensus standards wherever possible and to the greatest extent possible.

(2) QA programs should have independent review groups   1 P.C.
whose overview would constitute independent design
review.

Response: The study analysis of alternatives 13(b)(3) and (b)(5) and the conduct of the pilot program found that periodic independent audits of construction projects, of which QA would be one aspect examined, would be useful and would assist the NRC, as well as the licensee, in establishing an assurance of quality in the design and construction of the nuclear power plant. In the regulatory analysis preceding implementation of such a regulatory requirement, the NRC will consider establishing procedures and criteria that will allow certain of these independent audits to also serve the purpose of independent design review.

(3) Independent audits should be conducted at set stages   1 P.O.
of construction completion, such as 25%, 50%, and 75%.

Response: The recommendation for an independent review mentioned in the response to Comment 2 above would be conducted at predetermined stages throughout construction. The stages may be based on percent of plant construction and/or time (years). The study recommends that the independent reviews should occur about every two years. See Chapters 2 and 4.

(4) The independent auditing group should be selected by   1 P.O.
jurisdictional authority or by the NRC, not by the
utility.

Response: If independent audits are established through NRC regulation, it is anticipated that the auditors would have to meet criteria established by the NRC and would have to be approved by the NRC. However, the option of the utility to freely select and contract with an acceptable auditor should be preserved.

(5) The NRC should require independent audits that are   1 C.O.
based on the Preliminary Safety Analysis Report (PSAR).

Response: PSAR commitments that are not conditions of the construction permit currently are not enforceable. Chapter 6 addresses this problem and recommends that certain PSAR commitments should be made conditions of future construction permits. These commitments could then be included in the scope of coverage for the independent audits.

(6) INPO is a professional organization that   4 utilities
already provides such audits.   1 other
          1 P.O.

Response: Study staff observation and reviews of INPO evaluations have revealed areas that would need to be altered before the NRC could accept the results of INPO evaluations as part of the NRC inspection process. However, as noted under 10.2.2, Comment 4, assumption of such a quasi-regulatory role by INPO would damage INPO's ability to act as the industry's medium for improving and seeking excellence in industry performance. This study concluded that this loss would be irreplaceable, whereas the independent audit function can be performed by many organizations. See discussion of INPO in Chapters 2 and 6.

(7) INPO, the ASME, and the National Board   2 P.O.
are organizations of professionals that   1 other
already provide evaluations.   1 utility

Response: The study has interpreted this alternative [Alternative b(3)] to represent Congress' desire to examine whether audits currently performed by the above-mentioned organizations or other professional associations could be expanded, thereby increasing their effectiveness. In its review, including polling of the organizations, the study found that only INPO evaluations covered more than one professional discipline. Although all of these programs were capable of expansion within that discipline, and the NRC would welcome their doing so, the INPO program was studied most extensively because of its wide scope and closest correspondence to what the study believed to be Congressional intent of Alternative b(3). Therefore, the study did not believe that the above-listed organizations already provided the evaluations that Congress had envisioned. See Chapter 6.

(8) An additional layer of inspection will not solve the   1 utility
industry's quality problems.

Response: The study agrees. The focus should be on better inspections, not on more inspections. The pilot program study demonstrated that the independent audits were not an additional layer of inspection but provided inspection coverage in areas such as design, for which little inspection has been accomplished in the past. The independent audits were also shown to be an augmentation of NRC's regular inspection program rather than an additional layer. The independent audits would allow NRC to expand its current areas of inspections without a commensurate increase in NRC staff and would allow the NRC more flexibility in prioritizing its inspection effort to areas that appear to warrant increased coverage.

(9) There are too many yearly audits of component suppliers   1 other
already. The ratio of product costs to audit costs is
way out of line. Audit frequency should be a function of
product complexity.

Response: The study agrees that inspections and audits should be a function of project/product complexity. NRC has initiated two separate research projects directed toward developing greater guidance for acceptable ways to implement a QA program whose requirements are commensurate with the product's safety significance. This area has not received the attention it should have received in the past and is now a priority research effort. See Chapters 2 and 7.

10.2.4 Section 13(b)(4): Improvements to NRC's Programs, Methods and Organization

Comment	Commenter *
(1) NRC puts too much emphasis on procedures and documentation and not enough on implementation and product quality.	3 P.C., 2 C.O. 2 P.O. 2 other 1 utility

Response: The study found that this has been a significant shortcoming in the NRC inspection program. The NRC has started to shift emphasis of the inspection program away from the QA process and documentation, and toward implementation and product quality. This subject is covered in considerable detail in the report (especially Chapters 2 and 7). The large number of comments in this area made it of particular concern during the conduct of the study.

Comment	Commenter *
(2) Although serious problems have been overlooked under current NRC programs, existing programs can be effective. The NRC should concentrate on enforcing existing programs.	1 utility 1 C.O.

Response: NRC's existing programs have not detected some problems at a sufficiently early stage. Considerable effort was devoted during the conduct of this study to determine if quality-related problems in design and construction were due to a basic fault in the NRC licensing and inspection programs or to licensee's implementation of the programs. The study concluded that the problem has been in both rather than a basic fault in either program. This study outlines a course of action intended to strengthen each area. See Chapters 2, 3, and 7.

Comment	Commenter *
(3) NRC should inspect both safety-related and nonsafety-related work.	1 P.C. 2 C.O.

Response: The NRC currently has research efforts under way to address whether a change is required in NRC's methods for selecting which structures, systems, and components should be inspected and the appropriate quality assurance measures that should be applied. The research is directed toward providing better guidance on the prioritization of quality assurance measures to be applied to structures, systems, and components that are considered either "safety-related" or "important to safety". This is a priority issue and more guidance should be available when research is complete. See Chapters 2 and 7.

Comment	Commenter *
(4) The NRC has not exercised sufficient control to prevent danger to public health and safety by assuring quality of plant design and construction. The NRC should increase the number of resident inspectors to 20-30 per site and should be responsible for plant quality.	3 P.C. 5 C.O.

---

*For abbreviations and codes used to identify commenters, see Section 10.2.

Response: Having the NRC directly responsible for plant quality rather than the licensee would represent a fundamental change from the private owner-ship and responsibility course the U.S. embarked on in 1954. Such a change would represent a fundamental change in this country's public policy and in NRC's regulatory policy and would require legislative changes, as well as a substantial increase in Congressionally authorized funding and personnel limits. As this comment correctly observes, such change would require very significant increases in the numbers of NRC inspectors. It is doubted that even 20-30 per site would be enough. The study does not conclude that current circumstances warrant such a fundamental change.

(5) NRC fines are hidden in future rate increases   1 P.C.
     and therefore are not an incentive for improved   1 C.O.
     performance.

Response: Most Public Utility Commissions (PUCs) do not allow fines to be passed through to the ratepayer. However, this study did examine the effects of state PUCs on the NRC mission. This study concluded that some PUC actions have had a previously unappreciated impact on NRC's mission and are likely to have a different but equally significant impact in the future. This area is discussed in Chapter 8 and Appendix C. Further study is under way.

(6) The NRC does not properly investigate or evaluate   2 C.O.
     allegations.

Response: The NRC has included additional resources in its budget for investigating allegations and has formed a new Office of Investigations to handle many of the allegations received. Also, in 1982, the NRC developed a computerized tracking system to track the receipt and disposition of all allegations received by the NRC to ensure their proper treatment. Use of third-party audits for increased coverage of NRC's regular inspection program is expected to allow more staff time for reactive inspections such as those arising from allegations.

(7) The NRC should use statistically based sampling   1 P.C.
     techniques.   1 C.O.

Response: Statistically based sampling techniques are used when it is not possible to inspect all of or a substantial part of some activity. A licensee's QA/QC programs essentially provide for 100% inspection coverage of safety-related structures, systems and components, so a "sampling" process is not applicable to the licensee's QA/QC process. Because the NRC is not responsible for direct inspection, as is the licensee, but for auditing the licensee's compliance with the QA/QC program, the NRC performs only a very limited sampling of the licensee's efforts. Current estimates are that the NRC directly verifies no more than 1 to 2% of a licensee's activities. A sampling of such small size does not lend itself well to a statistically derived process.

For direct NRC inspection, the NRC inspection program has historically used engineering judgment to select items that best indicate the total process or that will have the greatest impact on plant safety. Also, the NRC wants to hear any allegations of improper practices from plant workers. The NRC inspection program continues to be based on the belief that the application of engineering judgment is superior to a mechanistic sampling technique. However, the study concludes that NRC's methods for implementing engineering judgment should be improved, and research is under way to improve current methods for classifying structures, systems and components according to their safety significance. See Chapter 2 and Section 7.4.

(8) NRC inspectors are technical specialists with little   1 utility
    QA knowledge and have problems relating findings to   1 P.O.
    QA program weaknesses.

Response: This comment highlights a finding of this report. Inspectors require technical training in QA, just as technical training is required in other technical and scientific fields. In 1983 the NRC instituted formal QA training courses for its inspectors in construction, operations, and modifications. This study has identified additional topics that appear to be needed for inspector training, including improving inspector and supervisor skills in relating inspection findings to QA program weaknesses and project management shortcomings. The study believes that a technical specialist appropriately trained in QA and management disciplines can be more effective for NRC inspection purposes than a QA professional without appropriate specialty training. See Chapters 2 and 7.

10.2.5 Section 13(b)(5): Audits by Third-Party Independent Inspectors

Comment	Commenter *

(1) Independent inspectors should be independent of   2 P.C.
    the nuclear industry, not just the utility, to   2 C.O.
    avoid conflict of interest.

Response: Independent inspectors must have the necessary qualifications to perform the audit. The best qualified personnel would be those with nuclear industry experience. The study recommends that the NRC establish criteria that include measures of the objectivity of the independent inspectors. See Chapters 2 and 4.

(2) Independent auditors and QA personnel should report   2 P.C.
    to the NRC Resident Inspector.   1 C.O.

Response: The NRC should not assume the licensee's responsibility for quality of construction. The study recommends that the NRC review the efforts and results of the proposed independent audits. If the NRC were not satisfied with the independent audit, appropriate actions would have to be taken.

---

*For abbreviations and codes used to identify commenters, see Section 10.2.

(3) The NRC should increase the use of independent inspectors.

2 C.O.
2 P.C.

Response: As noted under the comments for 13(b)(3) (Section 10.2.3) and the above comments, the study has found that independent auditors/inspectors can be a useful addition to NRC's program for assurance of quality and recommends their use be increased.

(4) More independent inspectors will not solve the problem of poor quality and quality assurance . in the nuclear industry.

1 P.O.
1 utility

Response: It is not necessarily a question of more or fewer inspectors, but how resources are used. A major finding of this study has been that some utility managements have not implemented an effective QA/QC program. Where this lack of implementation has resulted from a lack of commitment to an effective QA/QC program, the study agrees that more inspectors will not solve the problem of poor QA/QC in the nuclear industry. Whatever the case, the NRC must be capable of detecting that a QA/QC program is not being effectively implemented. Given NRC's current inspection resources and increasing reactive inspection workload, its detection capability is questionable. This problem becomes particularly evident when it appears that team inspections must be increased to solve the "threshold" problem. Use of independent inspectors to augment NRC's regular inspection program is considered to be the most feasible way to address this problem. Until a third-party program can be put in place, the study recommends an expanded Construction Appraisal Team (CAT) program. See Chapters 2 and 7.

(5) INPO programs provide sufficient independent inspectors.

2 utilities
2 other

Response: See response to Comment 6 under 13(b)(3) (Section 10.2.3) and Comment 4 under 13(b)(2) (Section 10.2.2).

(6) INPO, ASME, and the NRC currently provide sufficient independent inspectors.

3 utilities

Response: See response to Comment 7 under 13(b)(3) (Section 10.2.3) and response to Comment 4 in this section.

10.2.6 Section 13(c): Pilot Programs

Comment	Commenter *
(1) Pilot programs are of questionable value because people provide quality and people at each site are different.	1 other

---

*For abbreviations and codes used to identify commenters, see Section 10.2.

Response: The study agrees that qualified, capable people are crucial to achieving quality. One of the major points that the case studies found to be important was the quality of the people associated with the project, particularly the management. One of the major differences between apparently successful projects and unsuccessful projects was the capability and experience of a few key members of management and their commitment to and knowledge of achieving quality in construction. The pilot programs were intended to test the concept of using independent auditors as part of NRC's process of confirming that licensees have implemented an effective QA/QC program. As a test of the concept, the pilot programs were of considerable value. As would be expected in any test program, strengths and weaknesses of individuals and organizations resulted in differences among the pilot programs, but the test of the concept was successful.

As a result of the pilot programs, the study concluded that independent audits are a useful addition to the NRC program and that guidelines for what constitutes an effective audit can be developed. See Chapters 2, 3, and 4.

(2) A few revoked construction permits (CPs) would do   1 C.O.
more to emphasize quality than any pilot program.

Response: The NRC is required by law to issue licenses to qualified applicants, and it cannot arbitrarily revoke CPs for deficiencies or violations without giving the licenseee an opportunity to correct deficiencies and to conform to requirements. If a licensee is willing to upgrade QA/QC efforts to correct deficiencies so that the NRC's minimum criteria are met, there is no reason to revoke a CP. If the NRC's criteria are not met, the plant should not be issued a CP at the outset, and, if criteria are consistently not met after the the CP is issued, then revoking or suspending a CP may be the correct action.

(3) Midland, Zimmer, South Texas and Diablo Canyon   1 P.C.
should be included in the pilot program.   1 C.O.

Response: The Midland, South Texas, and Marble Hill projects were included in the pilot program. An independent audit of Zimmer was incorporated in the case studies, and Diablo Canyon was examined in the case studies. See Chapters 3 and 4.

(4) The pilot program emphasis should be on quality   1 other
control, including training and qualification.

Response: NRC's evaluation of the pilot programs puts great emphasis on the licensee's quality assurance program, including the quality control portion. Training and qualifications of quality control inspectors have been identified as areas of special NRC concern, and research is currently under way to assess the nature and extent of problems in these areas for possible NRC action. See Section 7.4. While training and certification of quality control inspectors are obviously of great importance to the NRC, the pilot programs were not viewed as the most effective place to solve the many questions that have arisen in this area, so they were not the area of greatest emphasis.

10.2.7 General Comments

Comment   Commenter *

(1) The study should recognize that three interrelated   1 utility
     parties are responsible for assuring quality and
     reliability: (a) owner/licensee, (b) A/E, constructor,
     contractors, labor unions, vendors, and (c) the NRC.

     Response: The owner/licensee is ultimately responsible for assuring
     quality and reliability of the plant. The A/E, constructor, contractors,
     labor unions and vendors are responsible to the licensee to the extent
     that the owner delegated responsibility. However, ultimate responsi-
     bility, even though delegated, is retained by the licensee. The NRC is
     responsible for the health and safety of the public. The NRC attitude
     in the past has been that if the licensee has not fulfilled its responsi-
     bility for building a safe plant, the NRC can still fulfill its responsi-
     bility by denying an operating license. Rather than passively relying on
     the possibility of such a strong action after a plant is essentially
     complete, the study has recommended a more active ongoing NRC program for
     ensuring quality in design and construction aimed first at prevention and
     second at detection. See Chapter 2.

(2) Quality is not bad at all nuclear plants.   1 utility
     Excessive emphasis has been placed on a
     few bad examples.

     Response: The study agrees that quality is not bad at all plants. NRC's
     programs set minimum standards for all plants. Plants that cannot meet
     those standards should receive additional emphasis as is required.
     Plants that have experienced general breakdowns of their quality assur-
     ance programs should receive even greater NRC inspection and enforcement
     attention. This attention is necessary for NRC to fulfill its responsi-
     bility to protect public health and safety, and it is not believed that
     this emphasis has been excessive. See Chapter 3.

(3) Quality would be better served by assuring competence   1 other
     of doers rather than overemphasis on competence of the
     verifier. The NRC should require certification of
     management, designers, field engineers, and others.

     Response: The study agrees that both the doer and the verifier have to be
     qualified to produce and assure quality work. For example, the NRC staff

*For abbreviations and codes used to identify commenters, see Section 10.2.

is currently working to establish a positive method of welder identification to ensure that only welders who have passed the qualification criteria actually perform the work. The staff is also working to establish revised standards for qualification of nondestructive examination personnel. Other such areas will also be examined. The study does not recommend establishing criteria for individual management positions or for the licensing of individual managers. However, this study has indicated that the NRC should develop criteria to evaluate the "management team". This area will receive much greater NRC attention in the future. See Sections 7.4.3 and 7.4.4.

(4) The Federal Register Notice did not contain specific   1 utility
proposals. These should be commented on, not just the   1 C.O.
Ford Amendment. The NRC should publish preliminary
results of studies after 9-12 months for public
comment.

Response: Because of the relatively short time available to prepare the report for Congress (15 months), time was not available to publish preliminary findings for public comment. Also in the interests of time, NRC elected to request public comments at the beginning of the study so that the comments could be used in devising the study plan. The NRC did not want to develop a study plan and discover, through the public comment process, that a significant item had been missed but could not be added because of time. The eventual study plan was presented at a public meeting of an ACRS Subcommittee in July 1983. To provide a broad range of expert advice and guidance in conducting the study, a review group of distinguished professionals from outside the NRC was established. See discussion of the review group and individual members' comments in Section 10.4.

(5) The NRC should provide incentive programs for   1 utility
craft workers to improve quality.

Response: Although the study agrees that incentive programs for craftsmen have the potential to improve quality, they are not something the NRC should mandate by regulation. Licensees may elect to use incentive programs if they so desire.

(6) The NRC should require a greater degree of mandatory   1 Utility
personnel qualification and requalification than
is presently the case.

Response: See response to Comment 4 under Pilot Programs (Section 10.2.6) and response to Comment 5 above. Also see Section 7.4.

(7) All personnel from management down to craft   1 utility
workers should receive training in quality.

Response: The study agrees. Training in the areas of quality, QA, and QC is important to foster a better understanding of how quality programs can benefit all involved, from managers to craftsmen. INPO is currently conducting periodic seminars covering these subject areas for senior utility management. NRC senior staff have participated and should continue to participate in these seminars. The study recommends more training in quality for NRC staff. See Chapters 2 and 7.

(8) Quality assurance would be improved if the   1 other
     industry's attitude were improved.

     Response: The study agrees. See the response to Comment 1 under
     Pilot Programs (Section 10.2.6), and Comments 3 and 7, above. See also
     Section 2.5.

(9) QA/QC is not always effectively used   1 other
     as a tool by management.

     Response: The study agrees. The case study analyses (Chapter 3) of
     nuclear power plants under construction examined the premise advanced by
     this comment in great detail and give examples where QA/QC was used
     effectively as a management tool and where it was not. The study has
     proposed revisions in NRC programs to correct this deficiency. See
     Section 2.4.1.

(10) The Congressional Amendment doesn't provide enough   1 other
     guidance. Studies are being undertaken without
     identifying the problem they are supposed to
     study.

     Response: The study does not agree. The study understands the problem to
     be a search for improved ways to ensure the achievement of quality in the
     design and construction of commercial nuclear power plants. The study
     activity resulting from the Ford Amendment was designed to identify the
     root causes of cases of both poor and good design and construction quality
     and to devise programs that would provide greater assurance that achieved
     design and construction quality is at least equal to NRC's mininum
     requirements.

10.3 ACRS COMMENTS

The ACRS Subcommittee on Quality and Quality Assurance in Design and Construction was briefed twice in open meetings by the NRC staff on the Ford
Amendment project, the study approach and preliminary results. These
briefings, which took place on July 18, 1983, and December 6, 1983, were
announced in advance in the Federal Register and were attended by members of
the public.

In addition, the ACRS Subcommittee reviewed and commented on an earlier
revision of this report. They provided oral comments to the NRC in a closed
briefing on February 24, 1984. The comments have been incorporated into this
report at the end of this chaper. The full ACRS reviewed an earlier version of
this report and was briefed on the study findings and recommendations in a
open briefing on March 15, 1984. The text of the ACRS letter to the NRC
Commissioners regarding this report follows.

March 21, 1984

The Honorable Nunzio J. Palladino
Chairman
U.S. Nuclear Regulatory Commission
Washington, D.C. 20555

Dear Dr. Palladino:

SUBJECT: ACRS REPORT ON DRAFT NRC REPORT TO CONGRESS ON IMPROVING
QUALITY AND THE ASSURANCE OF QUALITY IN THE DESIGN AND
CONSTRUCTION OF COMMERCIAL NUCLEAR POWER PLANTS

During its 287th meeting, March 15-17, 1984, the Advisory Committee on
Reactor Safeguards reviewed the draft NRC report to Congress, "Improving
Quality and the Assurance of Quality in the Design and Construction of
Commercial Nuclear Power Plants," Revision 3, dated March 13, 1984.
An earlier version of the draft report was considered during a meeting
of the ACRS Subcommittee on Quality and Quality Assurance in Design and
Construction held in Washington, D.C. on February 24, 1984. In addi-
tion, the Subcommittee reviewed and discussed the NRC's quality-related
initiatives during meetings held on July 18, 1983 and December 6, 1983.

The report is both useful and constructive. It gives thoughtful attention to the five alternatives which the Commission was required to
consider under Section 13(b) of the Ford Amendment (Public Law 97-415)
and reaches well-reasoned conclusions on each. Further, the results of
the pilot program mandated under Section 13(c) of the Ford Amendment
substantiate the conclusion that comprehensive audits of nuclear construction projects by qualified third parties can provide significant
additional preventive and detection capability as well as enhanced
assurance that nuclear plants are built in accordance with their design
and licensing commitments. The report is candid in conceding errors of
omission or commission on the part of the NRC which have contributed to
quality assurance deficiencies in the past.

During the Subcommittee's early review of the study, it suggested that
the Commission take advantage of the opportunity presented by Congress
and expand the scope of the study to address issues beyond those mandated. We are pleased that the report provides a more comprehensive
picture of the Commission's actions and initiatives.

Although the report is well written, it is voluminous and repetitious.
A concise executive summary would improve the report. This can be
accomplished without delaying the submission of the report to Congress.

The lessons learned from past problems in the design and construction of
commercial nuclear power plants are described. As indicated in the
report, little is said about the operation of plants although many of

the same observations and lessons should apply. Many of the problems described relate to the inability, or the difficulty, of assuring the quality of some plants as a result of shortcomings in quality assurance and/or quality control programs during design and construction. Not addressed is whether the QA/QC shortcomings had an effect on quality or had significant effect on public safety/risk.

The distinctions among quality, quality assurance and quality control, and their relationship to public safety/risk are, at times, not made clear in the report. This is compounded by the NRC's continued inability to clearly identify those systems and components for which quality is essential to public safety and thus for which programs to control and to assure quality are necessary. Probabilistic risk assessments (PRAs) could help in this regard. We encourage the NRC to expedite the collection of the data necessary to clarify these issues.

Further, although recommendations are based on the findings of the case studies, pilot programs, and other initiatives, it is not clear whether their implementation will actually improve quality or enhance public safety or whether they will merely improve the public's perception of safety. We recommend that the NRC Staff undertake to determine the relative risk significance of the various recommendations and proposed actions as well as determine whether safety would be enhanced by the proposed actions. The NRC should then concentrate its efforts on actions which will enhance public safety.

The report does not contain priorities or schedules for further development of the various recommendations or proposed actions. We believe that the NRC Staff needs to develop more specific recommendations following the submission of the report to Congress. We recommend that in forwarding this report to Congress, the Commission make clear its intention to develop a plan for achieving the assurance of quality in the design, construction, and operation of nuclear power plants. However, we caution that the development of a program plan should not be allowed to interfere with proceeding expeditiously with those actions found to improve public safety significantly.

The NRC Staff has identified management as a major factor affecting the success or failure in assuring the quality and safety of nuclear power plants. While we agree that a poor quality assurance program is an indication of poor management, an apparently good quality assurance program does not necessarily imply the presence of good management. We see the need for an organizationally independent quality assurance department that reports to senior management; however, we fear that the emphasis on independence has in some cases led to the belief that the assurance of quality is someone else's responsibility. To assure quality and public safety, a strong sense of the need for and the benefits from quality, the assurance of quality, and professionalism should permeate a licensee's and/or a vendor's entire organization. The

NRC should continue its efforts to stimulate this kind of professionalism in the nuclear industry.

One of the recommendations from the management analysis conducted by N. C. Kist & Associates, Inc. is to establish a quality assurance program within the NRC. Although noted in passing in the report, it remains a fallow recommendation. The report does contain a recommendation for performance audits of NRC QA activities. However, we believe that the relationship between QA and prudent management, as discussed in Section 3.5 of the report and in this letter above, is equally applicable to the entire NRC. Therefore, we suggest that the Commission give prompt and careful consideration to the recommendation that the NRC establish a program to assure the quality of its activities. We do not believe, however, that a formal QA program is necessary or desirable.

The recommendation that the NRC establish a body of experts to advise the Commission on the capability of applicants to effectively manage a nuclear construction project is worthy of further consideration. The ACRS currently does not contain extensive expertise of the types envisioned for the proposed advisory body, and to establish such expertise within the ACRS membership might sacrifice other requisite expertise. The report has also recommended that future construction permits be conditioned on a demonstration of the licensee's continuing ability to effectively manage the project. Those responsible for the development of these recommendations should consider the difficulties associated with judging such management capabilities.

The ACRS supports the NRC Staff's shift in inspection emphasis from looking at the content of quality assurance plans to looking at actual plant quality and at the implementation and effectiveness of programs to assure quality. However, we believe that the NRC Staff will experience difficulties implementing the modified inspection program until performance criteria are established.

Useful insights have been obtained from Integrated Design Inspections (IDIs), Independent Design Verification Programs (IDVPs), and Construction Appraisal Team (CAT) inspections. We recommend that for the present these inspection programs continue.

The concept of using designated representatives is worthy of further consideration. In addition to augmenting NRC resources, it may be a way of stimulating and rewarding professionalism and dedication to quality in the workplace. We would like to be kept apprised of the NRC Staff's efforts regarding the designated representative concept and other QA initiatives, and we would appreciate the opportunity to comment on them at a later date.

Additional comments by ACRS Member Glenn A. Reed are presented below.

Sincerely,

Jesse C. Ebersole
Chairman

Additional Comments by ACRS Member Glenn A. Reed

I consider the report to Congress to be deficient in its study of Alternative b(4), and I am concerned that Congress may continue endorsing regulatory approaches that are too similar in many ways to those that have in the past proved ineffective. I concur with this ACRS report concerning the above referenced report to Congress in most aspects, but in my opinion it does not go far enough, and is not critical enough with respect to the following:

1. The ACRS report states that the report to Congress "gives thoughtful attention to the five alternatives ...." I disagree that thoughtful or appropriate in-depth attention was given to Alternative b(4), which addresses improvements in the NRC's organization, methods, and programs for quality assurance.

2. The report to Congress recommends that a body of experts be established to advise the Commission on an applicant's management capabilities. The ACRS report states that this recommendation is worthy of further consideration. I disagree, and do not feel such expertise, with the time and objectivity, could be constituted to undertake this activity. Further, I disagree that such a body of experts is even desirable or necessary if a more astute study of Alternative b(4) is made.

3. What the ACRS report does not address, or recommend, is more in-depth consideration of the NRC's organizational structure, and what obstacles this present structure may place in the path of achieving quality in design and construction. The present NRC structure does not motivate professionalism and craftsmanship in the workplace. In my opinion, high quality can only be achieved by the enthusiastic and dedicated action of real professionals and crafts people who are motivated to standards of excellence by a regulatory structure that better recognizes human factors. I am aware of and have read an NRC Staff report which addresses the FAA designated representative (DR) system. In my opinion, the report to Congress should not have glossed over the FAA's DR program, but should have included a detailed study of that system and its potential for correcting the adversarial climate that is growing in the nuclear workplace. Given the current structures of the nuclear

industry and the NRC, the genuine professionals and crafts people are somewhat overwhelmed by top brass and regulations, yet the answers for real quality in this highly technical nuclear industry lie with those professionals.

In my opinion, the achievement of a high degree of design and construction quality can come from a modified version of the FAA system of DRs in design and architect-engineer organizations and in manufacturer and constructor shops. I would consider it appropriate for these DRs to be nominated by their peers, approved by their employers and perhaps the NRC, then established in a quasi-regulatory role while continuing their regular duties. Along similar lines, the NRC might consider structuring some licensed personnel in nuclear power plants into a DR system somewhat similar to the way in which the Massachusetts Department of Public Safety has incorporated licensed operating engineers into its regulatory structure.

References:

1. Public Law 97-415, NRC Authorization Act for Fiscal Years 1982 and 1983, Section 13 on Quality Assurance, dated January 4, 1983.

2. Draft NRC report to Congress, "Improving Quality and the Assurance of Quality in the Design and Construction of Commercial Nuclear Power Plants," Revision 3, dated March 13, 1984.

10.4 <u>REVIEW GROUP</u>

Expert review of the NRC study on quality assurance was provided by the individuals listed below. Two day meetings were held with NRC staff in June and September 1983 to provide comments on efforts to date and to help provide direction to future work. A third meeting was held in January 1984 to review the tentative conclusions and recommendations.

The meeting format was used as the most efficient way of informing the individuals of the information that was available and of receiving their oral comments. Individual comments were formally provided by individual members after each briefing. All of these individual comments were used to help guide the conduct of the study.

° Fred Albaugh
  Chairman

Independent Consultant, Past Director, Battelle Pacific Northwest Laboratories. Manhattan Project. GE.

° John Amaral

Corporate Manager of Quality Assurance, Bechtel Power Corporation. Former Chairman, Energy Division, American Society for Quality Control.

° Spencer Bush

Consultant, Review and Synthesis Associates. Former Consultant, Battelle, Pacific Northwest Laboratories. Member, Advisory Committee on Reactor Safeguards, 1966-77 (Chairman 1971). Manhattan Project.

° Thomas Cochran

Senior Staff Scientist, Natural Resources Defense Council.

° George Coulbourn

Director, Nuclear Power Systems, Boeing. Former Vice-President, Boeing Construction. Former Construction Manager, Indian Point #3.

° John Gray

Chairman, International Energy Associates Limited. Chairman, Energy Policy Committee, Atlantic Council of U.S. Former Manager, Shippingport. GE, Westinghouse.

° John Hansel

Independent Consultant. Former Project Manager, Gaseous Centrifuge Enrichment Plant, System Development Corporation. Former Director, Quality Assurance, Apollo Spacecraft, Space Shuttle Orbiter, and Launch Operations. President-Elect, American Society for Quality Control.

° Robert V. Laney

Independent Consultant. Retired Deputy Director, Argonne National Laboratory. Project Manager, Seawolf prototype. Bettis Laboratory. Former Vice-President, General Dynamics.

° Leland Bohl

Manager, Quality Assurance and Reliability, General Electric Nuclear Energy Group.

10.4.1 Review Group Comments on this Report

An earlier version of this report was reviewed by the members of the review group in a meeting with NRC staff on January 10-11, 1984. Each of the review group members provided oral comments on the report during the meeting. Six of the nine review group members also provided written comments. The report was revised to reflect many of the comments offered, both oral and written. The written comments follow.

POST MEETING
COMMENTARIES
OF
GROUP MEMBERS

THIRD REVIEW GROUP MEETING
JANUARY 11-12, 1984

Pacific Northwest Laboratories
P.O. Box 999
Richland, Washington U.S.A. 99352
Telephone (509) 375-2575

Telex 15-2874

January 16, 1984

Dr. W. D. Altman, Project Manager
Special Study on Nuclear QA
U.S. Nuclear Regulatory Commission
Washington, D.C. 20555

COMMENTS ON DRAFT REPORT - "IMPROVING QUALITY AND THE
ASSURANCE OF QUALITY IN THE DESIGN AND CONSTRUCTION OF
COMMERCIAL NUCLEAR POWER PLANTS"

These written comments will supplement the discussion and critique of
the subject draft report which took place at the meeting of the NRC's
QA Program Review Group in San Francisco on January 11 and 12. It does
not seem useful or practical at this point to attempt line-by-line
correction, but general comments will be made on the study as a whole
and on a number of topical items.

The report is, in this reviewer's opinion, a good report which, if
vigorously implemented, could substantially improve the regulation and
practice of assurance of quality of commercial power reactors. It is
responsive to the directives of Public Law 97-415 which mandated the
study. It reaches useful and generally unambiguous conclusions on which
of a number of possible measures to improve quality are likely to improve
assurance of quality and which are not likely to. It is, for the most
part, refreshingly candid in conceding errors of omission or commission
on the part of NRC that have contributed to quality assurance deficiencies
of the past. At the same time, it does not overplay its hand by offering
solutions to matters which, even if relevant to quality concerns, are
clearly beyond the normal role of the Inspection and Enforcement Division
to recommend or decide. Finally, subject to a major rewrite of one
section, the draft reviewed on January 11 and 12 promised a well-reasoned
and well-written report.

1. Management Commitment to Quality

The conclusion that the foremost requirement for a quality
reactor project is that utility top management and its project
management team be fully committed to quality and knowledgeable
of the methods and discipline required to achieve it is one with

which this reviewer completely agrees. Corollaries are that CP's have undoubtedly been granted in the past that should never have been granted, and that NRC should move diligently to identify such cases, to ascertain their present commitment and attitudes toward quality practices, and to bring about change as needed by persuasion, peer pressures, education, use of regulatory sanctions for observed transgressions and other means. However, this reviewer does not believe that NRC should have the authority to revoke a license, once given, on suspicion alone with the burden of proof placed upon the licensee to show that it should not be so. This could too easily lead to abuse of regulatory authority amounting to arbitrary confiscation of private property.

2. Audits and Inspections

A second major thrust of the study is that NRC enforcement of QA regulations will in the future be based less on monitoring compliance with approved prescriptive procedures and more on the results of audits of design quality and construction quality to be conducted by the licensee; by INPO, an industry organization; by NRC teams and by independent third parties. This is a major improvement on past practices, focusing as it does, on actual end-results observed in the field, rather than on mere statements of good intentions.

This panelist has one reservation about the plan for augmented inspection effort. Will less critical time-consuming QA reporting and test requirements imposed on licensees be eliminated or modified to compensate, at least in part, for the extra time required of the licensee in connection with the new and enlarged audit program? To this observer the draft is ambiguous on this matter and regulators are always more prone to add new regulations than to eliminate old ones of marginal value.

3. Standardized Designs and Requirements for High Percentage Completion of Design Before Construction

These two measures are closely related and both seek to avoid or minimize the need for major changes in design criteria or QA regulations after a project is underway. The record seems persuasive that such changes invite quality problems, project delays and cost overruns and the logic of this proposed requirement can scarcely be questioned. However, it is also true that the degree to which design criteria and regulations may be stabilized depends on the depth of technical understanding of reactor systems and of reactor health and safety hazards and also upon prevailing socio-political attitudes toward nuclear power.

Research and development studies, design development studies and
other relevant work is planned by NRC and the nuclear industry to
provide an improved basis for standardization and this is to be
commended. However, until these studies have provided the needed
information and NRC can thereby provide assurance against major
backfitting requirements no utility can be expected to commit the
funds needed for a complete design without the protection of a
Construction Permit in hand.

4. Development of a Prioritized System of QA Requirements

This a commendable program which inferentially acknowledges the
ad hoc or reactive nature of many QA regulatory practices of the
past and looks to future development of systematic, objective
criteria to guide QA priorities and the extent of QA requirements.
Like 3) above, its success will depend upon results obtained in
safety studies, design studies and the general advancement of
technical understanding of reactor systems. The program is necessary
for a stable, long-range future of commercial nuclear power.

5. Adversarial Attitudes

It is the opinion of this observer that adversarial attitudes that
have often prevailed between NRC and a licensee have substantially
contributed to quality problems by negating the possibility of
cooperative efforts to identify and solve problems of mutual concern
and for one of the two parties to understand and appreciate facets
of a problem peculiarly important to the other. Without belaboring
the origins of this situation, be it sufficient to say that it does
not have to be so. The report cites examples of cooperative yet
effective regulation in other nations and by other regulatory agencies
of the United States.

The QA report reviewed here chooses to say little about this issue
on the grounds that its effective correction is beyond the powers
of NRC/I&E to correct. Granted that this is true, the report does
offer an opportunity to call forceful attention to the issue, an
opportunity that should not be neglected.

6. More Prescriptive Regulation

The conclusion of the report that more prescriptive regulation is
not the key to elimination of quality problems is unequivocally
correct. No amount of prescribed QC testing, required procedures
and the like can ever eliminate the possibility of human error or

random mechanical failure. Rather it is the role of the regulator
to set basic performance criteria and then to audit the licensee
and his work to assure that his organization, personnel, designs,
methods and procedures provide appropriate safeguards against
threats to public health and safety.

F. W. Albaugh
Review Group,
Special Study on Nuclear QA

cc: JA Christensen

Review & Synthesis Associates

Spencer H. Bush, P.E. ● 630 Cedar / Richland, Washington 99352

January 16, 1984

Mr. James A. Christensen
Battelle-Northwest
P. O. Box 999
Richland, Washington 99352

Dear Jim:

COMMENTS ON DRAFT REPORT, "IMPROVING QUALITY AND THE ASSURANCE OF
QUALITY IN THE DESIGN AND CONSTRUCTION OF COMMERCIAL NUCLEAR POWER
PLANTS - A REPORT TO CONGRESS"

Executive Summary

I do not consider this an Executive Summary. It is a rearrange-
ment, principally of Chapter 2 with most paragraphs lifted in toto.
To me, an Executive Summary is a terse overview of the significant
content of the entire report, with emphasis on conclusions and
recommendations.

P. 6, III.A

Basically, this action item says nothing will be done. If not,
say so succinctly.

PP. 7-9, Re: Advisory Committee

I feel you would have a more viable committee if its scope were
broader and more vague. Also, I'm not so sure regarding the pro's
and con's of statutory vs. non-statutory.

P. 12, Line 4

This tends to look down on ASME/NB. I suggest replacing ",...and
that....continued" with "and they provide a valuable and continuing contribution."

P. 13, D.

"...prior nuclear construction experience...in its.... (insert word)

P. 18, Top of Page

What does "in-house inspection process" mean?

P. 24, IV.A

This is a very open-ended item entailing a great deal of effort and
it appears to be advanced with little or no justification.

Table of Contents

9) Other Agency and Foreign Quality Assurance Programs (not titles
of chapters).

P. 3, Chapter 1

Should Executive Summary mirror Item C, Compliance with the Pilot
Program, as being accomplished?

Chapter 2

Comments are primarily in the Executive Summary heading of this memo.

P. 37, 2.3.3.A

You use many words elsewhere to explain and justify a position.
Here, you casually suggest requirements that will cost millions.
If you wish to cite the movement of the NRC toward expanded use
PRA as a reason, I'll buy it.

Chapter 3

P. 3, 3.2

"....develop management criteria...". What does management indicate? Is it necessary? In the bottom two lines, "...having construction..."--do you mean QA/QC or construction or both?

Also, I'm not sure regarding your projection of management criteria;
e.g., experienced management. For example, Consumers Power/Midland.
(N.B. I agree experience is very important.)

P. 8, Top ¶, Last 6 Lines

I suggest deletion of Diablo. I doubt that any plant with a CP in
the same time frame and now operating could come through an in-depth
construction audit unscathed. I was there to approve an OL more
than ten years ago.

P. 9, Bottom

...quasi¢...

P. 14, ¶5

I realize this is a quote. Even so, I doubt that it was just seismic criteria. I thought it was siting criteria.

P. 36

These comments tend to be snide. Why go out of your way to make enemies?

Chapter 4

Much of this chapter could go into an appendix, a la Appendix A. Then you could highlight the important items.

PP. 1-3

Why do you flip back and forth between what's in this chapter and what's in other chapters? This, in my estimation, should be done in Chapter 1.

P. 5

"End-product" vs. "process" is not clear. This means different thi: to different people. Clarify.

P. 5, Bottom Item

You presume the PSAR's and FSAR's are O.K.--which isn't necessarily the case.

P. 6, Third Bullet

Wouldn't sequential sampling solve the scope problem?

P. 3, Middle ¶

Fifteen percent doesn't sound like much; however, large sums are committed in procurement, etc., that may not be recoverable.

P. 9, Item b

Unclear. I don't know what "reduced" and "promote" mean.

P. 20, Task D

Is concrete intended to be inside or outside ASME's scope? It will be different.

Chapter 5, ¶1

The paragraph is ambiguous. It talks of these activities not con-stituting audits, etc. If it still refers to ASME, it isn't true. If it refers to other standards, it would be clearer if a paragraph starts at "Applicable national standards.....".

Section 5.2.A, Last Line, ¶1

"...and (to?) the nuclear industry..."? You might indicate that

the broad audits could take credit for the narrow (and deeper)
ASME/NB audits in the delimited areas permitting a better focus
elsewhere.

P. 5, Middle ¶

In line 1, "qualit(f?)y"?

Chapter 6

P. 3, 6.2

"...either the NRC or the Licensee." This is significant and could
be strengthened to emphasize that it is essential for both parties
to be severely constrained if this is to work.

P. 8, Last ¶

What existed was a difference in the PSAR versus design drawings.
This doesn't come through.

Chapter 7

A fairly tight chapter. No basic comments.

Chapters 8 and 9

No comments.

Chapter 10

Did not review.

General

There is a generic class of QC/QA breakdown that you have not
touched on. For example, the low toughness support problem in my
estimation has more true safety implications than several others
discussed. This represents a breakdown across utilities rather
than within utilities.

A concerted effort should be made to condense the body by deletion,
shifting to appendices, etc. The repetition tends to dilute and
results in a loss of focus.

In the Executive Summary, it would be helpful to tie items to
appropriate chapters, sections and pages.

The ultimate text would gain with a more definitive chapter structure (regarding headings, subheadings, sub-subheadings, etc.) with
numbers to identify headings.

The report or sections thereof would be in better focus by graphing
or placing in tabular format such items as chronology, interactions,
key actions, etc.

Please, at a minimum, cite by reference the program planned for operating reactors. This will close the loop.

Incidentally, under 10.4, I'm not BMI-PNL. I'm Consultant, Review & Synthesis Associates, ex-BMI-PNL.

Very truly yours,

Spencer H. Bush, P.E., Ph.D.
Consultant
REVIEW & SYNTHESIS ASSOCIATES

SHB:dp

cc: Dr. W. D. Altman, USNRC/OI&E

BOEING ENGINEERING COMPANY

P.O. Box 3707
Seattle, Washington 98124-2207

January 17, 1984

Mr. James A. Christensen
Battelle
Pacific Northwest Laboratories
P.O. Box 999
Richland, WA 99352

Dear Mr. Christensen:

I would again like to express my appreciation for the work done by Battelle in arranging our third meeting. Whatever contribution we may have made was enhanced by your efforts.

As this is expected to be my final report on the NRC QA analysis required by the Ford Amendment, I have elected to provide several classes of comment. First, specific comments on the draft information provided during the process were provided verbally during our group review. These and other comments specific to your draft are provided directly on the enclosed document by means of marginal notes. Secondly, I have attached some general comments regarding the form and substance of your draft report (Attachment 1). Finally, attached are some more general thoughts concerning the current regulatory environment which are of a somewhat broader perspective (Attachment 2), but directly related to the scope of the study.

In retrospect, I feel that Congress has presented the NRC with an extraordinary opportunity. My condensation of the charge under the Ford Amendment takes the following form: "NRC, there has been a QA problem with some current nuclear power plant construction projects. You appear to be part of the problem. Examine the current organization relationships and practices and recommend improvements." I consider this an extraordinary opportunity because it provides a Congressional mandate for improvement; a convergence of recognized need, and requisite action with high visibility and transcending normal organizational inertia. I am hopeful that the NRC will respond with the bold initiatives clearly needed to restore the fundamental promise of safe and economical nuclear power to the nation.

Sincerely,

G. I. Coulbourn

Attachments

cc: W. D. Altman, NRC

FORMAT AND SUBSTANCE FOR THE NRC RESPONSE UNDER THE FORD AMENDMENT

1. The final form of the document should allow the reader to easily determine what is being recommended. It should provide direction into the body of the document to the findings and rationale behind the recommendations, to allow further study, if desired, by the reader. Copious detail should be placed in appendices.

2. An important portion of the Ford Amendement instructions involved examination of the effect organization relationships had on the achievement of quality. The draft document to date is weak addressing the effectiveness of the NRC organization, especially as it relates to its internal activities. It is also weak in analyzing the manner in which the various participants in the process organize, both to work with each other, and to accomplish their own responsibilities. As the draft document is currently written, it focuses primarily on two findings: (a) that utility management is weak, and (b) that more and better inspections are needed. As discussed elsewhere, the first is only one element of the problem, and then only in certain instances. I am apprehensive that failure to identify weaknesses in the NRC organization itself which impact quality assurance may reduce the credibility of the report. At this point, perhaps further study of the current organization structure is a means of addressing the issue.

3. NRC should step back and assess the total impact of the summation of all inspections, requests for information, and other technical and procedural interfaces between NRC and the utilities. It is likely that even a rudimentary cost-benefit analysis will disclose duplication and marginal use of resources. The opportunity to integrate multiple instructions should be examined prior to incorporating an additional overlay of inspection.

4. It should be clearly stated that the NRC cannot and should not be expected to achieve quality in design, construction, and operation directly. All the NRC can do is assure that quality is achieved by others by selective audit and followup. It appears that this fundamental precept is not generally recognized by others not involved in the process.

5. Careful attention is needed in establishing the acceptance criteria before third-party audits become a part of the overall process. There appears to be a dichotomy between the experience needed and the independence desired. For example, if a person has substantial experience in the nuclear power industry, in some quarters he is suspect as not being independent. On the other hand, one cannot expect competent inspection work to be performed by people without substantial experience.

6. The term, "principal AE criteria," is used in several instances in the document. This term is undefined and may not be clearly definable. If that is the case, then meaningful regulatory requirements relating to the definition of these criteria may be difficult to achieve. This appears to need further investigation.

7. It was observed that NRC has operated on the assumption that industry was uniformly competent, while considerable variation existed and the level of competency was changing (both improving and deteriorating). It is essential that NRC recognize that a variation in competence and approach exists within the various utilities building and operating nuclear power plants. It will clearly be counter-productive to the achievement of quality and to the cost-effective production of electricity to require all operating utilities to respond to the problems of the lowest common demoninator. Selectivity will be a very difficult goal to achieve, but should be adopted from the onset.

8. It was observed that some utilities do not appear to support quality assurance because, for them, it is a cost item versus a cost savings or management tool. If this allegation is correct, then it would seem to be essential to determine why quality assurance in some instances is merely a cost item versus a useful tool.

9. It is suggested that NRC management, in conducting its review of the final draft, critically assess the extent to which the method and manner of implementation of the actions identified is described. In many instances in earlier drafts, very attractive objectives and action items were identified, but the text lacked a description of how they were to be implemented. During the discussions of some of these action items, concerns regarding the manner of implementation were raised and left unanswered.

GENERAL OBSERVATIONS REGARDING THE CURRENT REGULATORY ENVIRONMENT

The laws of physics have not been repealed. The combination of relatively abundant, low-cost fuel with no higher use and very low environmental impact are inherent. The fundamental economic promise of nuclear power remains available to any society with the determination to achieve it. However, the management process in the United States has allowed the development of a regulatory environment wherein these benefits are currently no longer available. It is unlikely that any domestic utility will consider ordering a nuclear power plant so long as this environment persists. Following is my assessment of the fundamental problems within the context of the Ford Amendment study which appeared to warrant consideration.

1. Management of Change: There is a level of change action (technical, regulatory, and procedural) beyond which any program management structure can no longer prosecute its program. Utility management has consistently been faulted for quality assurance breakdowns. In some instances, the charge is well founded. However, in most instances, I believe the root cause is found in the circumstances which produced rampant, uncontrolled change. I submit that most of the utility management structures assembled to build the nuclear power plants of the past decade could have performed adequately in a more stable design and construction environment.

Accordingly, I recommend that NRC commission an examination of the change management process itself, both within NRC and in the other parts of the industry. This examination should focus on both the management of change as a discipline (elsewhere called configuration management) and upon the reduction of the volume of change. The latter can have numerous constituents; for example, higher percent design completion prior to start of construction, more restraint regarding in-process change, standardization, etc. All of these constituents require disciplined and consistent management.

2. Signal-to-Noise R io: The uncontrolled application of successive overlays of requirments and procedures has evolved into a data-management problem of major proportions. NRC can no longer quickly and consistently isolate important information from the trivial because of the mass of paperwork currently required. Frequently the system forces the paper to become the product, such that the correspondence between documentation and documented quality is lost.

This problem should be attacked from two perspectives. The volume of extraneous data currently required must be reduced by selectively culling that which is not clearly required. And, improved methods for isolating important information at an early stage must be implemented. I recommend that NRC commission an independent effort to examine these two objectives. This cannot be accomplished from within the organizations affected because of their limited perspectives and organizational inertia.

3. Long-Range Plan: I question whether an organization such as NRC can function effectively, lacking a comprehensive long-range plan. How can staffing plans be formed? What skills will be needed, when, and where? What levels? What are some of the technical and institutional trends likely to impact NRC obligations a few years hence? How should they be met?

At this time, a number of trends appear evident to me:

° There will be no new plant orders for some time.

° There will be further cancellations.

° Plants which encountered serious design and construction quality assurance will likely have difficulty in the operating phase.

° The source term issue must be addressed and benefits incorporated.

° The PUC's role is in transition.

° Utility operating groups appear inevitable.

I recommend that NRC develop and maintain a long-range plan.

4. <u>Misuse of the Regulatory Process</u>: Congress has been unable to establish and maintain a consistent energy policy for the nation. A subset of this failure is the abandonment of policy regarding the development and use of nuclear power. In the resulting void, the nuclear regulatory process is being used by individuals and special interest groups to formulate energy policy in accordance with their interests. An obvious example is the public hearing and public comment process. Public interaction processes are important and should be used to inform and enhance safety, but not to set energy policy. These conflicting uses create guarded, adversary relationships detrimental to information exchange and inevitably impacting quality itself.

I recommend that a part of the NRC's response under the Ford Amendment requirements suggest that Congress reexamine the fundamental role of the NRC respecting nuclear power, in recognition of the manner in which the regulatory process is currently being used. If a national debate is desired, then it should be decoupled from the regulatory process charged with preserving the health and safety of the public. The mission of the NRC should be restated. The purposes, forms, and mechanisms for the public interaction process should be clarified.

**INTERNATIONAL
ENERGY
ASSOCIATES
LIMITED**

January 16, 1984

Dr. Willard D. Altman
Project Manager
Special Study of Nuclear Quality Assurance
Office of Inspection and Enforcement
Mail: EWS-305A
U.S. Nuclear Regulatory Commission
Washington, D.C. 20555

Dear Dr. Altman,

As requested, I return herewith copies of the draft documents which we reviewed in San Francisco on January 10 and 11. I have not attempted to duplicate the comments I made during the meeting, having confidence in your capacity to glean from what was said by me and others which thoughts you may care to use.

My reaction to the documents was, and is:

1. The Executive Summary needs to be rewritten in its entirety and especially cries for up-front balance on the treatment of NRC/Industry culpability and credit.

2. Chapters 1 through 10 presented a complete and thoughtful analysis, findings, conclusions, and recommendations as well as some background material. Substantively, with editing and changes we discussed, you appear to have in hand a very useful and constructive product. You are aware of the organization, redundancy, and material changes suggested.

3. I gave to you at the meeting suggested simplification of page 27 - F. Project Ownership and Management Arrangements and page 30 - Possible Legislative Initiative - both in the Executive Summary. If those sections do not survive in the Summary, my suggestions may be carried into the body of the report.

2600 VIRGINIA AVENUE, N.W.
WASHINGTON, D.C. 20037
202 - 342 - 6700
Telex 89-2680 Cable IEAL WASHDC

We discussed the desirability of eliminating the "scrunity" of State PUCs on the premise that their impacts on utility decision making on QA could be dealt with in the prospective study of arrangements for utility ownership, financing, design, construction, and operation.

I subsequently recommended to you that any recommendations for a study such as the latter include provision for determining what desirable changes are possible "within the present system", and not requiring legislation, as well as determining what possibly desirable changes would require legislation.

Good luck as you proceed to wrap up this very interesting study. I've enjoyed working with you and the other key NRC staff, as well as with the other consultants to the Study.

Sincerely,

John E. Gray
Chairman

JEG:kd
enclosures (3)

104 Humbolt Ct.
Oak Ridge, TN 37830

Dr. J. A. Christensen
Sigma III Bldg./3000 Area
Batelle Pacific Northwest Laboratories
Post Office Box 999
Richland, WA 99352

Dear Jim:

I wish to thank you and other Battelle personnel for having had the opportunity
to serve on this panel. The panel's efforts have been productive and I believe
will prove to be beneficial to the NRC. I thank you for a fine job of coordinating our meetings and the arrangements - they have been superb.

As requested, I am providing you with written comments on the material reviewed
this week. I will not repeat a lot of the comments, but only expand on those
that I feel to be most significant from my standpoint. Dr. Altman kept a good
set of notes and should have all remaining comments and suggestions.

Comments:

1. The Executive Summary should be shortened and enhanced by some graphs or
    charts that tell the complete story in a few pages.

2. A better balance is required between the NRC and utilities. This statement
    applies to both good and bad aspects. Both are to blame for past problems,
    but both have also taken a lot of positive steps to improve the construction
    process. The report also needs to address which initiatives/actions can/
    will apply to the operational phase.

3. The report addresses a lot of planned and proposed actions/improvements.
    When studied closely, they have taken a giant step toward development of a
    quality system, this message does not come out loud and clear. Graphics
    would again be beneficial.

    Likewise, the plans also will create a need for future plans/studies, i.e.,
    classification of characteristics for NRC and licensees, and a true
    evaluation/definition of inspection needs. They will be in a better
    position to define the number of inspections, frequency, number and type
    of inspectors and training requirements.

4. I have been confused on the intent of IDI's. Perhaps others are also not aware of their significance. Report needs clarification and expansion.

5. A number of inspections, audits, units, reviews are suggested. I suggest a time line covering design, construction and start-up. Overlay this with all of the plans/checks to determine if a proper approach is being taken and is it balanced.

6. Drop discussion on graded approach or change terminology and add more clarification as to what you are trying to accomplish.

7. Chapter 2, page 16 is confusing. Split into at least (3) bullets.

    °   The level of detail in QA/QC requirements in procurement documents is extremely important.

    °   It is essential to follow up to evaluate subcontractors' performance against these requirements.

    °   Necessary to evaluate a contractor's ability to perform to these requirements prior to issuing a contract.

8. Better define the intent of readiness reviews and the fact that they will be tied to milestones.

9. Chapter 7, pages 4 and 5. Items 2 and 5 need to also be listed in the Executive Summary.

Hopefully the above comments will prove helpful. Please contact me by phone if required at (615)482-3981.

Sincerely,

John L. Hansel

cc: Willard D. Altman, NRC

Robert V. Laney

Consultant
Energy Project Management

24 Trout Farm Lane
Duxbury, Massachusetts 02332
Phone (617) 585-8912

January 16, 1984

Mr. William D. Altman
Project Manager
U. S. Nuclear Regulatory Commission
Office of Inspection and Enforcement, Room 305
4340 East West Highway
Bethesda, MD 20014

Dear Bill:

The draft NRC report to Congress on Improving Quality in the Design and Construction of Commercial Nuclear Power Plants, dated January 6, 1984, is returned herewith. You will find a number of comments written in the margins on those pages which have paper clips on them. These are largely editorial. More important comments are included in the body of this letter.

On the whole, I consider the report to be very good. You give thoughtful attention to the five possible courses of action which you were asked to consider, and you reach a well founded conclusion for each. Beyond this you provide the Congress a wider perspective on the status of the problem, its causes, and remedial actions. The case studies are especially useful inasmuch as they focus attention on real people and events and make us realize that the problems which led to this study are seldom as simple or one-dimensioned as sometimes portrayed.

The remainder of this letter is devoted to five subjects drawn from reading the draft report. For each I offer comments followed by a recommendation for improving the next draft.

Assessment of Blame

It is not a primary purpose of the report to assess blame for the quality problems which gave rise to the Ford Amendment and this study. However, in studying problem cases and alternative programs for improving quality, you have necessarily looked for root causes, made findings, and implied blame. To contribute to better understanding and to assure wider acceptance of the report, it is important that it be even handed in assessing blame. We could not have confidence in a program of improvement unless it derives from a recognition of all significant causes.

In Section 2.2.2, pages 10, 11, and 12, the study finds shortcomings in project and corporate management and in NRC's slowness in detecting these shortcomings and taking enforcement actions. Clearly these are root causes which contributed to the scale of the problem. Not mentioned, however, are, (1) the failure by the NRC to make a searching evaluation of licensee management competence before issuing a Construction Permit, and (2) failure by the NRC to foresee that even an otherwise adequate management could be swamped and demoralized by numerous regulatory and hardware changes mandated in the midst of construction.

> RECOMMENDATION. Without reducing or moderating the reported shortcomings of project and corporate management, point out the contributing effect of the two NRC shortcomings identified above. This will, incidentally, lay the groundwork for the action proposed in Section 2.2.3A(1) and (2) on page 14.

Relationship Between NRC and INPO

Fostering an effective relationship between the NRC and INPO, one which allows each to do that which it can do best, should continue to be a constant goal of both organizations. This consideration is most compelling during a period of changing roles and expanding activities, such as that described in the NRC study. It is desirable for the NRC to allow ample scope to the industry's move to improve construction quality represented by INPO's Reconstruction Projects Evaluations (CPE).

INPO is the central feature of industry's determined commitment to self-improvement and self-regulation. Simultaneously, INPO is the industry's chosen instrument for achieving rising standards of performance in all phases of nuclear power, including, most recently, design and construction. Thus it is particularly important that, when setting a new agenda for strengthening the quality of nuclear construction, all concerned should recognize that INPO is similarly engaged. In deciding what inspections, audits, or evaluations it will do, the NRC should encourage INPO to do those which INPO might do as well or better. If this requires modifying the scope or methods INPO now uses, as the CPE's, NRC should discuss this possibility with INPO, as an alternative to continuing both CAT's and CPE's.

The present study includes, in Section 5.2B, page 4-6, and 9.3.1, pages 8-13, excellent descriptions and discussions of the respective NRC and INPO roles in achieving construction quality. The study concludes that the present role differentiation should continue, with INPO in a "counseling and advisory role" and the NRC in its statutory role of setting standards and inspecting to assure that those standards are met. This may be the appropriate conclusion at the present time. However, in my opinion, this section of the report would be improved if it were amplified to recognize that there are circumstances which, in the future, might argue for adjusting the NRC/INPO interface and their respective inspection activities.

First, INPO is exploring ways by which it might exert pressure on member utilities to respond constructively to correct faults revealed by INPO's evaluations. In addition, INPO appears to be moving towards a performance "ranking" system which will provide a utility management with a specific measure of relative success in achieving rising standards. These and related INPO initiatives, as they mature, will benefit from NRC recognition and a willingness to consider role adjustment as appropriate.

Second, in concluding that NRC and INPO roles are, for the present at least, fixed and separate, the study accepts the indefinite imposition of two similar design and construction evaluation programs with resulting duplication of demands on licensees.

> RECOMMENDATION. This report is the appropriate place for the NRC to acknowledge that (1) INPO is developing into an effective industry instrument for raising the quality of operations and construction, and (2) since INPO's potential is not yet fully realized, the NRC should remain alert to future improvement in INPO's program which would justify the NRC's placing greater reliance on it.

On Being More Prescriptive

In Section 2.1, page 2, the study rejects the use of more prescriptive A/E criteria, observing that degree of prescriptiveness was not a contributing factor in any of the projects reviewed. Since prescriptiveness is a favorite remedy of many who do not understand the complexity of the design process, it will probably be proposed again and again as a remedy for some perceived regulatory or enforcement lack.

RECOMMENDATION. Use Section 2.1 to inform readers of
the study that there are two good reasons for the NRC
to avoid being more prescriptive. First, there is
more than one satisfactory way to perform most design
activities, and prescription would unreasonably limit
the licensee's choices. Second, too much prescription
of "how-to-do-it" by the NRC tends to put the government
into a management role where they do not belong and do
not want to be.

Assessment of Corporate and Project Management Capability

In commenting on Alternative b(3), Section 2.1, page 2,
and in Section 2.2.3A(1) and (2), page 14, the study advocates
substantive assessments of corporate and project experience
and management capability as a condition of issuing future
Construction Permits. The study proposes developing criteria
of management fitness to be used for this purpose, but does
not indicate whether or in what manner these would be applied
to present CP holders. NRC staff discussions with the Study
Group on January 10-11, 1984, indicated that, in some manner,
perhaps by third party audits as in Alternative b(5), the
same criteria would be used in a capability assessment.

RECOMMENDATION. The study proposes developing criteria
of management fitness for future CP applicants. It
should provide an answer to the obvious question concerning their applicability to present CP holders. I believe they must be written so as to be applicable to
present CP holders, for otherwise their present usefulness is quite small. If the criteria consist, as I
believe they should, of a small number of basic
principles of sound operation, if they are capable of
objective verification by experienced observers, and if
they are developed with the assistance of the industry,
it should be possible to test present CP holders against
them without significant claims of unfairness.

Use of Terms "Quality Assurance" and "Management"

These two terms are used throughout the report as
though their meanings are approximately the same. The
principal finding (page 2, Section IA of the Executive
Summary) recognizes the difference. Since management failure,
including QA failures, is reported to be the principal culprit,
the report would be clarified by more careful use of these
terms.

RECOMMENDATION.   As noted above.

I compliment you and the NRC/BNL staff for the thorough preparation and presentation of the material on which the Study Group was asked to comment.   The meetings were conducted with efficiency and you provided us with every reasonable opportunity to assist you.   I sincerely hope that my suggestions are useful.

Sincerely,

Bob —

Robert V. Laney

RVL:pb
enc
cc: Dr. Fred W. Albaugh
     Mr. James A. Christensen

APPENDIX A

CASE STUDIES OF QUALITY AND QUALITY ASSURANCE IN THE DESIGN
AND CONSTRUCTION OF NUCLEAR POWER PLANTS

PRINCIPAL CONTRIBUTORS

W. D. Altman (a)   M. G. Patrick (b)
H. Harty (b)   J. L. Heidenreich (c)
K. C. Carroll (d)   L. D. Kubicek (d)

PREPARED BY

H. Harty

March 1984

Prepared for
Division of Quality Assurance, Safeguards,
  and Inspection Programs
Office of Inspection and Enforcement
U.S. Nuclear Regulatory Commission
Washington, D.C. 20555

Pacific Northwest Laboratory
Richland, Washington 99352

(a) Nuclear Regulatory Commission
(b) Pacific Northwest Laboratory
(c) N. C. Kist and Associates
(d) EG&G Idaho, Inc.

EXECUTIVE SUMMARY

Quality-related problems in the construction and/or design of several nuclear power plant projects have received much publicity in recent years. Because of the perceived severity of the problems and the consequent publicity and expressed public concern, the Congress and the Nuclear Regulatory Commission (NRC) determined that the NRC staff needed to conduct a comprehensive study of alternative approaches to assuring quality of construction at nuclear plants.

This appendix describes the results of six case studies of nuclear power plant construction projects, three that experienced major quality-related problems and three that did not. The objective of the case studies is to identify the reasons (the root causes) why some plants had quality problems and some did not. Based upon the findings, the conclusions made are aimed at improving the assurance of quality of nuclear power plant construction.

The differences between nuclear power plant projects that have experienced major quality-related problems and those that have not are shown primarily in four factors: comparative levels of nuclear experience, extent of management control of a project, depth of commitment to quality, and the maturity of formalized procedures for controlling a project. Also, utilities experiencing quality-related problems failed to fully appreciate the complexity, special conditions, and requirements of nuclear construction. Utilities not experiencing quality-related problems may have had a stronger emphasis on plant reliability. The NRC (or predecessor agency) also contributed to quality failures, or failed to mitigate them, by inadequately evaluating whether a utility applicant had the necessary capabilities to undertake a nuclear project, or by not requiring an appropriate upgrading of capabilities before issuing a construction permit.

Because utility management could reasonably be expected to be aware of the importance of these factors, why then did some utilities experience severe quality problems? Several reasons stand out. Most nuclear utilities had a successful fossil background. Seeing that others had made the fossil-to-nuclear transition with apparent success, they naturally assumed they could do likewise. In evaluating their approach to nuclear plant construction, they appear to have assumed that nuclear plant construction was not significantly different than fossil plant construction. That belief would tend to influence several important decisions. For example, utilities felt they could rely upon their existing organizational structure, staffing, contracting methods, and construction methods for nuclear projects with only minor modification. One interviewee noted, "One might need a few additional nuclear experienced staff, but they could be fitted in." Also, utilities thought they could depend upon their traditional contractors and architect-engineers (A-Es) to construct the plants correctly and on schedule, even though they might not fully understand requirements of the nuclear industry.

In the case studies, utilities not experiencing major quality problems did several things differently. They separated the nuclear project(s) from their traditional power plant organization. A representative from one utility said, "The old system of discipline and functional responsibilities won't work for nuclear plants, not even for new fossil fuel plants now." They staffed their organizations with people who had appropriate nuclear experience and hired experienced A-Es and contractors. Conversely, some of the utilities experiencing problems began their programs at the height of the nuclear generation boom (mid-1970s), when there was a shortage of experienced personnel and firms. These utilities were understaffed and/or hired inexperienced firms, resulting in problems and/or inefficiencies.

Another factor affecting whether quality-related problems had occurred was the extent of management control of a project. In the projects not experiencing major quality problems (or ones that had turned around their quality-related problems), the utilities managed the projects, not only in name but in actuality. They were aggressively involved in the projects. There was no question of who was in control. They took the responsibility for licensing actions and for NRC relationships. They were deeply involved in cost, schedule and productivity considerations, as well as in quality. They made appropriate changes in the project organization and approaches as conditions warranted. They monitored their contractors closely or sometimes were the contractors themselves. These utilities set performance standards for the project rather than delegate this responsibility to their A-E or contractors. They approved key design drawings and established criteria and procedures or assured that their contractors did.

Utilities successful in constructing plants (in terms of project cost, schedule and quality) had a few (sometimes only two or three) leaders who controlled the project. During interviews, the utility staff, A-Es, and construction staff at all levels often singled out a vice-president in charge of the project, the project construction manager, or a team as being the source of quality for the project or the driving force behind the project. Sometimes multiple teams were being used. No similar figures or teams emerged in the projects experiencing major quality problems; leadership appeared to be diffuse. Aggressive leadership could also be clearly identified in those plants which, after early difficulties, had turned around their quality-related problems.

Utility control is not the only approach to successful nuclear plant construction. The "turnkey" projects of the 1960s where the vendors exercised unified control was an alternative method capable of success. However, fragmented approaches to management control as allowed by some utilities was a characteristic that correlated with whether major quality problems had occurred. When asked, "Why is it necessary for the utility to become so involved in nuclear projects?" one chief engineer said, "I don't really understand it. An A-E should be able to do the job once given direction by the utility--at least the better A-Es should. But they can't seem to do it anymore. They don't have the same interest and they don't have the drive."[a] Others had similar comments.

Commitment to quality was a third important factor. Utility managements involved in the six case studies said they were committed to quality in their plants and always had been. Almost without exception, management at all levels said quality and safety took priority over project cost and schedule. However, in view of the quality problems experienced, commitment to quality is reflected more in aggressive action to control the project than in verbal endorsement of quality.

Achieving quality did not derive from any unique organizational structures or quality assurance (QA) and quality control (QC). Rather, the six cases presented various organizational arrangements for these functions. Staffs varied greatly in size and reported at various levels in the utility/contractor organizations. Some had effective stop work authority and some did not. Some participated in construction activity planning and some were separated from the day-to-day activities. Some projects had multiple QA/QC programs and some had a single program for most of the construction activities. Some organizations had multiple QA layers that audited lower levels. No pattern of QA/QC organizational structure or delegation correlated whether plants had experienced major quality problems. However, a common characteristic of all successful projects was strong commitment to quality at all levels and elements of organization.

Two characteristics of licensees seemed to accompany quality problems. One characteristic was responding to NRC requirements not by objectively examining the substantive issue involved but by seeking to placate the NRC as easily as possible. Although hard to prove, cases of this attitude are certainly present. Conversely, an obvious characteristic of utilities without quality problems is an aggressive position in responding to NRC requirements and questions. On projects free of major quality problems, it was the licensee itself and not a contractor that primarily responded to and interfaced with the NRC.

The second characteristic of licensees with quality problems was failure to make plant reliability a major driving consideration in the plant design and construction. Two of the projects (and possibly three) that had not experienced major quality problems (and none that had) had a strong emphasis on plant reliability. In both cases, the utilities would be rewarded for improved operating efficiencies by their public utility commissions. Reliability is a more tangible goal than quality and appears to compete well with schedule and cost for management attention. Quality, on these projects, was not an end in itself; it flowed from other considerations, e.g., achieve reliability and quality goals will be met.

(a) Quotations are approximate but are believed to convey the intent.

A fourth factor affecting quality was the maturity of formalized procedures for controlling a project. The successes of those plants without major quality problems (and the failures of those with) can be attributed in part to having adequate (or inadequate) procedures for all aspects of the project which were rigorously adhered to (or ignored). All of the case studies substantiated this requirement. Nuclear power plants are both complex and subject to change during construction, requiring appropriate procedure-related disciplines. In response to the question, "Why has there been quality in the design?" one A-E's project engineer singled out, "A proceduralized approach was adopted early in the project for calculations, specifications and procurement" with "rigid internal audits." The six case studies revealed a wide range of sophistication in specifying, applying and auditing procedures used in QA, construction, design, procurement and other aspects of nuclear projects. It almost appeared that each project had to create and apply its own procedures. Greater sharing the state-of-the-art in this area among licensees would be beneficial to assurance of quality.

NRC actions, or inactions, also contributed to quality-related problems. The NRC's overriding failures were its inability to evaluate adequately whether a utility had the requisite capability to undertake construction of a nuclear power plant and its failure to require an appropriate upgrading of capabilities before issuing a construction permit. The NRC was also tardy in conveying its quality concerns to some licensees, giving them the perception that quality-related problems were not serious. Some licensees said in retrospect that the NRC waited too long to force suspension of construction.

The case studies reinforced the need for plant designs that are substantially completed before construction is started. Designs that are largely completed would help reduce the opportunities for quality deficiencies and would also help reduce inefficiencies in licensing and construction. Standardized plant designs can aid in achieving this objective and in reducing design and construction times, which would support another need--maintaining the "team" throughout the project. Extended project times lead to personnel turnover and increased opportunities for quality problems. Major gains in nuclear plant quality (and productivity) would also be achieved if the teams remained intact from project to project, rather than being dispersed and reformed into new groups for new projects.

Concurrent with the case studies, an independent evaluation of the Zimmer nuclear project was conducted by Torrey Pines Technology. The findings from the Zimmer evaluation closely parallel those of the six case studies.

Having described the salient features and practices of those projects that did and did not experience major quality problems in construction, it is important to note that neither group did all things right or all things wrong. The projects without major quality problems experienced quality failures and project inefficiencies; much of the work of the projects with major quality failures appears to have been of good quality. The former did not have experienced, dedicated personnel in every position, and their procedural controls were not flawless. It cannot be said that their projects are exempt from quality failures--only that the probability and extent of failure was less because of appropriate experience, management involvement, dedication to quality, and procedural controls.

The case studies have focused on what has happened in the past or is happening now, not on what is likely to happen in the future. The increased industry and NRC experience and the lessons learned greatly decrease the probability of major quality problems, especially in future generations of nuclear plants. However, there are several conditions under which major quality problems might recur. These include:

- a first-time utility with a staff or A-E/constructor that have inadequate nuclear experience.

- a very large growth in the number of nuclear plants being constructed that (again) overwhelms the industry's capabilities

- a long delay before nuclear plant construction activities start again, resulting in a dearth of experience in the industry

- regulatory actions at federal and state levels which undercut quality.

The NRC and the nuclear industry need to be aware of the implications for quality that these possibilities hold.

ACKNOWLEDGMENTS

The case studies documented in this report were performed as a cooperative effort by multidisciplinary teams led by the NRC project manager. The principal contributors to these efforts were: W. D. Altman, NRC Project Manager; H. Harty, PNL Project Manager; M. G. Patrick, PNL; J. Heidenreich, Kist Project Manager; K. C. Carroll, EG&G Project Manager; and L. D. Kubicek, EG&G. Others who contributed to these efforts in lesser but significant degrees include E. W. Brach (NRC), E. Bradford (EG&G), R. M. Kleckner (Kist), D. T. Ross (EG&G), G. T. Ankrum (NRC), J. A. Christensen (PNL), R. Dellon (Dellon Assoc.), J. E. Kennedy (NRC), H. J. Kirschenman (EG&G), M. McGuire (PNL), R. J. Sorenson (PNL), and M. E. Walsh (PNL).

This study would not have been possible without the willing cooperation of the six utilities involved:

Public Service of Indiana   Georgia Power Company
Houston Lighting and Power Company   Pacific Gas and Electric Company
Arizona Public Service Company   Florida Power and Light Company

Each of these utilities consented to sharing with its case study staff their experiences and the insights gained from them in the design and construction of nuclear power plants. They did so in order to benefit other similar projects and to provide a solid, factual foundation for improving quality and the assurance of quality in the nuclear industry. The purpose of this report is not to critize or compliment any specific utility. Rather, it is an attempt to evaluate the generic lessons arising from the experience and insights of the utilities involved.

CONTENTS

APPENDIX A

CASE STUDIES OF QUALITY AND QUALITY ASSURANCE IN THE DESIGN AND CONSTRUCTION OF NUCLEAR POWER PLANTS

A.1 INTRODUCTION

This appendix describes the results of six case studies of nuclear power plant construction projects in the United States. Three of the projects experienced major quality-related problems in design or construction and three did not. The root causes of the performance of each group are identified. Section A.1 presents introductory material on the study's background, purpose, and technical approach. Section A.2 presents conclusions and findings. Section A.3 describes the case study process and summarizes the major findings from each of the case studies. Results of an independent study of the Zimmer nuclear power plant construction project are included in Section A.3 for comparison purposes. References are provided in Section A.4.

A.1.1 Background

In recent years, there has been a series of well-publicized problems relating to the quality of construction and/or design at several nuclear power plant projects in the United States. It is important to understand what caused these problems and why some nuclear construction projects have been more successful in achieving quality than others. In an August 1982 paper to the Commission (NRC 1982), the NRC staff proposed a long-term review and study of the quality problems in the nuclear industry. A key feature of this long-term review was a series of analyses of representative nuclear construction projects having had varying degrees of success with respect to project quality to ascertain the underlying causal factors, or root causes of quality success or failure in nuclear construction projects. These analyses, which included site visits, were called case studies. They began in November 1982 and continued through August 1983. Six case studies were completed: three at projects that had experienced major quality-related problems and three at projects that had not. Three projects were in the range of 25-50% completed; three were recently completed or essentially completed. The projects were located in four of the five NRC regions.

An analysis of Cincinnati Gas and Electric's Zimmer plant was performed recently by Torrey Pines Technology (TPT). Because of the relevance of the TPT findings on Zimmer to the questions addressed by the NRC case studies, results of TPT's evaluation of Zimmer (TPT 1983) are also included in this appendix.

The case studies were not inspections, investigations, or audits, and thus were conducted outside the normal regulatory process of the NRC. For the most part they were limited to the construction sites and licensees' offices (and NRC regional offices); assurance of quality in the design and procurement processes was not examined in the same detail as the construction process.

A.1.2 Purpose

The purpose of the case studies was to determine the essential differences between nuclear power plant projects that had experienced major quality problems in design and/or construction and those that had not, and to highlight the lessons learned. These lessons provide a basis for considering changes to the NRC's activities supporting assurance of quality in nuclear power plant projects.

A.1.3 Technical Approach

Each case study had three phases: a pre-field activity, a field activity, and a post-field activity. The pre-field activity consisted of a general familiarization with the licensee and project, including the project quality assurance program and its history. Relevant NRC inspection and investigation reports and licensing documents were reviewed. Postulated root causes for successes or failures were developed to provide a framework for the subsequent interviewing process.

The field portion of the case studies typically commenced with briefings from the NRC regional offices and from licensee management to two of the case study team members prior to the full-team visit. Then the entire team would meet with the licensee's management at the start of the five-day site visit. During this meeting (and typically as a preface to each individual interview with licensee and contractor personnel), the purpose of the case studies was described. All were told that the case studies were not inspections or audits.

The case study teams were comprised of NRC and contractor personnel who collectively have experience in nuclear plant engineering and design, project management, construction management, operations, systems analysis, quality engineering, quality control, and quality management. Contractor personnel were selected from two national laboratories and from two consultant firms. To assure consistency in each case study, three team members, including the NRC case studies project manager, participated in all the case studies. Three others were assigned on a rotational basis. These six individuals comprised the core group for the case studies.

Twelve other staff from the national laboratories, NRC, and consulting firms participated in selected case studies to provide fresh ideas and perspectives. These individuals included management-level personnel qualified to critique the process. To further ensure that no key elements were being missed in the case study process, the results were periodically reviewed by a peer panel consisting of noted experts in their fields.

In the early case studies, three subteams of two personnel were used; one subteam concentrated on construction management and investigated the interfaces between the licensee and the contractors, paying particular attention to the quality control of construction. The second focused on project engineering and design processes and on interfaces with the architect-engineer, construction management, and quality assurance aspects of the project. The third concentrated on the quality assurance program, its organization, and personnel qualifications and training. A fourth subteam that was added later concentrated on corporate management's functions in the project and its approaches to the assurance of quality for the project.

Typically, 40 to 60 people, from top management to crafts and QC inspectors, were interviewed in each case study. The QA program was reviewed together with selected records, and a plant walk-through was conducted. The case studies did not include any technical review or evaluation of adequacy of plant design or construction. Apart from the plant walk-through, during which time team members were able to talk to additional craft workers, field engineers, and inspectors, no physical inspection of the plant was performed.

The field work concluded at the end of the week with an exit briefing for senior licensee management and staff. In a typical briefing, the case studies project manager presented the team's tentative findings regarding root causes of quality-related problems, or the absence of them, and related information. The NRC initiatives and the Congressional alternatives for improving quality were also discussed, as were the team's perspectives for the licensee to consider to further enhance its quality program. The licensee exit briefing afforded the opportunity to offer additional information, corrections of fact, and agreement or disagreement with the team's tentative findings and conclusions. The exit briefings were typically two to three hours long.

Post-field activity consisted of the preparation of a draft working paper. Subteams compiled individual reports which were incorporated into a case study draft working paper. The draft working papers served as resources for this study.

The information obtained through the interview process was taken at face value; however, several mechanisms for establishing confidence in cogent data were utilized. Generally, the findings and insights were corroborated by comparing information from more than one source:

- by interviewing personnel from a vertical cut of the project organization

- by extensive review of NRC file documents and other sources of data

- by interviews with regional and resident NRC personnel familiar with the project and its history

- by sharing and examining data at daily team caucus meetings.

Further confidence in the primary findings of the case studies can be gained from their similarity to those of the Torrey Pines Technology study conducted for the Zimmer nuclear plant (TPT 1983). The latter study used a different approach and was conducted in greater depth. It is summarized in Section A.3.1.

A.2 CONCLUSIONS AND FINDINGS

1. The single most important factor in assuring quality in nuclear power plant construction is prior nuclear construction experience (i.e., licensee experience in having constructed previous nuclear power plants, personnel who have learned how to construct them, experienced architect-engineers, experienced constructors, and experienced NRC inspectors).

    - This experience brings with it knowledge of the complexity of nuclear plant construction, understanding of regulatory-related quality requirements, the need for management leadership, and many other factors. These factors are poorly understood by those without experience and this lack of understanding leads to quality-related problems. Where licensees had marginal experience in critical areas (e.g., in the transition from construction to operation), they were prone to quality-related problems. The broader their inexperience, the more severe the problems were likely to be.

    - A high degree of design and engineering completion prior to construction, together with regulatory stability, might partially compensate for a lack of experience. There are some data showing that plants having higher design and engineering completion may have fewer construction-related quality problems arising from rework or extended project schedules. Standardization may produce comparable results.

2. A factor that ranks close to experience in importance is licensee management involvement in and control of the project. The project activities that compete with quality (i.e., cost and schedule) are not properly balanced without strong licensee management control and involvement.

    - Licensee contractors do not have the same overall responsibility that the licensee has nor do they have the same authority and resources to deal with quality-related problems. When a licensee abdicates its role, some aspect of quality, cost and/or schedule is likely to be compromised.

    - Licensees are also being forced to take more active roles in upgrading many aspects of the nuclear industry because of regulatory requirements--especially those aspects related to the quality of products or work from equipment suppliers and construction contractors. This has not been a role traditionally filled by licensees for their fossil-fuel (or other types of) plants. Where licensees have followed fossil-fuel practices and chosen not to be involved in supplier and contractor activities, quality-related problems were more likely to occur.

- Some licensees are now exercising the right to approve key A-E/C personnel for their projects to help assure quality and maintain project efficiency.

3. Another essential factor is a management commitment to quality. This is essential to facilitate activities that support quality of construction.

- All management claims to support quality, but verbal support is not sufficient. An understanding is required of why quality is important (e.g., as an important adjunct to achieve an acceptable level of safety, reliability, or scheduled completion) and how to obtain it. That understanding must be disseminated through the entire project team by training, personal contact, audit appraisals, support of QA/QC staff, incentives and other means.

- A commitment to quality seems encouraged by financial incentives. These may take the forms of an improved rate of return for high levels of operating efficiency, reduced maintenance costs, etc.--factors that may more than compensate for added construction costs incurred in the interest of quality or enhanced safety of operation. The role of public utility commissions in providing incentives for improved performance (a measure of quality) and adherence to NRC regulations needs to be considered.

- Safety by itself does not appear to be a sufficient motivation for ensuring good quality. For the most part, industry has been lagging the NRC with respect to assurance of quality. This is evidenced by the fact that industry does not appear to feel that greater attention to quality is needed. That situation is likely to change only when the utility industry focuses on an objective that is more meaningful to them--one that includes safety, perhaps reliability. Licensees seem to believe that their plants are (or will be) safer than the NRC credits them to be; thus, assurance of quality requirements often appear excessive to licensees.

4. Maintaining and documenting adequate quality requires appropriate procedures for all aspects of the project (i.e., construction, design, procurement, etc.). These procedures must be understood, rigorously applied, and adhered to at all levels of the project.

- There is a spectrum of assurance-of-quality practices ranging from outstanding to marginal in the nuclear industry. The superior practices appear slow to be propagated throughout the industry.

A.5

- The source of much contention about the adequacy of quality control in construction is the recordkeeping aspects of procedures. These need not be sophisticated, computer-based methods, but licensees with experience advocate the use of computer-based systems.

- The hiatus in new nuclear plant construction offers the NRC and the industry an opportunity to establish and disseminate improved practices with respect to procedures and records.

5. The case studies revealed several shortcomings in past or present NRC programs that have an effect on assurance of quality:

- The licensing focus with respect to assurance of quality has been on form, not substance; the NRC's inspection focus has tended to be on records rather than on quality of product.

- There has been little assessment of management capability as part of the construction permit review.

- The NRC's inspection presence at construction sites in the past has tended to be irregular and nonconstant and continues to be so in the initial stages of construction.

- The NRC's construction site resident inspection staff is too small to be expert in all phases of nuclear plant construction and construction management.

- The NRC has been slow to take action on management issues that are often at the root of quality-related problems.

- The NRC has failed to treat or sell QA as a management tool.

- Changing regulatory requirements have resulted in quality-related problems, and this factor has not been adequately addressed by the NRC.

- The NRC has done inadequate review or auditing in the past to verify quality in nuclear plant design processes.

6. Nuclear utilities are changing. Utility managements are becoming more aware of the special requirements for nuclear plants vis-a-vis other generating methods. Licensee nuclear staffs are increasing in size and capability. The utility industry seems to be assuming a larger role in the engineering services for operations. The transition from A-E to licensee for engineering services and the adequacy of the licensee to perform these services may need to be evaluated by the NRC.

7. The case study approach proved to be a useful tool for identifying and comparing assurance-of-quality practices. NRC regional and site personnel could benefit from case-type studies at other locations to gain insights

into alternative practices and to help avoid regional differences in approaches to quality.

A.3 DISCUSSION

This section describes the case study process and summarizes the major findings from each of the six case studies. The circumstances of each case study are described and the root causes of the quality-related problems--or lack of them--are identified. An independent study of the Zimmer nuclear power project is also summarized.

A.3.1 Case Studies

Case Study A

The licensee of Case Study A is constructing its first nuclear station, which consists of two large (>1000 MWe) units. Unit 1 is presently half completed; Unit 2 is about one-third completed. Construction permits (CPs) were issued in the late 1970s. Initial planning and site selection work commenced in the mid-1970s. Placement of safety-related concrete commenced in 1978.

The attitude of the licensee from the outset was one of confidence and adherence to practices that had worked in constructing previous fossil-fired plants. There was some recognition that nuclear projects would be different from fossil projects, but the differences were thought not to be great and could be largely overcome by hiring some managers and staff with prior nuclear experience. Also, the use of a nuclear plant design that was already well into construction at another location was a very positive factor. Completing the project on time and within budget was an important goal.

The licensee's prior construction experience consisted of about 20 fossil-fired plants. In some cases, the licensee had served as construction manager. The licensee had a construction department headed by a vice-president who was responsible for all utility construction. Over the years, the licensee had developed a close working relationship with, and confidence in, several major construction contractors who worked on its fossil-fired plants. The licensee's construction success for fossil-fired plants was a source of justifiable pride. Each plant had come on-line before schedule and within budget. The plants were of acceptable quality after the usual startup problems, and each plant operated safely and reliably. Quality was something put into the plant by the builders--there was no formal program for quality or the assurance of quality. To the licensee, quality was something that happened if you put good people on the project.

This licensee, in common with others in the industry, had a conservative management philosophy and was adverse to taking unnecessary risks. Contributing to this conservatism is the scrutiny of the public utility commission regarding how the licensee spends money and handles finances. These factors supported the licensee's cost and schedule consciousness.

Given the inherent conservatism of the licensee and the risks and uncertainties associated with nuclear power, why did it elect to build a nuclear plant? Many factors appear to have been involved, including projections of future energy demands, the price of oil and its future availability, the fact that other utilities, including first-timers, had built nuclear plants with apparent success, and analysis showing nuclear power to be not only cost effective but reasonably risk free. Going nuclear may have been a break with tradition, but it still represented a conservative decision.

The project was started under a Limited Work Authorization (LWA), which permitted non-safety-related work to be conducted prior to CP issuance. The licensee was the general contractor for the project. A firm experienced in the design and engineering of nuclear projects was retained as architect-engineer (A-E). A construction company that had previously participated in the construction of several fossil-fired plants for the utility was retained as the civil engineering contractor for the project. The civil contractor's nuclear experience was limited to providing workers for projects managed by other firms. It had never been the prime civil engineering contractor for a nuclear project. The licensee contracted with other firms for the mechanical, electrical and other aspects of the project. In the early phases of the project, the civil work fell behind schedule, and considerable pressure was applied by the Licensee to regain lost time.

About one year after CP issuance, the NRC identified deficiencies in the quality of the concrete work; e.g., severe cases of segregation and/or honeycombing. There had been many nonconformance reports filed regarding the concrete work since the start of the project. The utility agreed to upgrade its quality assurance program for concrete work and to determine through testing if previously poured concrete was adequate. About one month later, a former employee of the civil contractor alleged that surface defects in the concrete had been improperly patched. Concurrently, but independently, the National Board of Boiler and Pressure Vessel Inspectors confirmed code compliance problems with piping installations previously identified by a mechanical subcontractor.

The concrete deficiencies and the National Board findings led to an intensive NRC team inspection, which resulted in shutdown of all safety-related construction activities. The NRC determined there were programmatic questions concerning the licensee's project management, construction management, and quality assurance programs sufficient to warrant stoppage of safety-related construction work until they could be satisfactorily resolved.

The licensee retained a management consulting firm to perform an in-depth analysis of the project. The consulting firm confirmed the existence of, and helped identify, underlying programmatic deficiencies in the project. Their report outlined a 20-point plan to restructure and improve the project. Subsequent to that report, the licensee detailed to the NRC its effort to upgrade and implement its revised program for project and construction management and the assurance of quality.

A.8

To assure that the licensee's corrective actions were properly and effectively implemented, the NRC approved a five-step plan for gradual rescission of the shutdown order.

The licensee was permitted to resume receipt inspections of materials at the construction site about one year after the stop work order and after restructuring its project and construction management and quality assurance programs. Limited electrical and pipe installation work resumed six months later, followed by all remaining safety-related work, including concrete placement, in another four months. Unrestricted authority to continue the work was granted when the utility successfully demonstrated to the NRC that its revised project and construction management and quality assurance programs were implemented properly. The total time period from work stoppage to full resumption of all construction activity was about two and one-half years. Substantial non-safety-related civil work was completed while the stop work order was in effect.

During this period, the licensee substantially restructured its project management, construction management, and quality assurance programs (including records management). Substantial numbers of well-qualified people were hired. A nuclear division, whose sole responsibility was the nuclear construction project, was formed. The division manager, a senior vice-president, was located at the plant site. Morale improved considerably and team spirit and project determination pervaded the project.

Three years after the quality problems became so pervasive that all safety-related construction work was halted, the cognizant NRC regional office rated the licensee's QA program "outstanding" (the highest rating) on the annual NRC Systematic Assessment of Licensee Performance (SALP) review. The licensee received the rating of "outstanding" the subsequent year also.

The Case A study team identified the following root causes to be significant in contributing to the major quality failures experienced by the licensee.

The licensee's inexperience in nuclear power plant construction projects, and its failure to appreciate and understand the difference in difficulty and regulatory requirements between fossil and nuclear construction projects. The licensee had managed or overseen the construction of several successful fossil projects and it approached the nuclear project as an extension of the earlier fossil construction activity; i.e., to be managed, staffed, and contracted out in much the same ways as fossil projects. The licensee did not appreciate or understand the difference in complexity and regulation between fossil and nuclear projects and treated the nuclear project largely as just another construction project. The licensee's lack of experience in and understanding of nuclear construction requirements manifested itself as follows: lack of adequate staffing for the project, both in numbers, qualifications, and applicable nuclear experience; selection of contractors the licensee had used previously in building fossil plants, but who had very limited nuclear construction experience; over-reliance on these contractors for the management of the project and evaluation of its status and progress; use of fixed-price contracts where scope of work was inadequately defined; oversight of the project from corporate headquarters, with only minimal presence at the site; a lack of appreciation for the importance of ASME codes and other nuclear-related standards; a misunderstanding of the NRC, its practices, its authority, and its role in nuclear safety; and an inability to recognize that the piping and recurring concrete quality problems were merely manifestations or symptoms of much deeper underlying programmatic deficiencies in the management of the project.

The licensee's failure to understand and appreciate the potential merit of a formal program to assure quality. The licensee had built fossil units successfully in the past without having a formal program for the assurance of quality. For the nuclear project, NRC regulations required the establishment of a formal quality assurance (QA) program. The licensee viewed this requirement as just another government agency-imposed requirement necessary to obtain a license and treated it accordingly. The licensee inadequately staffed the QA function, in numbers, qualifications, and nuclear experience, and failed to listen to the QA organization when it reported quality problems and it (and other project components) asked for additional resources. Senior management was skeptical about formal QA programs; earlier, successful fossil projects had been completed without a QA program, and there were concerns about the QA organization trying to build an "empire." Quality, they felt, was something that came naturally to their projects.

The licensee's false sense of security in moving from fossil to nuclear construction. The licensee was unaware of the seriousness of the quality problem up to the issuance of the stop work order and had developed a false sense of security resulting in part from the following: past fossil plant successes; use of many of the same contractors who had worked on fossil units; believing the contractors when they indicated that the project had no major problems; believing that similar concrete placement practices and problems were common in nuclear construction; assuming that serious problems would not likely occur since the project's nuclear units were replicates of other plants being constructed by a more experienced utility; and believing that there were no major problems with the project since NRC inspection findings (until the inspection resulting in the stop work action) revealed none, having focused on details and minor problems. The licensee had little concept of the effect that regulatory changes were having on the "replicated" design.

The licensee's failure to adequately manage the project from the outset. This cause is related to the first cause; i.e., inexperience. In retrospect, the project was not being adequately managed by anyone. In the project structure, the role of project manager belonged to the licensee. The licensee acted as general contractor and construction manager, but managed the project more in an overview role. The licensee managed the project from corporate headquarters with minimal site presence and without effective control over its contractors. Accountability for the project was delegated among several organizations within the licensee's organization. The replication of design contributed in some degree to the failure to manage; the licensee felt that any major problem would develop first at the project being replicated, and it would have time to make adjustments on this project.

NRC licensing and inspection deficiencies. For construction permits, NRC licensing review is limited largely to technical and engineering issues. The NRC does not and did not in the case of the licensee, evaluate whether it and its contractors had the experience, knowledge, staffing, or ability to effectively manage and complete a project as complex as the construction of a nuclear reactor. Moreover, the NRC's inspection activity at the site was irregular and nonconstant, with several inspectors in different disciplines visiting individually for only short periods of time, and with no one (until the inspection resulting in the stop work action) recognizing that the reported deficiencies were symptoms of deep programmatic assurance of quality problems. The first resident inspector was not assigned to the site until four months after the stop work order. Just as the NRC, through its regional inspection program, was slow to put together the comments and evaluations coming from individual inspectors, so too was the licensee slow to recognize the extent of the programmatic quality assurance problems. Indeed, the licensee interpreted NRC's early narrow inspection findings as an indication that there were no major problems, and the licensee had some difficulty comprehending the stronger, more pervasively negative findings of the NRC inspection.

At the time of the Case Study A site review, the licensee had effectively implemented substantial modifications and improvements to the management of the project, and the project was regarded by cognizant regional NRC officials as having been turned around and as being something of a model project. The Case Study A team findings supported this assessment.

Case Study B

The licensee of Case Study B has one nuclear station in operation and a second under construction. Both consist of two large units (~1,000 MWe each). The former station has been in operation since the mid-1970s. The latter station is less than half completed. Its CPs were issued in the mid-1970s. Licensee fiscal problems required an approximate 18-month showdown in the construction of the station, so commercial operation is not anticipated until the latter half of this decade.

The licensee is the construction manager for the project. The major construction contractors--civil, mechanical, and electrical--all have had significant nuclear plant construction experience, as have many of the smaller contractors.

The A-E for the Case B nuclear station has had extensive experience in the design and construction of nuclear power plants. Some of the non-safety-related design is being done by the engineering staff of the licensee's holding company.

The licensee has experienced no major quality problems to date in the construction of this nuclear station (and, as far as the case study team knows, none occurred in the construction of the first station, either). There have been minor quality problems in the areas of engineering and construction, but the licensee has taken positive action to correct them. There has not been significant public intervention in the construction permit licensing or construction phases of the Case B nuclear station. No significant fines have been levied against the licensee for nonconformance violations or quality deficiencies.

The Case B study team identified the following root causes to be significant in contributing to the absence of major quality failures.

The licensee has an experienced design, construction, and construction management team. The licensee has had prior experience with a previous nuclear station, and many of the personnel who worked on it are now involved in the present project. This experience has given them an understanding and appreciation of the complexity of large nuclear station construction activities. Many of the staff have 5-15 years experience in nuclear work. The persons contacted, in general, had good qualifications for their assignments. There is a substantial training program and an overall impression of a high level of dedication and enthusiasm to the project. Early in the construction process, it was recognized that craft personnel available in the area needed further training on the special requirements of nuclear work, and this resulted in a comprehensive craft training program. The QA/QC staff is broad and deep in experience and qualifications.

The A-E has designed (and constructed) many nuclear power stations.

The major construction contractors (especially the mechanical and electrical contractors) and the smaller contractors have had previous experience in construction of nuclear projects.

The licensee has an orientation toward, and an attitude supportive of, quality in its nuclear project. The stated management philosophy of insisting on quality was not simply to satisfy the NRC, but to go beyond those requirements to have a reliable and safe operating plant. At higher levels in the management structure, the conviction appeared to prevail that public safety and company profitability demand assurance of quality in the construction (and operation) of nuclear plants, and that it is less expensive in the long run to "do the job right the first time." From the interviews conducted, both at the corporate offices and the site, it was evident that a sense of commitment to quality pervades the licensee's organization at all levels. The licensee volunteered to participate in the first Institute of Nuclear Power Operations (INPO) construction pilot audit and has expanded on it with its own self-initiated evaluation. The quality assurance staff has direct access to an executive vice-president. There was no indication from the interviews of cost/schedule overriding QA/QC. At lower levels, there was an expressed feeling that the company wants to do the job right. Employees at all levels appeared to have a constructive attitude toward the need for quality in general, and the proper application of quality assurance, in specific. A pro-company attitude and good morale on the part of the employees appear to exist.

The licensee manages the project, has clearly defined the responsibilities and authorities of the participants, and has provided adequate procedures to ensure compliance, especially at the interfaces. This is manifest most clearly in day-to-day activities at the site. The licensee is running the job. The licensee does not rely on the major contractors to perform overall management functions. There are limited and defined points of contact through which the licensee directs the work of its contractors. It is also manifest by the fact that the direction for the overall quality assurance program comes from the licensee and not from its subcontractors. Personnel within the licensee's and the major subcontractors' staffs were knowledgeable of their own, as well as others', responsibilities and authorities. (This, despite the fact that the organizational structure is quite complicated and not easily understood at first review; however, within the plant project team, the organizational structure is straightforward). Large geographical separation of some of the major organizations from the site; e.g., the A-E and the Nuclear Steam Supply System (NSSS) contractors' home offices, in particular, was seen to hamper communication.

The licensee supports its quality assurance program with adequate resources and backing. This is manifest by a Product Management Board comprised of senior utility management, senior project management, and senior A-E and NSSS representatives. The Board reviews the project, examines problems and maintains cognizance of nuclear matters. Quality does not seem to be sacrificed for schedule and cost considerations. (The case study team did not have occasion to evaluate schedule and cost pressures, however.) As previously mentioned, the licensee and contractors have good training programs for crafts and quality control personnel. The planning, scheduling, and budgeting activities appear to allow for adequate resources to do the job properly. Chronic delays were not evident. Procedure compliance was stressed at all levels and daily work schedules appear realistic enough to allow work to be completed in accordance with those procedures.

The licensee is proactive in looking for improvement in its assurance-of-quality practices. Key line managers were taken on a retreat by an executive vice-president to consider new approaches for assuring quality. This licensee volunteered to be the first to be evaluated under 10 CFR 50 Appendix B requirements in the early 1970s. Its own QA organization was asked by senior management to study the QA programs of other licensees for possible improvement as early as 1978. The licensee has been involved in one of the pilot studies for the INPO audits. It has also participated in self-initiated evaluations. There were numerous comments and indications in the interviews that problems, deficiencies, and areas of improvement can be surfaced without punitive actions.

The licensee's QA/QC function is active in reviewing, witnessing, and verifying contractors' work and in helping assure that corrective action is implemented. A well-staffed program with good procedures exists to ensure that construction conforms to the design. Licensee construction coordinators, many of whom have been quality control inspectors, do a preinspection of craft work prior to formal inspection by QC. There is feedback of lessons learned from earlier construction experience and from other projects. The licensee and its contractors have an effective corrective action program that brings about needed change. Design reviews by the licensee for constructability and operability were thorough. Licensee management interviewed indicated that they encouraged their staff to surface problems as soon as possible. In the long run, it was more beneficial and cost effective to do it earlier than later.

Case Study C

The licensee of the Case C study had established its own in-house engineering and construction management capability in the 1930s. During the late 1940s and early 1950s, outside A-E firms were used because of unusually large (post-WW II) system expansion requirements. In the mid-1950s, the licensee's earlier practice of doing its own engineering and construction management was resumed.

During the late 1950s and early 1960s, the licensee planned an ambitious program to construct several nuclear power stations. Nuclear power was recognized as a new technology and the licensee took actions to prepare itself for entry into this field. These actions included having observers at the construction sites of some early nuclear power plants, participating in the design of a test reactor, and studying A-Es' designs of proposed nuclear plants. The licensee decided to build its first nuclear plant--a small (<100 MWe) power reactor--through a "turnkey" contract for design and construction. The plant was completed in the early 1960s, and the licensee operated it successfully for about 15 years until it was retired. The licensee capitalized on the turnkey design and construction activity to familiarize its staff with nuclear activities and to enable it to engineer and construct subsequent nuclear plants. The licensee had been successful in engineering and construction activities on a variety of generating technologies and related electrical transmission systems.

During the early- and mid-1960s, the licensee announced plans for several nuclear plants. Environmental and/or seismic problems, coupled with intense intervention, resulted in all but the Case C nuclear station being canceled. These factors were also present in the Case C project, resulting in significant delays and cost increases.

The Case C nuclear station is comprised of two large (>1000 MWe) units. The licensee announced Units 1 and 2 in the mid-to-late 1960s. Construction permits were issued in the late 1960s and early 1970s. Unit 1 of the nuclear station was largely completed by the mid-1970s and fuel was received onsite for both units in 1975 and 1976.

Then a series of required modifications to the nuclear station delayed its completion. These were promulgated by NRC regulations such as pipe-break-outside-containment which necessitated, among other things, relocation of several conduits (1973-75); identification and/or reconsideration of a seismic fault which required such modifications as column stiffening, tank bracing, revised piping changes and equipment supports, diaphragm stiffening, buttressing and foundation changes (1978-79); the Brown's Ferry incident, which required modifications related to cable spreading, inerting atmosphere, new decking, and extensive concrete anchor bolt installation (1980); and the TMI accident, which required installation of extensive additional wiring, sub-cooling monitors, hydrogen recombiners, and other modifications (1981).

It is important to note that, over the time span of about eight years, one of the two units had been within a few months of being ready for fuel loading on several occasions. Thus far, Unit 1 has undergone three hot functional tests and three containment leak tests. Unit 2 has undergone one containment leak test.

In 1981, the licensee received a low-power license for Unit 1. It was suspended two months later following notification by the Licensee to the NRC that the diagrams used to locate the vertical seismic floor response spectra in the Unit 1 containment annulus area were in error. Briefly, the error occurred as follows. The licensee had transmitted to its seismic consultant a sketch with piping loads depicted from which the consultant was to determine the seismic response spectra. There was no indication on the sketch which unit the loadings applied to, although the consultant understood (correctly) that they were for Unit 2. The consultant thought that Unit 1 was a slidealong unit (instead of a mirror-image unit) and performed the analysis on Unit 1 based on that assumption. The information returned to the licensee was marked as "Unit 1." (In fact, the analysis applied to Unit 2, not Unit 1.) The licensee accepted the data at face value as being for Unit 1 and, because it knew the plants to be mirror-image plants, "flipped" the data so as to be applicable to Unit 2. (In fact, the data in the "flipped" condition were correct for Unit 1, not Unit 2.) The seismic response spectra were now incorrect for both Units 1 and 2.

Upon confirmation that wrong diagrams were used in developing Unit 1 design requirements, the licensee reanalyzed the design requirements for Unit 1 using the appropriate containment annulus frame orientation diagrams and determined that, as a result of the error, modifications were required to be made on several Unit 1 pipe supports. These modifications involved such actions as adding snubbers, changing the snubber size, adding braces, replacing structural members, and stiffening base plates.

In an inspection report of seismic-related errors, the NRC stated that the basic cause of this problem appeared to be the informal manner in which the subject data were developed by the licensee and transmitted to its seismic consultant, and the lack of independent review of the data within the licensee's organization prior to submittal to that consultant.

The licensee had been the architect-engineer/construction manager for the Case C nuclear power station. One of the major actions that the licensee took as a result of the aforementioned error was the formation of a Project Completion Team comprised of the licensee's engineering/construction personnel and personnel from a newly hired A-E firm.

An extensive Independent Design Verification Program (IDVP) was initiated in early 1982 in response to the seismic errors discovered in 1981. The Project Completion Team also conducted a concurrent design verification program.

As of January 1983, much of the design and construction required as a result of a wide range of reviews spawned by discovery of the seismic diagram error had been completed. The licensee had applied for reinstatement of the low-power operating license.

At the time of the case study, neither the IDVP nor the licensee's design verification program had revealed significant further deficiencies in the design or construction of the nuclear station. The design errors that were identified were not considered to have prevented the affected systems from performing their functions satisfactorily.

The Case C study team identified the following root causes to be significant in contributing to the quality failures experienced by the licensee.

The primary root cause of the design-related quality problem was the licensee's failure to plan, establish, and effectively implement a management system which provided adequate control and oversight over all aspects of the project. The licensee failed to fully control the flow of information across all the interfaces inherent in the engineering/design process and failed to provide appropriate reviews of the information transmitted.

Several factors appear to have contributed to this failure. Using the experience gained from their earlier turnkey plant and participation of the staff in other nuclear projects, the licensee, after considerable evaluation, assumed the role of A-E for this nuclear project. As previously stated, the licensee had good success with various types of generating projects it had engineered and managed over the years. The nuclear project was fitted into a design, engineering, and management system that may not have been adequately modified to handle all aspects of nuclear work, including the control of quality at design interfaces. Generally, it has been more difficult to apply QA to the engineering process than to the construction process; historically it has not been done effectively and the licensee had similar difficulty. Even though QA was apparently rigorously applied to the construction of the project in question (and growing in strength as NRC requirements and guidance evolved), the licensee did not implement NRC quality requirements for engineering as intensely as it did for construction. The licensee's attitude seemed to be that the engineering organization was comprised of professionals capable of doing what is right without overlaying a stringent formal quality assurance program beyond the normal controls considered part of good engineering practice.

Another factor in the problem of assuring quality in engineering dealt with changes in NRC requirements that occurred between the late-1960s and the late-1970s. It appears that the licensee did not completely understand the implications of the changes as they occurred; hence, an engineering QA program that the Atomic Energy Commission (the predecessor agency to the NRC) might have found acceptable early in the project might not pass NRC scrutiny in the late 1970s.

Secondary root causes also contributed to the quality failures. These included the following.

a. Failure to understand and appreciate the potential merit of a formal institutionalized QA program. This is suggested by the fact that the Project Completion Team adopted the A-E's quality assurance program, even though they were concerned about imposing a new system on the project at a late date. (The licensee's engineering procedures were maintained, however.) Examples of program deficiencies (drawn from various reports on the project and discussions with NRC inspectors) that had occurred during the project and the key indications of these deficiencies were as follows:

Design Control:

- The licensee's engineering staff did not always document important data transmitted to subcontractors.

- Design information was orally transferred to subcontractors.

- Assigned cognizant engineers were sometimes bypassed in the information or approval process.

- Adequate internal communications among the disciplines did not always exist within the licensee's organization.

- Requirements for independent reviews were not always followed.

Control of Instructions, Procedures, and Drawings/Document Control:

- The licensee's engineering management did not develop and/or implement formalized procedures to comply with early QA program requirements.

- In some cases, outdated drawings were used to establish seismic criteria.

- In some cases, diagrams in lieu of released drawings were used-- a contributing factor to the seismic problem.

Control of Service Contracts:

- Proceduralized activities for service contracts were lacking to control all interfaces with some subcontractors.

- Informal "letter-type" contracts and documents were used.

- Service contracts were not treated as formally as hardware contracts.

- Formal quality requirements were not placed on some subcontractors until the late 1970s.

b. NRC's failure to sell QA as a management tool. The NRC requirement for quality assurance seemed to come across as just another requirement. The emphasis from the NRC seemed to be on externals--the trappings of a QA program, rather than its substance: develop a QA manual, set up a QA organization, have the QA manager report high in the organization, etc. The NRC tended to lose sight of what it was trying to achieve and failed to provide adequate guidance on what a quality assurance program should be. The NRC failed to inspect against QA requirements in the engineering area to the extent they inspected against QA requirements for construction.

Case Study D

Early in the 1970s, the Case D licensee decided to construct nuclear generating plants. A possible natural gas shortage, the favorable economics of nuclear power, and public acceptance of nuclear power were reasons the nuclear option was deemed by the licensee to be a logical choice. Two projects were initiated, one in which the licensee would be sole owner (and which was later canceled) and the other a joint partnership with the licensee as project manager for all aspects of engineering design, construction, and operation. This latter project comprised two large (>1000 MWe) units. The first-unit operation was projected for the 1981-1982 timeframe, with second-unit operation to follow about two years later. Both have been delayed.

The licensee had no prior nuclear experience, but this was not seen as an insurmountable obstacle. Many other utilities were (or had been) in the same position, and the leaders in the industry were viewed as not having that much more experience.

In selecting an architect-engineer/construction manager/constructor (AE/CM/C), the licensee had compiled a candidate list that included the firm selected. Because many nuclear plants had been on order in the late 1960s and early 1970s, most A-E firms were committed and the licensee realized there would not be an opportunity to select from a large number of firms. It selected a large engineering and construction firm as both A-E, construction manager and constructor, one that had performed well for the licensee in non-nuclear projects. This firm was noted for its ability to complete large construction projects within cost and schedule. Its primary forte up to the early 1970s, however, had been in other than nuclear work. It did not have as extensive nuclear experience as many other A-E or constructor firms. This would be its first major nuclear engineering and design project, and the first nuclear project for which it was construction manager.

When the licensee applied for a construction permit in the mid-1970s, it was received about 6-8 months earlier than either the licensee or its A-E/C expected. While this may have been the result of a national emphasis to streamline the licensing process (a few years previously, the oil embargo had taken place and there was national concern over energy independence), it also had the effect of confirming, in the eyes of the licensee, the effectiveness of the AE/CM/C. The licensee maintained (during the site visit) that rapid licensing resulted in construction being started before an adequate amount of design and engineering (estimated at less than 25%) had been completed.

During the early phases of the project, the licensee was also staffing its own project management organization to fulfill its commitments to the project. Early in the project, the licensee used a matrix-type organization to manage the project. The approach was recognized to be embryonic, but thought capable of doing the job. The licensee recognized that managing a nuclear plant construction project would require a greater involvement than that required for a fossil plant. Project management rested on an organization that had responsibility for both nuclear and fossil projects.

In the course of the project activities, the licensee's staff had to issue some stop work orders to the A-E/C on specific tasks, e.g., work on concrete and on welding on the containment vessel liner. The licensee became concerned that the A-E/C was not accurate in its estimates of cost and schedule status of the project. Further, according to the licensee, the A-E/C was not demonstrating an adequate understanding of quality assurance or how it should be applied to a nuclear plant. The A-E/C wanted to do a good job, but it was not effectively balancing costs, schedule, and quality, according to the licensee. At about this time (mid-to-late 1970s), and perhaps coincident with the cancellation of many nuclear plants, the licensee believes there was a waning of interest by the A-E/C in the project, with a consequent loss of engineering and management resources.

In late-1978, the licensee initiated a six-month study of whether the A-E/C should be replaced. Consultations with other A-Es and constructors led the licensee to conclude that it would do best to support and improve the A-E/C organization and to become more involved in the design and construction activities. Thus, during the course of the project and up into the early 1980s, the licensee increased its involvement in the A-E/C activities. In 1978, following a consultant report that there was a high likelihood of both cost and schedule overruns, the licensee acted to strengthen its project management. It made the power plant engineering and construction manager the nuclear project manager and created a project management team reporting directly to him. About 30 experienced personnel were added from a consultant organization until the licensee could replace them with comparable personnel.

In 1979, the licensee expressed written concern about the A-E/C's performance and directed it to take several actions in the areas of construction supervision, planning, scheduling, control of construction work, labor productivity, and site housekeeping. The A-E/C agreed in large measure with the licensee's assessment and already had begun corrective measures to improve its performance. While some concerns were promptly resolved, others continued to require the attention of the licensee.

Thereupon, the licensee took a stronger stance by trying to help the A-E/C recognize its responsibilities and by injecting more licensee personnel into the contractor's realm of operation in an effort to compensate for the difficulties being experienced. It became more obvious to the licensee as time went on that the A-E/C's strength in this project was as a constructor, and not as an A-E because the engineering effort was not sufficiently leading the construction effort. Instead, construction was essentially driving the engineering portion of the project.

Symptoms of QA program breakdown gradually appeared at the construction site as the project became more involved in complex work. There were allegations of quality control (QC) inspectors having to rush through or overlook inspection functions, being intimidated and threatened by construction personnel, and lacking backing by their site supervisors. There were also allegations about bad construction practices, workmanship, and falsification of records.

Concurrently, the NRC initiated an investigation through its regional office in response to the allegations. Ten allegations were investigated from July 1977 to November 1979. The results of these investigations substantiated the allegations of harassment, intimidation, and lack of support of QC inspectors. The investigation demonstrated shortcomings in the project management and that the implementation of the QA/QC program did not meet the standards required to assure that the facility would be constructed to NRC requirements.

In April 1980, following a lengthy investigation of improper construction procedures and alleged inadequacies in construction and inspection, the NRC issued a Show Cause Order for safety-related sections of the plant. In total, 31 allegations of impropriety were made, and 19 were substantiated. The Show Cause Order findings are summarized in the following partial quotation:

"This investigation has determined through the examination of current work activities and interviews with over 100 personnel onsite that the QA/QC program at the project is impaired ... Allegations of harassment, threats and intimidations of QC inspectors by construction personnel that were common knowledge through rumors have been substantiated ... Difficulties in controlling structural concrete activities and quality problems in completed portions of structures have been continuing problems at the project since 1977 ... Procedures lacking in clarity, qualitative acceptance criteria, personnel with inadequate training, experience and/or education, production pressures, harassment, and intimidation have all contributed to this situation ..."

"That the project QA management may not fully recognize the requirements for QA/QC organizational freedom is evidenced by a January 4, 1980 lecture by the A-E/C project QA manager ..." "... strongly emphasized the fact that a A-E/C QC inspector's decisions are subject to question, challenge, and supervisory review and reversal ..."

"In the area of soil foundations, serious questions remain as to whether the implaced compacted backfill has met the required densities ..."

"Although safety-related welding activities are at an early stage at the project, serious problems were identified in the areas of welder qualification, welder process controls, and NDE performance interpretation ..."

"Further, although not reviewed during this investigation, Licensee personnel indicated significant problems relative to the storage and maintenance of equipment and processing of quality records ..."

In July 1980, the licensee responded to the Show Cause Order in which it claimed that it had undertaken, along with the A-E/C, a comprehensive examination of their organizations with the intent of enhancing their combined capability to design and construct the plant to conform with all applicable standards and commitments. Both had undertaken major changes in organization, personnel, and procedures to meet this objective.

The licensee contended that these improvements by itself and the A-E/C revitalized the project's QA program.

In spite of efforts to reconcile differences and to establish a credible program, the relationship between the licensee and the A-E/C continued to deteriorate. This culminated in the termination of the A-E and construction management parts of the A-E/C's contract by the licensee in the fall of 1980. According to the licensee, the A-E/C subsequently terminated its construction contract as well.

In September 1981, the licensee replaced the A-E/C with another A-E/C and an independent constructor. The latter was given responsibility for QC and QA activities, reporting directly to the constructor's offsite corporate headquarters. QA/QC activities on the part of the constructor were to be monitored by an independent QA department maintained by the A-E/C. An overview of all QA/QC activities was to be maintained by the licensee. This management system was intended to provide checks and balances to avoid a recurrence of the types of problems that had occurred previously.

Within the licensee's organization, additional changes were made to strengthen the project and to improve oversight over both the A-E/C and the constructor.

Safety-related work resumed in the fall of 1982. Construction completion goal dates for the two units were rescheduled to the mid-to-late 1980s.

The Case D study team identified the following root causes to be significant in contributing to the major quality and quality assurance problems experienced by this project.

The primary root cause for the construction difficulties was the inexperience of the project team. While the licensee had extensive experience in constructing and operating fossil fuel-fired plants, it had not been involved with constructing a nuclear plant. It apparently failed to appreciate the difference in scope and complexity between the two, as reflected in the management methods and procedures applied to the project by both itself and the prime contractor.

The licensee's lack of nuclear experience was further aggravated by the lack of experience of key individuals involved with the construction project. This project was the first nuclear project for the project manager, project engineering manager, and the quality assurance manager. The licensee was organized by technical discipline into a matrixed fossil-nuclear organization. Personnel were shuffled from fossil to nuclear and vice-versa as the need for a particular discipline arose. As a consequence, a requisite core of full-time professionals was slow in developing. The licensee did hire some staff with nuclear experience; however, they were not sufficient to provide the necessary core of competence.

Another problem resulted from the three management levels between the site quality assurance organization and the executive vice-president responsible for the project. The delay and filtration of information caused by this managerial superstructure contributed to incomplete understanding at the executive level of the problems that were developing.

Historically, the licensee had depended upon its contractors to do the bulk of the planning and execution of fossil plant construction jobs. The licensee assumed that this same approach would be appropriate for the nuclear project and, consequently, placed too much reliance on the prime contractor.

While not adequately involved at higher levels of management, in some respects the licensee became too involved at lower levels. Licensee personnel found themselves directly in the approval chain for A-E/C design approvals and other documents. This had the effect of unduly restricting work flow. Everyone in the chain had veto authority, and everyone had to agree. Toward the end of the A-E/C's tenure, the licensee assumed nearly all of the contractor's responsibility in an intensive but vain effort to help the contractor's effectiveness. In effect, the engineering work that was performed was the product of the A-E and the licensee instead of the product of the A-E with licensee overview.

The A-E/C, like the licensee, had inadequate nuclear experience. As a consequence, according to the licensee, the A-E/C did not understand the complexity of nuclear plant design and construction and did not bring to bear the necessary technical and management skills. These problems were aggravated by the earlier-than-expected approval of the construction permit and, therefore, the A-E/C did not have the planned time to come up to speed on design and personnel competence.

Design work proceeded slowly and specifications and procedures were inadequate and formatted in complex ways. There appears to have been insufficient engineering support for design and construction. The capabilities that the A-E/C did have were channeled into those areas in which it had experience, to the neglect of other equally important areas, according to the licensee. Engineering efforts were scheduled based upon dictates from construction. This led to unrealistic demands on the engineering groups. Quality assurance and quality control were also dominated by construction. There were many conflicts between QA/QC and construction in which construction generally prevailed. Project management did not have an adequate understanding of the interfaces and responsibilities for such functions as QA/QC, engineering, design, and construction. As a result, the constructor did not react in a timely, effective way to problems and did not employ proper management systems to reveal the causes of problems and to prevent them from recurring.

There was inadequate management support of quality. Neither the licensee nor its A-E/C appeared to have had a full understanding of quality and quality

A.22

assurance concepts as they applied to nuclear plant construction. Although
both made verbal commitments to quality, these were not actualized in the construction process. The licensee was not appropriately involved in monitoring
the total scope and details of activities and did not know how to take effective corrective action to prevent recurrence of problems. The A-E/C did not
sufficiently insulate QA/QC from cost and schedule demands, nor shield them
from intimidation or harassment. Consequently, construction supervisors dominated the QA/QC functions, both in the field and in the form of published
policy, which emphasized minimizing cost and maintaining schedule. The long
chain of command filtered information and introduced inefficiencies into the
decisionmaking and implementing processes. To further compound these problems,
the licensee had none of its own QA inspectors at the site until 1980. This
gave low visibility to management support of quality, which may have been
interpreted as a lack of backing from top management for quality.

There was an insufficient review by NRC of the licensee's (and its A-E/C)
experience in nuclear plant construction, and an inadequate involvement in the
inspection process in the early phases of construction. A recurrent theme was
that the NRC licensing process did not adequately address the ability and
experience of the project management, nor was there adequate evaluation of
whether the nuclear industry had over-extended itself at the time this plant
was contracted. The inspection process also tended to ignore management
issues. The irregular presence of NRC inspectors at the site early in the
project was cited as a contributing factor. The process used by NRC in identifying and dealing with problems was cumbersome and required excessive amounts
of time. In effect, the NRC approach was one of allowing troublesome situations to progress to the point that a case could be built for taking the dras-
tic action represented by a Show Cause Order. Some of the problems involving
the NRC required up to two years to resolve.

The changing environment of the nuclear industry was a contributing factor
to quality-related problems. The rapid proliferation of regulations during the
mid-1970s was cited as particularly troublesome, especially since the design of
this particular plant was probably less than 25% complete when construction
began in 1975 and proceeded more slowly than it should have in relation to construction activities. Regulatory changes from the TMI and Brown's Ferry
incidents were also a severe blow to the project, according to the licensee.

Declining energy projections and increasing interest rates made funding
plant construction more difficult. Incidents within the industry, such as TMI
and Brown's Ferry, reflected into changed design requirements. All of these
changes coming in rapid succession further complicated the task for the relatively inexperienced nuclear staff of the A-E/C and its A-E/C licensee-
constructor.

Case Study E

The licensee had previously constructed two large (~700 MWe) turnkey nuclear projects in the late 1960s. CPs were issued in 1967 for both units, with one unit achieving commercial operation in 1972 and the other in 1973. The licensee assigned a small group of its own engineers to the project to begin to accumulate a nuclear experience base.

The licensee's next nuclear project was the construction of two 810-MWe units of similar design. Construction of the first unit began in 1969 and commercial operation was achieved in 1976. Unit 2 (Case Study E) was announced in 1971, but major construction did not commence until 1977. The licensee contracted design and construction management (including QA/QC) on Unit 1 to an A-E firm with considerable experience in design and construction of nuclear projects. The licensee performed a project overview function.

The rapid rise in oil prices brought on by the Arab oil crisis in the early 1970s motivated the licensee to restart construction of Unit 2. In 1976, construction of Unit 2 was proceeding under a Limited Work Authorization (LWA); however, work was halted for 15 months by court injunction. After this injunction was resolved, the NRC issued a CP in June 1977 and major construction commenced.

The 15-month delay had advantages. During this period, the integrated management team was structured, a detailed master project schedule was developed, design completion was advanced, procurement of engineered components was continued, and a much more detailed level of planning was achieved. These factors were identified by the licensee as major contributors to the project's success. The licensee recognized that it had the talent to assume full management of the project and made the decision to do so. An integrated management organization using personnel from the licensee and A-E for key positions was established. The integrated management concept worked well and a spirit of teamwork, commitment, and loyalty to the project was achieved.

Advancement of the design was a particularly significant item. The design was approximately 75% complete when construction resumed. Vendor drawings on equipment were available, and construction drawings reflected correct equipment installation details. Some nuclear projects have experienced significant problems because designs were not sufficiently advanced for construction to proceed efficiently. Typically, construction begins with designs about 50% completed, sometimes less.

During the 15-month delay, the licensee had its field engineering work force develop many of the construction activities in considerable detail. This information was used in preparing procedures and was integrated into the design. The licensee also used the time to prepare effective procedures to control the project, including refinement of its own QC procedures.

The licensee had decided to continue procurement of engineered materials during the 15-month delay. This decision resulted in vendor drawings being available to the A-E and to field forces well in advance of equipment installation or construction-related activities.

As a result of its experience and these factors, the project achieved a 59-month time span from start of concrete to completion of cold hydro, static testing, 35 months better than current industry averages.

The licensee experienced no major quality problems during construction; however, on several occasions during construction, extensive reinspection efforts were required because adequate inspection records were not available. For example, an NRC inspection resulted in 12,000 socket welds having to be reworked and reinspected. Other quality-related problems typical of large construction projects also occurred. The licensee provided its QA organization with the following authorities as a check on its QC operations, which reported directly to the construction organization:

- QA held the "N" Stamp for the Licensee, which strengthened its overview function through access to records and the authorized nuclear inspector (ANI).

- QA performed daily surveillance of construction work, including formal audits of the entire project function.

- QA was responsible for the records vault, and through this activity monitored QC inspections.

The licensee stated that having QC report to construction permitted a better working relationship between crafts and QC, and thus better project results. While this action resulted in a more-or-less adversarial relationship between QA and QC, management's message on quality was "do it right the first time." This message supported the licensee's effort to stay on schedule.

The licensee identified what it thought to be the ten most important factors in completing the plant essentially on schedule, within cost, and without major quality-related problems:
1. management commitment
2. a realistic and firm schedule
3. clear decision-making authority
4. flexible project control tools
5. teamwork
6. maintaining engineering ahead of construction
7. early startup involvement
8. organizational flexibility
9. ongoing critique of the project
10. close coordination with the NRC.

Apart from the initial 15-month licensing delay there were no other significant licensing delays. No significant public intervention occurred in the construction phases of the Case Study E nuclear station. No fines were levied against the licensee for nonconformance violations or quality deficiencies during construction and startup of the project.

The Case E study team identified the following to be significant in contributing to the absence of major quality failures.

The licensee had an experienced design, construction and construction management team. A major factor in the project's success was the nuclear experience of the licensee and its staff and contractors. The Case E project had a seasoned group of managers and a tried, although evolving, set of project controls. The A-E commented that an estimated 75% of the skilled labor force carried over from the Unit 1 project to Unit 2. Early in the previous project, an extensive training program was instituted to develop additional craft persons, a factor important to achieving quality. An estimated 50% of the A-E supervision also continued from the previous project.

The licensee recognized the need for effective planning and implemented it. During the 15-month licensing delay, effort was redirected towards developing detailed plans and schedules to facilitate the construction phase. The requirements were integrated into the design and procurement process to minimize disruption of the construction process later. A realistic and firm schedule resulted from the planning process.

The licensee exercised control of the project through strong owner involvement, commitment of resources, and an effective integrated organization. An important root cause for this project's success was the licensee's firm management, including providing all of the onsite quality control and quality assurance functions. Clear decision-making authority was placed at the proper level. The licensee established a matrix organization comprised of its staff and of the A-E/constructor staff that created an environment of affiliation and loyalty to the project.

The project became the priority project for the licensee, who committed the necessary resources for the project. The licensee committed to a project schedule of 65 months (from first concrete to core load) and consistently invested additional resources to maintain or recover schedule whenever needed.

Recognition of the need for early startup involvement. The Case E construction plan had startup logic involved in it with the decision to involve operating personnel at an early time. That decision reflected into certain innovative construction approaches on the project. In previous projects, operations personnel were not involved until the project, or at least major systems, were essentially completed. The operations involvement took place over an 18-month period and included about 60 personnel. There were 494 turnover packages. Early turnover helped resolve problems, including quality-related problems early on.

Problem identification and solution was an important part of the licensee's management philosophy. The licensee followed a policy of resolving problems at the earliest possible time. As problems or changed conditions confronted the licensee, it formed teams to resolve them in a timely manner. An independent engineering verification program was instituted about a year before fuel loading. The A-E maintained a larger field force than on previous plants to process field change requests, nonconformance reports, etc., more rapidly.

The licensee had task forces examining how impending changes might impact construction. To help circumvent delays that might arise from regulatory matters, it maintained three engineering personnel at NRC Headquarters, some at the utility's engineering office, and three at the site to interface with NRC personnel and provide timely responses to licensing questions. It avoided adversarial relationships. On the NRC side, its inspection surfaced problems early, thereby avoiding major issues that might continue long into the construction period.

The licensee achieved a high level of teamwork on the project. In discussing teamwork, the licensee stated that all of the participants in the project worked to meet the project objectives, not their own (sub)-objectives. Heavy emphasis was placed on integrating work with NRC, EPA, trade councils, etc. The entire state Congressional delegation supported the licensing schedule. There were quarterly labor-management meetings (there has been no work stoppage of significance since 1980). Labor was involved in improving productivity.

Interviews with personnel on this project revealed a positive orientation toward the project. They were proud of what they had accomplished--they identified with the project. The reduced number of individual contractors on the job may have been a factor in achieving the strong team effort.

The licensee recognized the merits of an institutionalized QA program and was innovative in structuring and implementing its QA program. This root cause was manifest as follows:

a. The licensee established a single QA/QC program for the project. A single program reduced confusion through fewer interfaces and uniformity of requirements. The A-E made the comment that a single QA/QC program was an asset that avoided gaps in the program with the increased possibility of things falling through the cracks.

b. The licensee had a corporate commitment to quality. The licensee extended its QA program to programs other than nuclear, which indicates its recognition of QA as an effective management tool. It is involved in a program with eight other utilities that audit one another's QA program.

    The licensee's QA organization became the ASME "N" Stamp holder for Case E, which permits greater control of the inspection process and a different perspective than provided by NRC inspection.

c. The licensee balanced schedule and QA commitments. The licensee responded to several setbacks by defining solutions and applying whatever resources it took to resolve each problem and recover schedule. Emphasis on maintaining schedule did not compromise quality. Good management practices can produce quality amidst commitment to schedule.

Case Study F

The Case F nuclear power project was organized in mid-1974 and an application for a construction permit was filed with the NRC for three 1,270-MWe generating units. The construction permit was granted in the spring of 1976 and construction began in June of that year. At the time of the case study (August 1983), the status of the three plants was as follows: hot functional tests were performed on Unit 1 in mid-1983. During these tests, reactor coolant pump problems developed, and cracking was noted in the control element assembly guide tubes, as discussed later. Unit 2 hot functionals are scheduled for early 1984. Unit 3 is about three-fourths completed, with commercial operation expected in 1986.

Several utilities participate in the project with ownerships ranging from about 10-30%. One of them was selected to be the licensee who would manage the project and operate the plant on behalf of the others. The utility selected had no prior nuclear construction experience, but at least one of the presently participating utilities had constructed and operated nuclear plants.

Information provided by the licensee showed that the project was conceived in the early 1970s. The initial planning was done by the licensee with a small staff it had assembled for that purpose, all of whom were experienced in the nuclear field. This staff analyzed what had gone wrong at other nuclear projects and arrived at findings that played an important role in organizing and carrying out the Case F project. They felt that a long-term commitment of qualified people to the project was important, both from the licensee as well as its contractors. They noted that utilities typically tended to do the wrong things and get involved in the wrong places in nuclear projects, such as wanting to approve everything. Utilities often believed they knew more about all aspects of the projects than their contractors or the regulator. It was found that utilities were often very untimely in their actions and decisions, which caused costly delays. Finally, they perceived that utilities often have the wrong type of organization. For nuclear projects, they found that the organization must be both management and detail oriented.

Based on these general findings, the licensee's staff came up with some recommendations that formed the basis for its project organization. First, there should be a strong project concept, both within the licensee's and A-E's organizations--but with a singleness of purpose. Second, the licensee should manage the interfaces. Third, there should be single points of entry for all correspondence to each organization, and the communication channels should be monitored to ensure effectiveness. Fourth, clearly written design criteria should be established and maintained current as changes occurred. Fifth, the licensee should establish which documents produced by the A-E and others it would review. Sixth, the licensee should be responsible for obtaining all project permits and licenses. Seventh, purchasing and construction work should be controlled through administrative procedures (such as having standard terms and conditions for contracts and purchase orders), a qualified bidders list, and work initiation procedures. Eighth, safety and quality must come ahead of schedule and cost, not only for the licensee, but for its contractors, also.

These priorities must also be conveyed to the project regulators. Ninth, adequate systems and procedures must be established to monitor the project.

Based on discussions with the licensee, it was determined that these early recommendations were implemented as follows:

1. An A-E construction manager/constructor experienced in nuclear construction was hired.

2. Contracts with major contractors required a long-term commitment of key personnel.

3. Interfaces were defined and procedures were developed to ensure the proper flow and interpretation of information and to permit monitoring.

4. Frequent meetings were held with the major contractors' senior management to discuss project problems and to facilitate decisions.

5. Contractor responsibilities were defined for design, specifications, purchasing materials, and hiring and managing labor forces.

6. The licensee set up a strong project organization with staff hired from other utilities, architect-engineers, vendors, and the NRC. The head of the licensee's nuclear project had considerable experience with designing and constructing commercial nuclear reactor projects. The licensee's organization actively overviewed and closely monitored its contractors. Construction input was provided early in the design effort. Operations input occurred early in the design effort, also.

7. Licensing activities were assigned to executive levels to help ensure prompt decision making. It was the licensee's philosophy to be responsive to the regulators.

In the context of these recommendations, the Case F project was implemented.

To date, no major quality problems have arisen in the construction of the Case F nuclear power station. Also, no significant public intervention has occurred in the licensing or construction phases of the station. As previously mentioned, significant primary pump problems have occurred in startup operations, and other startup problems have surfaced as well.

The licensee has experienced construction problems typical of large construction projects. Poor communication about project completion existed between the licensee and its contractors, and the licensee took the necessary steps to reorganize the scheduling function. Poor productivity had to be overcome, and the licensee insisted on changes in personnel and organization of its A-E/C. The turnover rate was considered high for the field engineering staff, and the licensee found it difficult to retain a good staff. System walkdowns revealed quality deficiencies that required rework. Unit 1 experienced major problems in the transition from construction to operations. The good management practices that led to construction success were not applied equally to the transition from construction to operation. Unit 1 hot functionals have revealed the primary pump deficiencies. While major quality-related problems have not been experienced in construction, there is a strong possibility that the design verification process supporting new components, such as the primary pumps, was not adequately explored.

The Case F study team identified the following root causes to be significant in contributing to the absence of major quality failures in construction.

The licensee determined in advance the important factors in constructing a nuclear project and took the necessary actions to achieve them, including hiring key personnel with nuclear experience, retaining an experienced A-E/C, and creating an organization appropriate for the project. The licensee recognized from the outset that construction of a nuclear power plant would be different from previous projects it had undertaken. This realization resulted in several key decisions that were strong contributors in avoiding significant quality problems in construction. First was the recognition that fossil fuel plant experience alone would not be adequate for the project staff; selective recruiting of personnel for key positions with nuclear plant construction experience was essential. Second was the licensee's action in retaining an experienced architect-engineer/construction manager/constructor for the project. Third was the recognition that it would not be appropriate for the nuclear construction project to be fitted into an existing organization component; a separate, strong project organization would be required--one which could closely monitor and actively overview the management of the project. The combined experience resulted in many actions appropriate for controlling and monitoring the project. One action that was singled out by the licensee and its A-E was the development of a detailed scale design model (costing several million dollars) of the plant to supplement design drawings as a basis for configuration control. This model, together with a design that was estimated at about 60% complete when construction started, was credited by the licensee and its contractors as being instrumental in facilitating the construction activities and avoiding many problems experienced in other nuclear projects.

On the other hand, a lack of experience has led to confusion and inefficiencies in the startup testing program on Unit 1. That activity does not appear to reflect the same degree of understanding, planning, and preparation that was applied to the construction phases. The startup testing program has been restructured more than once in the past several years. There appears to have been a lack of appreciation that nuclear is a more complex startup process than fossil plant startup, and that turnover from a strong construction manager requires a well thought out transition plan and startup program. The transition was to have the constructor do the prerequisite tests and the operations staff perform the preoperational tests. That did not work satisfactorily for several reasons, but probably primarily due to a lack of a well thought out plan. Startup of the subsequent units involved an operations/construction team involved in both prerequisites/preoperational tests at an earlier stage, with a greater focus on completion of systems (versus the area concept of completion).

During hot functional tests on Unit 1, the pump problems occurred because of a clearance problem between the pump impeller and diffuser, and perhaps compounded by flow conditions that existed during the tests. Previous factory tests of shorter duration had not revealed a similar problem. Cracking in the control element assembly tubes, which occurred during the initial startup tests, also appeared to be associated with the flow conditions. First-of-a-kind equipment, such as the pumps, frequently require modification in the start-up phases of operation. The licensee stated that it had been relying on prior orders for similar components by other licensees to work out the "bugs" on the first-of-a-kind equipment. Nuclear plant delays or cancellations invalidated this approach. Experience would have suggested that a revised approach to permit more extensive design verification testing of such equipment prior to installation would be prudent.

The licensee pursued several management practices, especially the working involvement of upper management, which permitted the project team to function effectively. The working involvement of upper management was important in many respects. They were sufficiently involved that when corrective action was needed, it could be taken in a timely and decisive manner. They set the tone for the project's orientation toward quality, expressed in several ways, but importantly in terms of high plant reliability goals, as well as maintaining quality standards for non-safety systems and for temporary construction. They established a philosophy of good public relations and a nonadversarial working relationship with the NRC. They arranged for appropriate contracting practices and labor relations. They minimized the number of contractors on the project, clearly defined responsibilities of the participants, and established sound procedures for design and construction activities. Finally, they helped assure uninterrupted financial resources for the project. Good management practices appropriate to nuclear projects were clearly another root cause in avoiding significant quality problems in construction.

A relatively high design completion at the start of construction of Unit 1 and the replication of the design for the two subsequent units permitted problems experienced in Unit 1 to be corrected in advance in Units 2 and 3. An example is the transition from construction to operation described previously. The design completion was estimated at about 60% when construction was started. The use of the model as a design model also helped to reduce interferences and resulting field changes. Construction planning activities were enhanced. The licensee adjusted well to the changing regulatory environment over the life of the project.

The responsibility for quality was placed at the working level. Field engineers were required to sign off on inspection hold points before involving the QC inspector. This also helped preclude QA/QC personnel directing the work through the inspection process. The licensee established its QA requirements sufficiently broader than NRC requirements (though with appropriate cognizance) so if the latter were changed, the former would remain unchanged.

The attitude of senior project management that the NRC could help the project helped avoid unnecessary confrontations that were counterproductive. The licensee, as a matter of policy, established a constructive nonadversarial working relationship with the NRC. The vice-president of the Nuclear Department has been the licensee's prime contact in licensing matters and has set the criteria and guidelines for interactions with the NRC. The independent design reviews conducted by the licensee has NRC staff as listeners/observers.

The proceduralized approach to design and construction was an important contributor in avoiding major quality-related problems. The project had workable procedures to control calculations, specifications, procurement, and other facets of the construction process that had been adopted early in the project. The licensee established design criteria for the project in conjunction with the A-E, and they have been the governing guidelines for the project. The document specifying the criteria has been the control document for the life of the project, which extends into operation.

The A-E's resident engineer said that the basis for quality at the project was that the quality control procedures were specific. "It is an expensive process," he said, "but it works."

The constructor prepares work planning procedures/quality control instructions, which control safety-related work for non-safety balance of plant items, though less inspection is applied.

The Zimmer Case Study

Concurrent with the case studies, an independent analysis was made of Cincinnati Gas and Electric Company's (CG&E) Zimmer Unit 1 nuclear project (TPT 1983). The study was mandated by the NRC in a Show Cause Order to CG&E in November 1982. One of the provisions of the Show Cause Order was the requirement that CG&E have a qualified consultant conduct an independent review of the project management of the Zimmer project. Torrey Pines Technology (TPT) was retained by CG&E to conduct this independent review, including CG&E's quality assurance program and its quality confirmation program. The review was to identify the organizational changes needed to ensure that construction of the Zimmer 1 plant can be completed in conformance with the NRC regulations and the construction permit. This section of the case studies report summarizes the TPT findings. The summary is intended to provide additional information from a second perspective on the root causes of quality-related failures in nuclear plant construction.

The selection of TPT was subject to the NRC approval of its independence and capability to perform the review. Several public meetings were held, and the NRC reviewed TPT's proposed program plan for conducting the review. As a result of the program plan review, greater emphasis was placed on evaluating CG&E's management of the Zimmer project and less on a detailed review of procedures, specifications, records, etc. The program plan was also revised to include the evaluation of the Zimmer project management from the inception of the project to the present. The revised program plan was approved by the NRC in a public meeting with TPT on May 26, 1983, with a provision that TPT include an evaluation of the relationship between CG&E and Reactor Controls, Inc., one of the contractors.

The basic approach used in the TPT study was to separately examine key characteristics and aspects of the Zimmer project management and QA programs. As a cross-check, selected "case studies"[a] were also examined to assess the collective role and behavior of management in response to specific problems and/or series of events. The specific areas reviewed were as follows:

- CG&E management attitude toward "whistle blowers"
- structural steel in the control room
- 2400 feet of small-bore piping
- welder qualifications.

TPT reviewed the organizational structure, policies and procedures, and QA activities of CG&E, including its interfaces with its contractors: Sargent and Lundy (S&L), Henry J. Kaiser Company (HJK), General Electric Company (GE), Catalytic Incorporated (CI), and Reactor Controls, Incorporated (RCI). The review was divided into four periods: 1) project inception to the assumption of increased construction responsibilities by CG&E in 1976, 2) from 1976 to the Immediate Action Letter in early 1981, 3) from the Immediate Action Letter to the Show Cause Order in November 1982, and 4) subsequent to the Show Cause Order.

Information was obtained by interviewing CG&E's Zimmer project management staff, representatives of contractor organizations, and representatives of related organizations such as the NRC, National Board of Inspectors, and intervenor groups. The interviews included past and present management and other individuals having information pertinent to this review. Selected records and files were examined to obtain relevant documents/information to supplement and verify the information obtained in the interviews. The interviewees and the supplemental documents were selected on the basis of TPT's professional judgment.

The total program effort was approximately 60 man-months; over 3200 documents were reviewed; and approximately 100 people were interviewed, several more than once. The investigation did not include any technical review or evaluation of the adequacy of the Zimmer plant design and construction. No physical inspection of the plant was performed.

The TPT study of the Zimmer project (TPT 1983) showed the following factors to be the important causes of the quality-related problems experienced.

<u>The Licensee and its constructor lacked prior nuclear experience.</u> The TPT report states:

"... CG&E and, to a large extent, its constructor HJK, lacked prior experience in its assigned roles in this nuclear power plant project. Although in the early 1970s numerous other utilities also lacked prior nuclear experience, the constructor (HJK) of the Zimmer

---

(a) Term used by TPT to identify a portion of its review requiring in-depth documentation review, which should not be confused with the NRC "case studies."

project was unique from the standpoint that it did not have, nor did it later obtain, any additional commercial nuclear power plant prime construction contracts. Consequently, it appears that neither CG&E nor HJK had sufficient experience or the external interactions necessary in order to respond in a timely and effective manner to the rapidly evolving, more stringent interpretations of NRC requirements. As a result, it was not recognized until very far along in the Zimmer project that a much more formalized, rigorous approach was needed to control and document the quality of the design and construction of a nuclear plant than that required for the design and construction of a fossil fuel plant. This was probably the single, most significant factor contributing to the present situation at the Zimmer plant ..." (Vol. 1, p. 4).

"... CG&E attempted to use a project management approach at Zimmer that had been previously used successfully in the construction of fossil fuel plants. The approach, which was not unusual at that time, was to rely on a small, dedicated management team using relatively informal management systems and techniques ..." (Vol. 1, pp. 6-7).

The Licensee did not have an adequately sized staff, nor one with adequate experience. The TPT report states:

"... In comparison with other nuclear utility companies, staffing of both CG&E and the subcontractor organizations was inadequate throughout the 1970s. The CG&E management and professional staff was of inadequate size and had insufficient experience and training in the design and construction of nuclear power plants. After the IAL[a] in April 1981, additional staff was recruited, including a large proportion of temporary employees--some in management positions. A small number of CG&E personnel with prior nuclear experience has been added to the staff since the SCO,[b] but it still remains understaffed, and this situation needs to be corrected ..." (Vol. 1, pp. 7-8).

The Licensee failed to manage the project. The TPT report states:

"... Key managers and professional staff were not dedicated solely to the Zimmer project. Several key managers had conflicting responsibilities that detracted from their management overview of Zimmer. Except for short periods of time, the CG&E manager responsible for the entire Zimmer project was not located at the site. These conditions, coupled with the lack of an integrated project management system, contributed to the creation of informal autonomous organizations within the project with lines of communication that were not always consistent with the published project organization charts. Also, there was a too-heavy reliance on contractors for project

---

(a) Immediate Action Letter.
(b) Show Cause Order.

management and control. The CG&E policy of delegating the responsibility of major elements of the work to reputable experienced contractors is not inconsistent with the approach taken by other utilities in the construction of nuclear power plants; however, CG&E does not have the management system, implementing procedures, and staff required to control the work performed by its subcontractors. The net result was to impair the visibility of the project to CG&E top management ..." (Vol. 1, p. 8).

"... CG&E top management appeared to lack an adequate degree of involvement in, and commitment toward, QA at Zimmer. Up until 1981, the president of CG&E appeared to be insulated from an accurate picture of the status and inadequacies of the Zimmer QA program. The CG&E project organization provided minimal executive summary information to management on overall quality problems, status, and QA program effectiveness. Executive reports generally addressed details and highlighted 'brush-fires,' rather than providing a management perspective ..." (Vol. 1, p. 9).

"... Up to 1981, CG&E lacked effective control over the design function. More audit emphasis should have been placed by CG&E on field design control procedures. This could have helped to identify and correct, in a timely manner, the design control problems experienced at Zimmer. CG&E initiated an intensive effort after the SCO to get this system back on track ..." (Vol. 1, p. 11).

"... CG&E did not provide sufficient direction and support for the establishment of a comprehensive audit program executed in accordance with the requirements and intent of 10CFR50, Appendix B. Consequently, the CG&E QA audit program appeared to be ineffective. Individual problems were attacked, but the magnitude and extent of problems apparently remained largely undetected. Many noncompliances detected by outside audit groups should have been found by the CG&E QA audit group ..." (Vol. 1, pp. 11-12).

"... In general, review of subcontractor's activities appears to have occurred aggressively only between CG&E and HJK. There is little evidence that S&L, RCI, or CI activities were effectively reviewed, monitored, audited, or critiqued by CG&E. This CG&E policy of delegating the responsibility for major elements of the work to reputable experienced contractors is not inconsistent with the approach taken by other utilities for the construction of nuclear power plants; however, CG&E does not have the management system, implementing procedures, and staff required to control the work performed by its subcontractors ..." (Vol. 1, p. 18).

The Licensee failed to elevate its commitment to quality and quality assurance to an equal status with cost and schedule. The TPT report states:

"... CG&E had a corporate fiscal policy that minimized expenditures. Such a policy, taken in the proper perspective, benefits both the ratepayers and the stockholders of the Company; however, this emphasis completely dominated other important priorities, such as quality and quality assurance. Cost reduction and schedule mainten- ance was encouraged to the extent that construction forces worked only to compliance with the minimum NRC standards and regulations. This approach, combined with the rapidly evolving and more stringent interpretation of these regulations over the years, contributed sig- nificantly to the current problems at the Zimmer project ..." (Vol. 1, p. 4).

"... The emphasis was on getting the plant built on schedule, at the minimum cost ..." (Vol. 1, p. 7).

"... Management at Zimmer had not done an adequate job in highlight- ing the QA program as one of the key elements in the successful con- struction of a nuclear power plant, or in providing the appropriate level of support that would ensure effective program implementation.

The level and status of the CG&E QA organization through the years was generally inadequate to provide an effective nuclear QA pro- gram. The major shortcomings in this area are the small and inexpe- rienced CG&E QA staff, cost and schedule pressures on the QA organi- zations, and failure to effectively correct and prevent recurrence of problems. CG&E management generally did not establish definitive policies, verbal or written, concerning QA at Zimmer, and no strong message by CG&E management in support of quality and quality assur- ance was evident. Instead, CG&E management policy insisted that all concerned (CG&E and subcontractors) minimize the time and money spent on QA programs ..." (Vol. 1, p. 9).

"... There exists no effective assurance that documents to be main- tained as records are complete, accurate, valid, or readily retriev- able. It would also appear that management did not take effective action early enough in the construction project to ensure the valid- ity and availability of these documents. A centralized records cen- ter was set up after the IAL, and the turnover of documents from other site locations is in progress. However, progress is slow and it is not being accomplished in a thorough manner.

From the beginning of construction until the present, the corrective action system was generally not effective in assuring that identified discrepancies in material/systems/procedures were investigated in a timely manner, analyzed to determine root causes, and corrected by priority actions to prevent recurrence. Standard management tools to collect relevant data, analyze the data relating to the problem, pro- pose alternatives on the basis of analyzed data and the operating environment, and select solutions were available, but were apparently not utilized or, at the least, were not effective. In addition, there is little evidence to indicate that management established an effective system to track 'open' items to assure their completion ..." (Vol. 1, p. 12).

The NRC failed to impress on the Licensee the importance of quality. The TPT report states:

"... Quality problems existed during the early stages of construction which remained uncorrected during that period due, in part, to a lack of attention and follow-through on a corrective action course by the NRC. Although CG&E QA was generally responsive to NRC concerns, these concerns were neither extensively nor aggressively pursued by the NRC. Consequently, CG&E management failed to recognize the underlying message in the Inspection Reports (IRs) relating to the problems that existed at Zimmer. As a result, corrective action was not taken in an effective or timely manner. CG&E was allowed to continue construction while being lulled into a false sense of satisfactory performance until the late 1970s and early 1980s ..." (Vol. 1, p. 5).

The Licensee did not have adequate procedures to control the project, nor were those in effect adequately implemented. The TPT report states:

"... CG&E established an Owners Project Procedures (OPP) Manual for the Zimmer project in 1972 which delineated the project organization, including reporting lines within CG&E, for the major subcontractors (HJK, S&L, GE); defined the responsibilities and authority of the various positions; and named the personnel who would act in those positions. These formal overall project policies concerning responsibility and authority over the functions at Zimmer appear to have been adequate, but they were not implemented adequately by project personnel.

CG&E did not have an integrated, comprehensive set of project management procedures documented and implemented to ensure that all elements of the project (e.g., Construction, Engineering, Quality Assurance, Licensing, Cost, Scheduling, etc.) were coordinated. This impaired communication between departments and, in some instances, resulted in conflicting requirements and/or a duplication of effort ..." (Vol. 1, p. 7).

"... CG&E project management and control systems, including performance measurement and document control, were inadequate. The systems utilized did not integrate the planning and scheduling of various project management activities such as construction, QA, engineering, and, subsequently, the transition to operations. Management reporting systems were also poor ..." (Vol. 1, p. 8).

"CG&E's control of the process of developing, maintaining, and implementing subtier procedures, instructions for work, and inspections that affect quality has been less than effective from the start of construction to the present. There are many instances of inadequate control over design documents, design document changes, welding

forms, inspection methods/procedures, documentation of work accomplished, conformance to work procedures, and QA procedures ..."
(Vol. 1, pp. 10-11).

A.3.2 Summary of Case Study Findings

Although no single factor distinguishes nuclear plants that have experienced major quality-related problems from those that have not, a combination of
utility/contractor experience and/or personnel experience in nuclear plant construction provides the greatest assurance that quality-related problems will be
avoided. Based on the six case studies (and substantiated by the independent
review of the Zimmer project), if a utility had constructed previous nuclear
plants or if it hired experienced personnel for its own staff and had an
experienced A-E/C, it tended to avoid major quality problems. Where the utility depended on non-nuclear, e.g., fossil experience of its staff and its
A-E/constructor, it was prone to experience major quality problems. Experience
by itself may not preclude major quality problems, e.g., Midland, with an
experienced utility and architect-engineer, still it is probably the greatest
assurance factor in achieving quality of construction.

Because this study was limited to six case studies (and a review of a
seventh), it did not evaluate in-depth a larger grouping of other utilities
that have built nuclear plants without apparent major quality problems, and as
first-time ventures. However, experience would be expected to be a significant
factor, especially in the timeframe of the projects studied for this report.

A second important factor that emerges is the importance of the licensee
actively managing the construction project. Projects experiencing major quality problems placed too much reliance on their contractors or their own
in-house capability, a reliance that was not justified based on previous experience. The licensee failed to effectively implement a management system that
provided adequate control and oversight over all project aspects. Those projects not experiencing major quality problems "ran the project." They were
deeply involved in planning, establishing criteria and procedures, approving
important drawings and specifications, overviewing their contractors' activities, and identifying and solving problems. They clearly defined the responsibilities and authorities of the participants and monitored the interfaces to
assure that responsibilities were being properly discharged. There was often a
working involvement of upper management or, failing that, a good understanding
of the project's needs in terms of finances, manpower, autonomy (from prevailing practices, etc.) which were provided. As with experience, some nuclear
plants may have been constructed where the licensee did not actively "manage"
the project (such as in the turnkey projects), but in the present timeframe and
based on the case studies, licensee management involvement and control are
important factors in constructing plants without major quality problems. Constructing nuclear plants in the 1970s with the many regulatory changes that
occurred, with a supporting nuclear industry in which the most experienced
A-Es, constructors, and major contractors were stretched to capacity, with the
increased complexity of the larger plants, and with rapidly escalating costs,
was in a far different environment than those constructed earlier. The
requirements that assured success in nuclear construction also escalated.

A.38

Success in constructing fossil-fuel plants, together with a little nuclear experience, was no longer sufficient to guarantee that the nuclear project could be completed without major problems (in quality, cost, or schedule). Utilities with only those qualifications, but which avoided major problems, probably had good fortune and astute management who were able to discern impending difficulties and compensate for them accordingly.

Active management of the construction project was clearly shown in those projects that had experienced major quality-related problems, but that had since turned them around. Those which had not completely turned their projects around tended to maintain that "we always managed the project," "it was someone else's fault," or, "it was a fluke." With those which had, there was a change in management involvement. Project leadership emerged clearly. There was no question who was providing project leadership. Senior management was often relieved of other utility responsibilities to devote sole attention to the project; they often moved to the construction site; their project staffs were divorced from those responsible for more traditional generating plants; and substantial additional experienced personnel were hired. Procedures were strengthened and enforced. There was more active involvement with the NRC, especially by upper management. Where necessary, modifications were made to contracting methods to give the licensee more control over quality (fixed-price contracts were often converted to cost-type contracts); there was substantial strengthening of the QA function within contractor organizations. These types of actions are also manifest by those utilities that have not had major quality-related problems.

As utilities gain experience in nuclear project management and develop a core of experienced personnel, they appear able to delegate the project management successfully to lower management without lessening the utility's active involvement in the project. As regulatory stability is achieved and plant designs become more standardized (or plants are more fully designed before construction commences), the more experienced A-E/C probably can assume an increasing degree of responsibility for plant construction without such intense utility involvement.

A third important factor relates to the licensee's commitment to quality, or perhaps to plant reliability. While all (or almost all) personnel interviewed believed quality was important, ranking with safety, those licensees (and their contractors) that had experienced quality-related problems tended not to appreciate that quality assurance was a management tool that would help assure quality in nuclear plant construction. Those who had not experienced major quality-related problems supported their quality assurance programs with adequate resources and backing, and tended to take proactive roles in seeking improvements in their quality assurance programs. They were prone to put responsibility (and authority) at the level where the know-how was and emphasized doing it right the first time. They seemed to place more stress on plant reliability than those who had experienced major quality problems. That is, plant reliability was a more evident concern that surfaced frequently in discussions with the licensees' staffs.

There did not seem to be a particular QA/QC organizational structure, function, authority, etc., that characterized plants with or without quality-related problems. Rather, the six cases presented a variety of organizational arrangements for these functions. Staffs varied in size over a considerable range. QA staffs reported at various levels in the utility/contractor organizations. Some had stop work authority and some did not. Some participated in construction activity planning and some were separated from the day-to-day activities. Some projects had multiple QA/QC programs and some had a single program for most of the construction activities. Some organizations had multiple layers of QA which audited lower levels. No pattern of QA/QC organizational structure or delegation correlated with whether plants had experienced major quality problems or not. Projects having major quality-related problems in construction, when once turned around, tended to establish a strong QA function as their main line of defense against (further) quality failures, whereas those not having major quality-related problems in construction (perhaps arising from their greater experience) emphasized craftsmanship responsibility for quality as their main line of defense.

A fourth important factor relates to procedures. All licensees have procedures, but there was a difference in what was done with them. For those with quality-related problems, one or more of the following conditions existed: they lacked adequate procedures; they had procedures, but did not rigorously follow them; or they relied on them to do what is a management function of overview and control. Those without major quality-related problems spoke of the use of detailed procedures for design, procurement, construction, and inspection activities, and of the need for adherence to them. They seemed to have a better appreciation of their value and limitations than those with quality-related problems. In the six case studies, there was a wide range of sophistication in specifying, using, and auditing procedures. Some quality-related problems could probably be avoided by helping licensees appreciate state-of-the-art applications of procedures and related controls.

The presence of four factors--experience, management control, commitment to quality, and properly implemented procedures--should be sufficient to avoid quality-related problems in nuclear plant construction. There is overlap in the factors; that is, experience will tend to assure that the other factors are appropriately implemented. Evaluation of these factors by the NRC will be difficult because there is a large subjective aspect to them. One senior vice-president suggested that the NRC might appoint a panel of experienced nuclear utility executives to evaluate whether a first-time licensee has the requisite capability to construct a nuclear plant successfully. Their own experience should permit them to adequately evaluate the subjective aspects.

Not all licensees and their contractors fully appreciated the requirements of the regulatory process, nor coped well with the changing regulatory environment. All of the nuclear plants considered in the case studies were under construction during a period of considerable regulatory change (the 1970s). One difference between projects with and without major quality problems lay in how much they relied upon the NRC (or other bodies such as ASME or INPO) as an indicator of construction quality at their projects, and how they related to the NRC. As a class, those with problems seemed to assume that (a) a lack of

NRC prompting or (b) a lack of dramatic action by the NRC on quality-related matters meant no significant quality problems existed. Those without problems were proactive in resolving regulatory matters and anticipating regulatory changes. Those licensees represented themselves (i.e., they were the spokesmen) in dealings with the NRC. They took the initiative and the lead in licensing matters--not their contractors. They understood the implications of the impending decisions. Some licensees stationed personnel at NRC offices to promote rapid resolution of regulatory problems. Those without problems had two other characteristics: they had non-adversarial relationships with the NRC, and they attempted to anticipate the effects of impending regulatory changes on their construction projects. All aspects of this factor could be attributed to one or both of the first two factors--experience and management control.

A factor that does not seem to be recognized by either the licensees or the NRC is that the longer the construction period, the greater the risk of a quality failure, and the greater the need for additional attention to quality matters. The problem has two interrelated facets. First, the longer the construction period, the more regulatory change a licensee will have to cope with. Regulatory changes often result in design changes and rework. These changes and rework are often made under less than optimum conditions, conditions not conducive to quality workmanship. Second, the personnel making the changes may not be the ones who did the (design or construction) work the first time and may not understand all of the assumptions, interactions, or special conditions considered initially.[a] Also, there is some impact on morale from having to make changes, especially if thought to be marginal ones. For projects that have been under construction for an extended period (perhaps eight to ten years or more), special attention to quality matters may be appropriate.

NRC actions, or inactions, also contributed to quality problems that licensees experienced. Quality problems arising from regulatory changes have already been discussed. The NRC's failure (or inability) to adequately evaluate a licensee's management capability to undertake a nuclear project and its understanding of its required role is a major shortcoming in the licensing process. Clearly, some licensees should not have been granted a license under the prevailing environment and conditions, nor with the teams they assembled for their projects.

The NRC's failure to take action with the licensees on a more timely and firm basis allowed poor quality practices to exist and possibly proliferate in the industry while the licensees assumed NRC inaction meant the practices were approved or at least not sufficiently bad to make a big issue of them. In at least some cases, the NRC's presence at the site in the early phases of construction was sufficiently sporadic that the developing poor quality practices were not fully appreciated. Licensees generally believed that an NRC presence was needed at the site continuously from the start of construction.

_____

(a) Modifications to operating reactors may be made under even less optimum circumstances, but the special conditions are less unexpected and potentially better planned for.

Other factors may have played a role in assurance of quality matters, such as failure to appreciate that quality failures were symptomatic of major problems, communication problems, the level of intervention, use of innovative practices, use of detailed design models, etc., but they are considered secondary to those cited earlier. These secondary causes may be useful as indicators that primary root problems may exist. For instance, failure to appreciate that quality failures may be symptomatic of greater deficiencies might point to management failure to understand the merit of a formal institutionalized QA program, or to NRC's failure to convince the licensee that QA is an important management tool. A long period between the inception of construction and operation may be indicative of a failure to manage the project. A failure to respond effectively to NRC quality-related findings may be indicative that the licensee has developed a false sense of security.

The case studies presented a wide variety of approaches and techniques for constructing nuclear power plants. Some of these have been described previously. No single project combined all of the most advanced, efficient approaches and techniques. Some had highly computerized methods for tracking all design, specification, and record information. Others used largely manual tracking systems. Some had a large (100-150) contingent of A-E staff on site to facilitate engineering support; others had a small (~25) contingent (although the trend was towards larger on-site design staffs).

Some projects had detailed, highly computerized, systems-oriented approaches to measuring cost and schedule status. Others used more traditional "bulk" methods or manual systems for tracking. Some had innovative approaches to construction; e.g., concrete placement or sequencing. Other used more commonplace approaches. The effect of those various approaches in achieving quality in nuclear plant construction could not be evaluated. At lower levels of management and at working levels, innovative and efficient approaches and techniques used at other nuclear sites are not generally known. The fast pace and required commitment to a single construction project seems to impede dissemination of good practices. Better dissemination of the more effective procedures and methods is needed in the industry to promote quality.

It appears to take considerable time for a project team to finally "get it all together," unless the project is an immediate follow-on from a similar plant. At least most licensees stated that "things got better" as the project continued. In the latter phases of the project, working relationships were well understood and construction activities tended to flow more efficiently. Unfortunately, when a project is completed, most "teams" are disbanded. A new team has to form when the next project is initiated and, again, it takes time to "get it all together." Much more cost-effective plants, and probably plants with better quality, could be constructed if there was more continuity in the whole nuclear construction (design/construction/startup) process.

The case studies focused on quality-related problems that had occurred in the past and on present practices at projects that had not experienced quality-related problems. The NRC and the nuclear construction industry has changed and is changing. Promulgation of new regulations to address the root causes of what has happened in the past, but which are judged unlikely to occur again, may be counter-productive. In this context, it might be useful to consider the conditions under which major quality problems might recur. These could include the following:

- a first-time utility with a staff or A-E/constructor that have inadequate nuclear experience. This could result in a replay of some of the case studies reported here. NRC attention to licensee's experience, the experience of the licensee's team, and the other important factors identified in the case studies would preclude this situation from recurring.

- a very large growth in the number of nuclear plants being constructed/modified that (again) overwhelms the industry's capabilities. Sufficient data are available to estimate the industry's capabilities at present. These estimates can be adjusted to account for the effects of a nuclear hiatus, retirements, etc. If the capabilities appear to be exceeded, special care can be taken in granting additional construction permits.

- a long delay before nuclear plant construction activities start again, resulting in a dearth of experience in the industry. This situation is similar in nature to that described in the preceding case.

- regulatory actions at federal and state levels which undercut quality. Possible actions range from excessive ratcheting of NRC regulations to state regulatory utility commission actions which are counterproductive to quality. In addition to being evaluated for their effect on cost or safety, regulatory actions should be evaluated for their effect on quality.

The NRC and the nuclear industry need to be aware of the implications for quality that these and similar possibilities hold.

A.4 REFERENCES

Torrey Pines Technology (TPT). 1983. Independent Review of Zimmer Project Management. General Atomic Project 2474, Torrey Pines Technology, San Diego, California.

U.S. Nuclear Regulatory Commission (NRC). 1982. Assurance of Quality. SECY 82-352, Washington, D.C.

APPENDIX B

MANAGEMENT ANALYSIS: U.S. NUCLEAR REGULATORY COMMISSION
PROGRAMS FOR ASSURANCE OF QUALITY IN DESIGN AND CONSTRUCTION
OF NUCLEAR POWER PLANTS

PRINCIPAL CONTRIBUTORS

J. L. Heidenreich

December 1983

Prepared for
Division of Quality Assurance, Safeguards,
  and Inspection Programs
Office of Inspection and Enforcement
U.S. Nuclear Regulatory Commission
Washington, D.C. 20555

N. C. Kist and Associates, Inc.
Napierville, Illinois 60540

EXECUTIVE SUMMARY

A series of problems in the quality of construction at commercial nuclear power plants has resulted in Congress requiring the NRC to conduct a study of existing and alternative programs for improving quality assurance and quality control during design and construction. Included in NRC initiatives to assure quality in design and construction was a review of NRC quality assurance activities.

This management analysis was performed in conjunction with other NRC activities related to the Congressional legislation and NRC initiatives and included a review of implementation of management practices, past and present programs for assurance of quality in design and construction, organizational relationships between the Office of Inspection and Enforcement and Regional Offices, and a determination of root causes of the NRC's inability to prevent problems and slowness to identify and act on problems at the Diablo Canyon, Marble Hill, Midland, South Texas, and Zimmer nuclear power plants. The analysis was performed by reviewing literature provided by the Quality Assurance Branch of the Office of Inspection and Enforcement and by limited interviews with NRC personnel at the Office of Inspection and Enforcement and Regions II, III, IV and V. Time did not permit a visit to Region I.

The AEC/NRC have made the commercial nuclear power plant industry responsible for assuring the safety of its operations and have monitored the industry on a limited sampling basis. The construction of nuclear power plants has been a learning process for the AEC/NRC and the nuclear industry. NRC programs for assurance of quality during design and construction have evolved along with the nuclear industry and in response to adverse industry events. Although 10 CFR 50 became a regulation in 1954, it was not until 1967 that Appendix A of 10 CFR 50, containing the first mention of a Quality Assurance Program requirement, was published for comment and 1970 that Appendix B of 10 CFR 50, defining criteria of Quality Assurance Programs, was issued. From 1970 to about 1975, guidance documents for establishing and implementing quality assurance programs and AEC/NRC programs for assurance of quality were developed and implemented. Over the years, experience and adverse industry events, such as the Browns Ferry fire and the accident at Three Mile Island, have resulted in efforts to increase the safety of plants under construction and in operation. Instability in the regulatory process, caused by imposition of additional regulations and guidance, has contributed to longer construction times and increased opportunities for errors. Better preventive action and planning of programs would minimize the instability.

Regulations pertaining to quality assurance have not been sufficiently prescriptive or definitive to assure their clear understanding. As a result, many guidance documents have been developed. However, guidance documents have been neither mandatory nor sufficiently prescriptive or definitive to assure their understanding. The original intent of Appendix B of 10 CFR 50 applying to all aspects of a reactor without separate classes of applicability for safety-related items and items important to safety has not been fulfilled and regulations have not adequately defined safety-related items, items important to safety, and applicability of quality program requirements. Regulations should be more prescriptive and definitive in elements of control. Better regulations would eliminate the need for many guidance documents.

Licensing programs have been deficient in reviews of quality assurance programs prior to issuance of authorizations and Construction Permits and in evaluation of licensee and contractor experience, attitude, and management capability. Quality assurance programs have not been a condition of authorizations and Construction Permits and there was no requirement for submittal of program changes for NRC approval until 1983.

AEC/NRC monitoring of design and construction activities on too limited of a basis has caused inability to prevent problems and slowness to identify and act on problems. Little inspection was performed during construction prior to 1968. The direct inspection effort of the regionally based inspection program used until 1980 was about 16 days a year at each plant. Inspection orientation was towards documentation and records review until about 1979 when it changed towards hardware and results. A mindset existed that there was no immediate threat to the health and safety of the public until a nuclear power plant became operational and that plants would not be licensed until ready for operation as determined by pre-operational and startup tests. A Resident Inspector was assigned to each construction site starting in 1980. For multiple plant sites, one Resident Inspector covers all of the plants. An average of 1.5 man-years/unit is devoted to inspection during design and construction.

Budget and manpower restraints have precluded implementation of programs. Approximately 1.0% of NRC personnel are Resident Inspectors assigned to construction sites and 0.6% conduct the Licensee Contractor and Vendor Inspection Program. About 12% of the NRC budget is allocated to Inspection and Enforcement, of which inspection of design and construction is but a small part. The current inspection program is being rewritten with a goal of reducing it by 40% in recognition of budget and manpower restraints. Team inspections (PAT, CAT and IDI) are limited to a small number of plants because of budget and manpower constraints. Inspection programs appear to be designed around available resources. Inspection programs need to be designed around what must be done and the necessary resources to implement the programs need to be provided. The use of licensee inspection plans and establishment of hold points should be included.

Inspection procedures and modules have been intended as guidance and reliance has been placed on the engineering judgment of the inspector and Regional Office management for proper implementation. The degree of inspection program implementation has varied across the Regions dependent upon management's approach to regulations and the capability of personnel. Inspector experience has decreased over the years and it appears that training in quality assurance and performing inspections has been insufficient. Salaries have not been competitive with the industry, which has resulted in the loss of trained and experienced personnel to the industry and difficulty in attracting and keeping personnel. Inspection modules need to identify mandatory requirements and inspectors should receive additional training in quality assurance and in performing inspections. Less reliance on individual engineering judgment results in greater uniformity of implementation.

The NRC assumed part of licensee responsibility for evaluation of vendors through implementation of the Licensee Contractor and Vendor Inspection Program. The legal base for direct NRC inspection of vendors and any resultant enforcement action is not clearly addressed in regulations. The LCVIP does not include material manufacturers or material suppliers, sources of many material related problems during construction. The NRC has been slow to respond to findings and recommendations of previous studies of the LCVIP. Regulations should be changed to permit industry organizations to evaluate vendors with NRC overview or to establish licensing or certification programs for vendors, including material manufacturers and suppliers.

Enforcement programs have not been aggressively implemented and have not encouraged conformance to commitments. Early enforcement action consisted of "jawboning" sessions and issuance of routine enforcement letters. A mindset existed that there was no immediate threat to the health and safety of the public until a plant became operational. Programs have tended to result in categorization of nonconformances to the lowest action levels. The action point system, categorizing of nonconformances to lower action levels, and limited monitoring of design and construction activities resulted in inability to raise problems to thresholds of stronger enforcement action. The AEC/NRC have placed insufficient importance on procedural matters and have had a tendency to accept a fix to a specific problem without requiring determination of the magnitude of the problem and correction of the root cause. They have had difficulty in recognizing the significance, magnitude, and complexity of problems and did not consistently require expeditious handling of corrective action. Management has been hesitant to take strong enforcement action.

The deficiencies previously discussed were causes of NRC inability to prevent problems and slowness to identify and act on problems at the Diablo Canyon, Marble Hill, Midland, South Texas and Zimmer nuclear plants. In particular, the root cause at Diablo Canyon was insufficient attention in the area of design; the root causes at Marble Hill were inadequate review of experience and management capability, irregular NRC presence, and inability to recognize the significance and magnitude of problems; the root causes at Midland were irregular presence, reluctance to take enforcement action, and the mindset that the plant would not be licensed until ready for operation as determined by pre-operational and startup tests; the root causes at South Texas were inadequate review of experience and management capability, irregular presence, inability to recognize the significance and magnitude of problems, and the mindset that the plant would not be licensed until ready for operation as determined by pre-operational and startup tests; and the root causes at Zimmer were inadequate review of experience and management capability, failure to require licensee review of problems to determine their magnitude and correct their root cause, inability to recognize the significance and magnitude of problems, loss of inspection experience in the Region, and the mindset that the plant would not be licensed until ready for operation as determined by pre-operational and startup tests.

Commercial nuclear power plants under construction have been built during a period of learning and understanding the beneficial effects of an effective quality assurance program. Caution must be used in judging design and construction activities of the past against the standards of today. The next generation of nuclear plants will have the benefit of many man years of construction quality assurance experience. It is vital that the knowledge and understanding gained to date be properly incorporated in the NRC requirements for future nuclear installations.

TABLE OF CONTENTS

MANAGEMENT ANALYSIS
OF
U.S. NUCLEAR REGULATORY COMMISSION

1.0 INTRODUCTION

1.1 PURPOSE OF MANAGEMENT ANALYSIS

To determine shortcomings in the U.S. Nuclear Regulatory Commission (NRC) policies and programs for assurance of quality in the design and construction of commercial nuclear power plants and improvements that could be made.

1.2 SCOPE

The scope of this management analysis of NRC policies and programs for assurance of quality in the design and construction of commercial nuclear power plants was:

- to review and analyze implementation of management practices

- to review and analyze past and present programs for assurance of quality in design and construction

- to review and analyze organizational relationships between the Office of Inspection and Enforcement and Regional Offices

- to determine root causes of the NRC's inability to prevent problems and slowness to identify and act on problems at the Diablo Canyon, Marble Hill, Midland, South Texas and Zimmer nuclear plants.

1.3 BACKGROUND

During the past several years, there have been a series of well publicized problems in the quality of construction of commercial nuclear power plants.

At Midland, excessive settlement of the diesel generating building was observed in 1978. Investigation revealed that the settlement was a result of inadequate and poorly compacted soil and that other safety-related systems and structures were affected. Design and construction specifications for placement of soil fill materials had not been followed and there was insufficient control and supervision of soil placement activities by the utility and its contractors. In 1979, a civil penalty of $38,000 was issued for HVAC problems and in 1982, a civil penalty of $120,000 was issued for breakdown of the Quality Assurance Program.

At Marble Hill, all safety-related work was halted in 1979 because of concrete consolidation problems, improper repair of the imperfections, inadequate or nonexistent records traceable to the repairs, inadequate training and supervision of personnel responsible for the repairs, and insufficient awareness of the problems and control by the licensee.

At South Texas, safety-related work was halted in 1980 because of problems with concrete placement, welding, procedural violations, records falsification, personnel qualification, harassment and intimidation of inspectors, and insufficient design work. NRC investigations revealed shortcomings in management and implementation of the QA/QC Program.

At Zimmer, construction was nearly completed when in 1981 allegations prompted investigation of quality problems. Following the investigations, the NRC issued a $200,000 fine for quality assurance breakdowns. In 1982, safety-related work was halted. The major problems were identified as QC documentation, procedure violations, inadequate nonconformance reporting system, deficiencies in drawings, specifications, instructions and procedures, material control, and licensee audits and corrective action. Additional investigations reported inadequate management controls and inadequacies in administration of the Quality Assurance Program.

At Diablo Canyon, the NRC issued an Operating License in September of 1981 and revoked it two months later following licensee identification of errors in the seismic design of some piping and equipment restraints. NRC investigations revealed that proper quality assurance controls were not implemented in technical and procurement communications with service-type contractors and document control was inadequate to assure ready access to the most recent information available.

This series of problems in the quality of construction resulted in Congress requiring the NRC to conduct a study of existing and alternative programs for improving quality assurance and quality control in the construction of commercial nuclear power plants (U.S. Congress 1983).

In recognition of the problems and in anticipation of the Congressional mandate, the NRC established a series of initiatives designed to assure quality in design and construction of nuclear power plants and the NRC's ability to monitor and evaluate it (NRC 1982). Included in the initiatives was a review of NRC quality assurance activities to determine shortcomings and improvements that could be made.

This management analysis of NRC programs for assurance of quality in design and construction of commercial nuclear power plants was performed in response to the Congressional legislation and the NRC initiative.

1.4 TECHNICAL APPROACH

This management analysis was performed by reviewing literature pertaining to past and present AEC/NRC programs for assurance of quality in design and construction of commercial nuclear power plants and previous studies of those programs, and by limited interviews with the staff of the Office of Inspection and Enforcement in Bethesda, Maryland, Region II offices in Atlanta, Georgia, Region III offices in Glen Ellyn, Illinois, Region IV offices in Arlington, Texas and Region V offices in Walnut Creek, California. Personnel interviewed at the Office of Inspection and Enforcement were:

- Deputy Director, Division of Quality Assurance, Safeguards, and Inspection Programs
- Chief, Construction Inspection Branch
- Chief, Operating Reactor Programs Section
- Chief, Construction, Vendor and Special Programs Section
- Chief, Licensing Section of Quality Assurance Branch

From 12 to 16 personnel were interviewed at the Regional Offices. Personnel interviewed had the following job titles:

- Regional Administrator

- Deputy Regional Administrator

- Director
  - Division of Project and Resident Programs
  - Division of Engineering
  - Division of Vendor and Technical Programs
  - Division of Resident, Reactor Project and Engineering Programs
  - Division of Reactor Safety and Reactor Projects
  - Enforcement

- Branch Chief
  - Engineering Programs
  - Reactor Projects
  - Construction
  - Vendor

- Section Chief
  - Management Programs
  - Plant Systems
  - Materials and Mechanical
  - Reactor Projects
  - Reactor Systems
  - Reactive and Components
  - Program Support
  - Project Operations

- Enforcement Officer

- Engineer
  - Nuclear (Reactor Licensing)
  - Reactor
  - Project

- Inspector
  - Reactor
  - Project
  - Electrical Construction

The analysis includes licensee, contractor, and NRC Resident Inspector perceptions of problems with the NRC and suggestions for improvement obtained during NRC Site Assessment Case Studies performed in response to the Congressional legislation (U.S. Congress 1973).

1.5 LIMITATIONS OF THE MANAGEMENT ANALYSIS

The management analysis has been limited to NRC programs for assurance of quality in design and construction of commercial nuclear power plants and does not include other NRC programs.

The analysis has been based solely upon literature reviewed and information obtained during interviews. N.C. Kist & Associates, Inc. has not performed activities to authenticate the information obtained and makes no representations to this effect.

The study of NRC programs has been performed in conjunction with and not independent from the NRC. The Quality Assurance Branch of the Division of Quality Assurance, Safeguards, and Inspection Programs of the NRC Office of Inspection and Enforcement provided the literature reviewed, scheduled trips and interviews, and participated in the trips and interviews. The NRC did not, however, participate in the analysis of the information obtained or in the preparation of this report.

Limited interviews of personnel were performed. Two days were spent at the Office of Inspection and Enforcement in Bethesda, Maryland and two days were spent at each of the Regional Offices visited.

N.C. Kist & Associates, Inc. expended approximately two man-months of effort in performing the analysis.

2.0 SUMMARY AND CONCLUSIONS

2.1 General

This analysis of implementation of basic management practices, past and present programs for assurance of quality in design and construction of commercial nuclear power plants, organizational relationships between the Office of Inspection and Enforcement and Regional Offices, and root causes of the NRC's inability to prevent problems and slowness to identify and act on problems has revealed the following shortcomings in NRC policies and programs.

2.2 Organization

- Allocated resources have been insufficient for effective implementation of programs.

- Several functions of the Quality Assurance Branch of the Office of Inspection and Enforcement appear to duplicate functions of the Reactor Programs Construction Branch.

  - developing inspection procedures

  - performing assessments of inspection program inplementation

  - coordinating with industry the development of overview programs.

- Organizationally, there is no single overview of Resident Inspector and Speciality Inspector activities below the level of Deputy. Administrator or Administrator in Regions I, II, and III, which may create a potential for inadequate consolidation of inspection information.

- Differences exist in Regional Office organizational structures and job titles for personnel assigned similar positions, which may lead to differences in job descriptions and understanding of responsibilities.

2.3 Management Practices

- The following basic management practices have not been effectively implemented:

  - clearly defining objectives to assure their understanding

  - providing clear and constant direction

  - establishing a firm and expeditious decision-making process

  - providing adequate resources

  - performing meaningful regular assessments of the adequacy and effectiveness of NRC activities

- taking prompt, forceful corrective action in response to problems and deficiencies.

2.4 Standards Program for Assurance of Quality

. The development and application of quality assurance standards have evolved with the growth of the nuclear industry and there has been insufficient preventive action and planning.

. The original intent of Appendix B of 10 CFR 50 being applicable to all aspects of a reactor, without separate classes of applicability for items important to safety and safety-related items, has not been fulfilled.

. Regulations have not adequately defined safety-related items, items important to safety, and the applicability of quality program requirements and have not been sufficiently prescriptive or definitive to assure their clear understanding. .

. Guidance documents have been neither mandatory nor sufficiently prescriptive to assure their understanding.

. Instability in the regulatory process has resulted in longer construction times and more opportunities for error.

2.5 Licensing Program for Assurance of Quality

. Licensing Programs have been insufficient to help assure quality during design and construction.

- prior to 1970, there was no documented guidance for review of Quality Assurance Programs before issuance of permits

- from 1970 until 1975, guidance documents for review of Quality Assurance Program descriptions did not require a description of the complete program nor a detailed description of how the commitments were to be implemented

- reviews of Quality Assurance Program descriptions have emphasized completeness in addressing requirements of Appendix B of 10 CFR 50 without detailed evaluation of how the program would be implemented

- regional personnel responsible for reviewing QA Manuals were not trained in reviewing manuals

- the Quality Assurance Program has not been a condition of authorizations and Construction Permits

- until 1983, submittal of PSAR changes for NRC approval was not required

- design work and procurement of major components has been permitted prior to submittal of Quality Assurance Program descriptions

B.6

- approval of Quality Assurance Program descriptions has been heavily based on reviewer judgment as opposed to clearly defined acceptance criteria

- there has been inadequate evaluation of licensee and contractor experience, attitude and management capability.

2.6 Inspection Program for Assurance of Quality

. There has been insufficient AEC/NRC inspection during design and construction

- little inspection was performed prior to 1968

- there was irregular and non-constant presence until 1980 (a minimum of six inspections a year were to be performed and inspections were performed by regional personnel of varying disciplines)

- the annual direct inspection effort of the regionally based inspection program was about 16 days at each plant

- until 1979, inspection orientation was towards documentation and records review

- GAO concluded in 1978 that the NRC's inspection program cannot independently assure that nuclear power plants are constructed adequately

- inadequate attention has been given to design activities

- a national average of 1.5 man-years/unit is devoted to inspection during design and construction.

- approximately 1% of all NRC personnel are Resident Inspectors assigned to construction and 0.6% are in the Licensee Contractor and Vendor Inspection Program

- for multiple plant sites, one Resident Inspector covers all the plants during construction.

. Budget and manpower restraints have precluded complete implementation of inspection programs.

- the regionally-based inspection program

- the resident inspection program

- Performance Appraisal Teams

- Construction Assessment Teams

- Independent Design Inspections

- the Licensee Contractor and Vendor Inspection Program

- the current inspection program is being rewritten with a goal of a 40% reduction in recognition of budget and manpower restraints

- diverting inspection personnel to investigate allegations and team inspection findings has resulted in missing inspection "windows of opportunity" and inability to complete inspection modules

- Pre-Construction Permit activities have been insufficient in their:

  - attention to design activities

  - review of Quality Assurance Programs and their implementation

  - evaluation of licensee and contractor experience, attitude, and management capability.

- Inspection Programs appear to have been designed around available resources instead of determining what must be done and obtaining the resources to do it.

- Regional and resident inspection programs have been intended as guidance, not as mandatory requirements, and have been based upon the use of individual engineering judgment regarding the adequacy of activities performed.

- The degree of implementation of inspection programs has varied across the Regions dependent upon management's approach to regulations and the capability of personnel.

- Insufficient attention has been paid to personnel matters.

  - Inspector experience has decreased over the years and it appears training in quality assurance and performing inspections has been inadequate

  - NRC salaries have not remained competitive with the industry.

- Inspection Programs have not included hold points designating activities requiring NRC inspection.

2.7 Licensee Contractor and Vendor Inspection Program

- The Licensee Contractor and Vendor Inspection Program LCVIP evolved as a result of the learning process and of licensee inability to assure the quality of items and services supplied by their vendors.

  - Prior to 1969, vendor qualification and monitoring was viewed as the licensee's responsibility.

- In 1970, regional inspectors evaluated licensee vendor inspection programs.

- In 1973, a trial vendor inspection program was initiated for fuel fabricators.

- In 1974, a trial LCVIP was initiated because 63% of construction and operation problems were traceable to vendor errors in design or fabrication.

- In 1974, a task force recommended expansion of the LCVIP as a result of increases in vendor-related problems.

- In 1977, electrical equipment was added to the program.

- In 1978, the effectiveness of vendor design programs began being evaluated.

- In 1979, inspections became reactionary as a result of Three Mile Island.

• The legal base for direct NRC inspections of vendors and resultant enforcement action is not clearly addressed in regulations.

• The NRC has been slow to respond to findings and recommendations of previous studies of the LCVIP.

• The NRC has assumed licensee responsibility for evaluation of vendors through the LCVIP and has not taken sufficient enforcement action with licensees to force them to fulfill their responsibilities.

• The LCVIP does not apply to material manufacturers and suppliers.

2.8 Enforcement Program for Assurance of Quality

• Enforcement Programs have not been aggressively implemented.

- early enforcement action consisted of "jawboning" sessions and routine enforcement letters

- mindset existed that there was no immediate threat to the health and safety of the public until a nuclear power plant became operational

- tendency of nonconformances in design and construction to be categorized to the lower action levels since the safety function or integrity could not be clearly shown to be impaired or lost

- the action point system, categorizing of nonconformances to the lower action levels, and periodic nature of inspections resulted in inability to raise problems to thresholds of stronger enforcement action

- failure to recognize the significance, magnitude and complexity of problems

- tendency to accept a fix to a specific problem without requiring a determination of the magnitude of the problem and correction of the root cause

- failure to force expeditious handling of corrective action

- AEC/NRC management hesitancy to take action.

. Enforcement Programs have not encouraged conformance to commitments.

- Failure to conform to commitments, such as PSAR, Regulatory Guides, etc., when lack of conformance did not constitute an item of noncompliance, was considered a deviation, the lowest level of categorization. Commitments are not regulatory requirements and have not been binding. NRC approval has not been required to cancel or change commitments.

. The AEC/NRC have had difficulty in recognizing Quality Assurance Program breakdowns because of:

- the periodic nature of inspections

- categorizing of noncompliances to lower action levels

- low level of attention afforded commitments

- insufficient significance attached to procedural matters.

2.9 NRC Inability to Prevent Problems and Slowness to Identify and Act on Problems

The root causes of NRC inability to prevent problems and slowness to identify and act on problems at Diablo Canyon, Marble Hill, Midland, South Texas and Zimmer nuclear plants follows.

. Diablo Canyon

- insufficient attention in the area of design.

. Marble Hill

- inadequate review of licensee and contractor experience and capability to manage construction of a nuclear power plant

- irregular, non-constant presence
- inability to recognize the significance and magnitude of problems

. Midland

- irregular, non-constant presence

- reluctance to take enforcement action

- loss of inspection experience in the Region

- mindset that it was the licensee's responsibility to properly construct the plant and it would not be licensed until ready for operation as determined by pre-operational and startup tests.

. South Texas

- inadequate review of licensee and contractor experience and capability to manage construction of a nuclear power plant

- irregular, non-constant presence

- inability to recognize the significance and magnitude of the problems

- mindset that it was the licensee's responsibility to properly construct the plant and it would not be licensed until ready for operation as determined by pre-operational and startup tests.

. Zimmer

- inadequate review of licensee and contractor experience and ability to manage construction of a nuclear power plant

- failure to require licensee reviews of problems to determine their extent and to take corrective action regarding the cause of the problem

- inability to recognize the significance and magnitude of the problems

- loss of inspection experience in the Region

- mindset that it was the licensee's responsibility to properly construct the plant and it would not be licensed until ready for operation as determined by pre-operational and startup tests.

The following improvements could be made in NRC policies and programs for assurance of quality during design and construction of commercial nuclear power plants.

. Stabilize the regulatory process through more preventive action and planning.

. Streamline regulations and guidance documents and make them more prescriptive and definitive in terms of required elements of control without specifying how the elements of control must be implemented. Regulations that can stand on their own would eliminate the need for many guidance documents. Clearly define the applicability of

B.11

quality program requirements, safety-related items and items important to safety.

. Make the Quality Assurance Program and licensee commitments a condition of authorizations and permits.

. Replace Licensing review of the Quality Assurance Program description as presented in the Preliminary Safety Analysis Report with a Licensing or Office of Inspection and Enforcement review of the licensee Quality Assurance Manual and require the Manual to detail how the Quality Assurance Program shall be implemented. Require Licensing or Office of Inspection and Enforcement approval of Quality Assurance Manual changes. Establish definitive acceptance criteria for Manual reviews specifying required elements of control but not methods of accomplishing them. Do not permit work to be performed until approval of the Quality Assurance Manual.

. Evaluate licensee and contractor experience, attitude and management capability prior to issuance of authorizations and permits. Establish parameters and acceptance criteria.

. Require demonstration of capability to implement the Quality Assurance Program prior to issuance of authorizations or permits.

. Devote greater attention to design activities.

. Develop programs based upon what must be done and then obtain the necessary resources to implement the programs.

. Establish mandatory requirements in inspection programs and reduce dependency upon individual engineering judgement.

. Require an Inspection Plan of licensees and contractors and establish NRC hold points.

. Reevaluate personnel practices, including salaries.

. Change regulations to permit industry organizations to evaluate vendors instead of individual licensees and monitor their activities or establish licensing or certification programs for vendors. Extend the program to include material manufacturers and material suppliers.

. Take stronger enforcement action. Require expeditious handling of corrective action, including determination of the magnitude of problems and correction of their root causes.

. Perform detailed annual audits of licensee Quality Assurance Program implementation

. Review functions to be performed by the Quality Assurance Branch and Construction Programs Branches of the Office of Inspection and Enforcement to assure efforts are not duplicated.

- Eliminate differences in basic Regional Office structures and job titles to assure uniformity of functional responsibilities.

- Increase the training of inspectors in the areas of quality assurance, auditing, and implementation of inspection modules. Broaden the capabilities of inspectors to encompass all disciplines or provide additional support.

- Establish an audit program of NRC activities utilizing qualified personnel not having responsibility in the areas audited.

- Establish a Quality Assurance Program within the NRC.

3.0 MAIN DISCUSSION

3.1 ENABLING LEGISLATION

3.1.1 Description

3.1.1.1 Atomic Energy Act of 1946

The Atomic Energy Act of 1946 created the Atomic Energy Commission (AEC), empowered it to control all aspects of atomic energy, and forbade private ownership of nuclear materials. The AEC's primary activities related to the control of nuclear weapons.

3.1.1.2 Atomic Energy Act of 1954

The Atomic Energy Act of 1954 empowered and directed the AEC to promote nuclear energy and to regulate the nuclear industry. Among the provisions of the Act were to issue licenses to private companies to build and operate commercial nuclear power stations and to adopt whatever regulations it deemed necessary to protect the health and safety of the public.

3.1.1.3 Energy Reorganization Act of 1974

The Energy Reorganization Act of 1974 abolished the AEC and eliminated the conflict of interest of promoting and regulating nuclear energy by creating the Energy Research and Development Administration (ERDA) and the Nuclear Regulatory Commission (NRC).

ERDA was responsible to bring together and direct Federal activities relating to research and development of various sources of energy, to increase the efficiency and reliability in the use of energy, and to carry out the performance of other functions, including but not limited to AEC's military and production activities and its general basic research activities (U.S. Congress 1974).

NRC was responsible for all the licensing and related regulatory functions of the AEC and the functions of the Atomic Safety and Licensing Board Panel and the Atomic Safety and Licensing Appeal Board (U.S. Congress 1974).

NRC's licensing and related regulatory authority (U.S. Congress 1974) extend to:

- Demonstration Liquid Metal Fast Breeder reactors when operated as part of the power generation facilities of an electric utility system or when operated to demonstrate the suitability for commercial application of such a reactor.

- Other demonstration nuclear reactors except those in existence before the effective date of the Energy Reorganization Act of 1974 when operated as stated above.

- Facilities used primarily for the receipt and storage of high-level radioactive wastes resulting from activities licensed.

- Retrievable Surface Storage facilities and other facilities authorized for the express purpose of subsequent long-term storage of high-level radioactive waste generated by the Administration, which are not used for, or part of, research and development activities.

The Energy Reorganization Act of 1974 (U.S. Congress 1974) established the organization of the Commission and Offices of Nuclear Reactor Regulation, Nuclear Material Safety and Safeguards, and Nuclear Regulatory Rsearch.

The Commission is composed of five members appointed by the President, by and with advice and consent of the Senate. Appointments must be made in such a manner that not more than 3 members are of the same political party. Each member serves a 5-year term with terms expiring in consecutive years. The President designates one member as chairman to serve during the pleasure of the President and any member may be removed by the President for inefficiency, neglect of duty or malfeasance in office. Each member has equal responsibility and authority in all decisions and actions, has full access to all information relating to his duties or responsibilities, and has 1 vote. Action of the NRC is determined by a majority vote of members present.

The Office of Nuclear Reactor Regulation is responsible for such functions as the NRC delegates (U.S. Congress 1974) including:

- principal licensing and regulation of all facilities and materials licensed under the Atomic Energy Act of 1954, as amended

- review of the safety and safeguards of all such facilities, materials and activities, including but not limited to monitoring, testing and recommending upgrading of systems designed to prevent substantial health or safety hazards and evaluating methods of transporting nuclear materials and transporting and storing high level radioactive wastes

- recommending research necessary for the discharge of the functions of the NRC.

The Office of Nuclear Material Safety and Safeguards is responsible for such functions as the NRC delegates (U.S. Congress 1974) including:

- principal licensing and regulation involving all facilities and materials, licensed under the Atomic Energy Act of 1954, as amended, associated with the processing, transport, and handling of nuclear materials, including the provision and maintenance of safeguards against threats, thefts, and sabotage of such licensed facilities, and materials

- review of safety and safeguards of all such facilities and materials, including, but not limited to monitoring, testing, and recommending upgrading of internal accounting systems for special nuclear and other nuclear materials and developing contingency plans for dealing with threats, thefts, and sabotage relating to special nuclear materials, high-level radioactive wastes and nuclear facilities

. recommending research to enable the NRC to more effectively perform its functions.

The Office of Nuclear Regulatory Research is responsible for such functions as the NRC delegates (U.S. Congress 1974) including:

. developing recommendations for research deemed necessary for performance by the Commission of its licensing and related regulatory functions

. engaging in or contracting for research which the Commission deems necessary for the performance of its licensing and related regulatory functions.

The Energy Reorganization Act of 1974 also included a survey to locate and identify possible nuclear energy center sites, quarterly submittal by the Commission to Congress of a report listing abnormal occurrences at or associated with any facility licensed or regulated and dissemination of such information to the public within fifteen days of Commission receipt of such information, development of a plan for the specification and analysis of unresolved safety issues, employee protection against discharge and discrimination because the employee commenced or participated in a proceeding under the Act or the Atomic Energy Act of 1954, including investigation of such charges by the Secretary of Labor, annual authorization of appropriations to the Commission which reflect the need for effective licensing and other regulation of the nuclear power industry in relation to the growth of such industry, and Comptroller General of the United States audit, review and evaluation of implementation of the provisions of the Act pertaining to the Nuclear Regulatory Commission not later than 60 days after the effective date of the Act.

3.1.1.4 Additional Enabling Legislation

Congress provides additional enabling legislation as part of its annual authorization of appropriations.

3.1.2 Analysis

Until the Energy Reorganization Act of 1974, the AEC was empowered and directed to promote nuclear energy and to regulate the nuclear industry. This dual responsibility created an inherent conflict of interest which resulted in widespread criticism of the AEC emphasizing their promoting role at the expense of their regulating role.

The Energy Reorganization Act of 1974 eliminated the inherent conflict of interest by making the NRC responsible for all licensing and related regulatory functions and ERDA responsible for directing Federal activities related to research and development of various sources of energy.

Enabling legislation has provided the AEC/NRC adequate authority for fulfilling its responsibilities. This study has not identified a need for changes to enabling legislation.

3.2 ORGANIZATION

3.2.1 Description

3.2.1.1 General

The Energy Reorganization Act of 1974 (U.S. Congress 1974) transferred to NRC the Chairman and members of the AEC, the General Counsel, and other officers and components of the Commission except functions, officers, components, and personnel transferred to ERDA.

The Commission is responsible for licensing and regulating nuclear facilities and materials and for conducting research in support of the licensing and regulatory process, including protecting public health and safety, protecting the environment, protecting and safeguarding materials and plants in the interest of national security, and assuring conformity with antitrust laws (NRC 1983). To fulfill its responsibilities, the Commission has used; standards setting and rule making; technical reviews and studies; conduct of public hearings; issuance of authorizations, permits and licenses; inspection, investigation, and enforcement; evaluation of operating experience; and confirmatory research. The Commissioners are described under Enabling Legislation in this section of the report. Reporting to the Commissioners are the:

- Office of Public Affairs
- Office of Congressional Affairs
- Atomic Safety and Licensing Board Panel
- Atomic Safety and Licensing Appeal Panel
- Advisory Committee on Reactor Safeguards
- Office of Investigations
- Office of Inspector and Auditor
- Office of Policy Evaluation
- Office of the General Counsel
- Office of the Secretary
- Executive Director for Operations.

The Executive Director of Operations (EDO) performs functions as the Chairman or Commission directs and is governed by policies and decisions of the Commission (NRC 1983). Reporting to the Executive Director for Operations are the:

- Office of Administration
- Office of the Executive Legal Director
- Office of Resource Management
- Office of Small and Disadvantaged Business Utilization and Civil Rights
- Office for Analysis and Evaluation of Operational Data
- Office of International Programs
- Office of State Programs
- Regional Offices
- Office of Nuclear Material Safety and Safeguards

- Office of Nuclear Reactor Regulation
- Office of Nuclear Regulatory Research
- Office of Inspection and Enforcement.

The NRC has operated with a budget ranging from 333 million dollars to 513 million dollars over the past five years. The annual Congressional authorization specifies the amounts that will be used for specific activities. The following summary shows the average allocation of funding to each specified area of activity during the last five year period (U.S. Congress 1983 et al).

Area of Activity	Average % of Total Appropriation
Nuclear Regulatory Research	49.7
Nuclear Reactor Regulation	15.4
Inspection and Enforcement	12.0
Nuclear Material Safety and Safeguards	8.4
Program Direction and Administration	8.3
Technical Support	4.2
*Standards	4.1

*1979 and 1980 only. Not listed as a separate category after 1980.

The elements of the organization primarily involved in NRC programs for assurance of quality during the design and construction of nuclear power plants are the Office of Nuclear Reactor Regulation, the Office of Nuclear Regulatory Research, the Office of Inspection and Enforcement and the Regional Offices.

The Office of Nuclear Reactor Regulation develops and administers regulations, policies and procedures. The Division of Licensing directs and administers the licensing process for all utilization facilities including safety and environmental evaluations of reactors required to be licensed for operation. It directs and supervises the processing of applications and petitions for license amendments and issues, denies, and amends all limited work authorizations, permits and licenses for reactors, administers the Standardization Program, and serves as NRR coordinator with the Office of Inspection and Enforcement (NRC 1983).

The Office of Nuclear Regulatory Research plans, recommends and implements the programs of Nuclear regulatory research necessary for performance of licensing and related regulatory functions. The Division of Engineering Technology plans, develops, and directs research programs and develops standards for the design, qualification, construction, inspection, testing, operations and decommissioning of nuclear power plants (NRC 1983).

3.2.1.2 Office of Inspection and Enforcement

The Office of Inspection and Enforcement was formed by the Commission during the Energy Reorganization Act of 1974. Its function (NRC 1983) is to develop policies and programs for enforcement and inspection of licensees, applicants, and their contractors and suppliers to:

- ascertain whether they are complying with NRC regulations, rules, orders, and license conditions

- identify conditions that may adversely affect public health and safety, the environment, or the safeguarding of nuclear materials and facilities

- provide a basis for recommending issuance or denial of an authorization, permit or license

- determine whether quality assurance programs meet NRC criteria

- recommend or take appropriate action regarding incidents or accidents

- develop policies and implement a program of enforcement action

- direct emergency preparedness activities

- provide guidance to Regional Offices on program matters

- appraise program performance in terms of effectiveness and uniformity.

In January 1983, several organizational and functional changes were made in the Office of Inspection and Enforcement because of their expanded role in quality assurance.

1) The Division of Reactor Programs was redesignated the Division of Quality Assurance, Safeguards, and Inspection Programs. Primary emphasis continues to be placed on quality assurance while integrating quality assurance concerns and principles into the reactor construction and operating reactor inspection programs.

2) The Division of Engineering and Quality Assurance was redesignated the Division of Emergency Preparedness and Engineering Response.

3) The Reactor Training Center became the Technical Training Center.

Quality assurance functions of Nuclear Reactor Regulation and Nuclear Reactor Research were transferred to the Quality Assurance Branch of the Office of Inspection and Enforcement. Within the Division of Quality Assurance, Safeguards, and Inspection Programs, the Quality Assurance Branch consists of 12 personnel and performs the following functions (De Young 1982):

- Develops a comprehensive NRC program for Quality Assurance of licensee facilities to be applied to design, fabrication, construction, testing and operation. This encompasses licensees, vendors, architect-engineers, constructors, and other licensee agents.

- Develops requirements and standards based upon regulatory experience and industry coordination.

- Reviews existing requirements and standards to clarify and optimize the effectiveness of QA requirements and standards.

- Reviews existing office programs to optimize the effectiveness of QA activities.

- Responsible for developing QA-related inspection procedures and for performing assessments of QA inspection program implementation by the regional offices.

- Develops and coordinates with the regional and other headquarters offices, NRC initiatives to confirm the management effectiveness of licensees in assuring the quality of licensee and contractor activities during design, fabrication, construction, testing and operation.

- Develops and coordinates with the regional and other headquarters offices, NRC initiatives to independently verify the quality of construction at selected utilities.

- Coordinates with industry the development of overview programs for improving the effectiveness of QA programs and their implementation.

The Reactor Construction Programs Branch consists of 21 personnel and performs the following functions (De Young 1982):

- Develops the NRC inspection policies and programs for reactor projects from the time of an application for a construction authorization or permit to the time the operating license is issued. Includes inspection programs for associated nuclear steam suppliers, architect-engineers, constructors and component vendors. Excluded from the branch responsibilities are the preoperational preparations that do not pertain to the actual construction of the plant. The policies, strategies, and programs will be revised principally to improve staff resource effectiveness by integrating applicable licensing procedures and experiences with those from the office inspection activities.

- Based on the results of assessments of program implementation and on recommendations from regional offices, NRR, and NMSS, revises established programs, as necessary, to increase their effectiveness to better coordinate inspection activity with licensing policy and objectives, and to tailor the programs to anticipated resources.

- Develops estimates of resources that are needed to perform the various elements of the programs which have been established, or which are under consideration.

- Represents NRC to outside agencies and technical organizations such as INPO, ASME, and IEEE, in order to further the development of integrated construction inspection programs and to make best use of available resources of NRC, licensees, and associated organizations.

- Develops and maintains the Construction Appraisal Team (CAT) programs for reactors under construction including assessment of regional office implementation. Conducts CAT team inspections at licensee facilities.

• Assesses regional office implementation of established inspection programs to determine to what degree program requirements are being met. Assesses the effectiveness of each established program and determines whether the regions are implementing the programs in a technically adequate and consistent manner. This process will include field observations and examinations at licensee sites and at licensee and regional offices. Provides guidance to the regions regarding areas of program implementation which need improved performance and areas where the program can be cut back to better fit available resources.

3.2.1.3 Regional Offices

The Regional Offices execute established NRC policies and assigned programs relating to inspection, enforcement, licensing, state agreements, state liaison, and emergency response within Regional boundaries. Regional Office activities include project and resident inspection programs, engineering, radiological safety, emergency preparedness, and materials safety programs. Region IV is responsible for implementation of the Licensee Contractor and Vendor Inspection Program. In 1980, the NRC began to expand the scope of functions of Regional Offices to create an agencywide regional operation which includes licensing as well as inspection and enforcement functions.

Regional Administrators have managerial and supervisory responsibility for all functions and personnel assigned to their Region. Regional organizations include an Administrator; Deputy Administrator; Enforcement Director, Coordinator, or Specialist; Counsel or Attorney; Public Affairs Officer; Division Directors; Branch Chiefs, Section Chiefs; Resident Inspectors; Specialty Inspectors and support personnel.

The two Regional Office groups of major interest to this study are the Division of Project and Resident Inspector Programs, which administers assigned project and resident inspectors, and the Division of Engineering, which provides technical or speciality inspectors to perform work such as quality assurance reviews or nondestructive examinations.

In Regions IV and V, inspection responsibilities of these two divisions is consolidated in the Division of Resident Reactor Project and Engineering Programs.

Each division is comprised of two branches supervised by Branch Chiefs who are responsible for providing management of the division's functions for assigned facilities within the Region. Each branch is comprised of sections supervised by Section Chiefs who are responsible for providing management of functions at from three to five nuclear power plant sites. Within sections are Project Inspectors who are responsible for overseeing implementation of the inspection program at one or more sites and helping to coordinate regional activity at the sites. Also within sections are Resident Inspectors who are responsible for implementation of the inspection program at their assigned site.

Of the 3300 employees working for the NRC, 890 are located at the Regional Offices and of these, 460 are classified as inspectors. Of the 460 inspectors, 32 are Resident Inspectors for construction and 22 are involved in the Licensee Contractor & Vendor Inspection Program.

A more detailed breakdown of inspector personnel by Region follows: (Blaha 1983):

Region	Total Personnel		Total Inspectors		Resident Inspectors-Construction	
	Actual	Budgeted	Actual	Budgeted	Actual	Budgeted
I	211	218	118	124.5	6	7
II	213	222	118	125	5	7
III	222.3	216	116	118	14	10
IV	151	139	64	67.5	4	4
V	93	92	44	47.5	3	4
TOTALS	890.3	887	460	482.5	32	32

3.2.2 Analysis

3.2.2.1 General

The scope of this study limited organizational analysis to relationships between the Office of Inspection and Enforcement and Regional Offices. Additional study of other NRC offices involved in the assurance of quality during design and construction of nuclear power plants is warranted.

Communications between NRC headquarters and Regional Offices appear to be adequate, although more personal unscheduled meetings in handling problems and suggestions should be encouraged. Complaints were heard that by the time a suggestion travels from a Resident Inspector upwards through the Regional Office and then downward in the I and E chain to the QA Division, much of its effectiveness is lost. Regular meetings of individuals involved with standards and inspection modules would be beneficial.

Over the past five years, an average of 12 percent of the NRC budget has been allocated to all inspection and enforcement activities. The portion of the 12 percent assigned to inspection of design and construction activities was not readily obtainable, but would be small. Allocated resources have been insufficient for effective implementation of programs and is discussed under the programs in this report. In order to assure quality during design and construction, additional budget allocations to inspection activities appears necessary.

3.2.2.2 Office of Inspection and Enforcement

Recent changes in the Bethesda, MD headquarters organization have shifted the quality assurance functions from other offices to the Office of Inspection and Enforcement Division of Quality Assurance, Safeguards, and Inspection Programs. The consolidation of these functions within a central group should provide more effective management of the functions. The functions to be performed by the Quality Assurance Branch and Construction Programs Branch appear adequate to assure quality in design and construction of commercial nuclear power plants. However, the effectiveness of the NRC will be dependent upon the implementation of the functions. Additional guidance, describing in more detail the implementation of each function, appears to be necessary.

Some of the functions appear to duplicate efforts. The Quality Assurance Branch is responsible for developing QA-related inspection procedures and for performing assessments of QA inspection program implementation by the Regional Offices. The Reactor Programs Construction Branch is responsible to develop the NRC inspection policies and programs and to assess Regional Office implementation of established inspection programs. The Quality Assurance Branch is to coordinate with industry the development of overview programs for improving the effectiveness of QA programs and their implementation. The Reactor Construction Programs Branch is to represent the NRC to outside agencies and technical organizations such as INPO, ASME, and IEEE, in order to further the development of integrated construction inspection programs and to make best use of available resources of NRC, licensees, and associated organizations. These functions should be reviewed to assure efforts are not being duplicated.

3.2.2.3 Regional Offices

Each Region does not have the same organizational structure and job titles for personnel assigned similar positions. For example, Regions I, II and III, have two separate divisions for Project and Resident Programs and Engineering while these activities are combined into one division in Regions IV and V. The responsibility of implementing enforcement policies and procedures is held by an Enforcement Specialist in Region I, a Director of Enforcement in Regions II and IV, an Enforcement Coordinator in Region III, and an Enforcement Officer in Region V. Differences in organization and job titles may lead to differences in job descriptions and misunderstandings of responsibilities. Regional Office organizational structures should be standardized for identical functions.

In Regions I, II, and III, project and resident inspectors are part of the Division of Project and Resident Programs and specialty inspectors are part of the Division of Engineering. Organizationally, there is no single overview of all inspector activities below the level of Deputy Administrator or Administrator. Functionally, interaction between the inspection personnel, Section Chiefs, Branch Chiefs, and Division Directors may provide overview of all inspector activities, but the organization would indicate a potential for inadequate consolidation of inspection information. Consideration should be given to providing a single overview of all inspector activities below the level of Deputy Administrator or Administrator.

The organizational structure of Regions results in four levels of supervision between an inspector and the Regional Administrator, which could result in attenuation of information. Regional personnel indicated there was little attenuation of information between the inspectors and the Administrator on anything of significance. A formal NRC policy was placed into effect following Three Mile Island to permit submittal of differing professional opinions to the Commission over Regional management.

Personnel allocation to inspection activities during design and construction of Nuclear power plants has been insufficient. Of the total number of NRC personnel, approximately 1.0 percent are Resident Inspectors assigned to construction sites and approximately 0.6 percent are involved in the Licensee Contractor and Vendor Inspection Program. Current NRC headquarters personnel estimates of manpower performing inspections during design and construction is 1.5 man-years/unit. At the time of pre-operational activities, inspection effort increases to about five to seven man-years.

The EDO stated in SECY-82-352:

"Although a resident inspector is now assigned to every site at which construction is more than 15 percent complete, the NRC is limited in its ability to assure compliance with all NRC requirements because of the limited inspection resources."

Inability to fully implement past and present programs as a result of budget and manpower restraints has been a contributing factor to the AEC/NRC inability to prevent problems and slowness to identify and act on problems. Additional discussion of resource allocation pertaining to past and present programs may be found in other sections of this report. In order to assure quality during design and construction, additional allocation of personnel to inspection activities is necessary.

3.3 MANAGEMENT PRACTICES

3.3.1 General

To effectively regulate and control the commercial nuclear power plant industry in the United States, it is necessary for the NRC to implement basic management practices, such as:

- clearly defining objectives and philosophy

- assuring clear understanding of objectives and philosophy

- defining organizational structure, functional responsibilities, authorities, and interfaces

- defining a detailed approach towards accomplishing objectives in instructions, procedures and other documents which may be easily understood

- providing clear and constant direction

- establishing a firm and expeditious decision-making process

- assuring good communications

- providing adequate resources

- performing meaningful, regular assessments of the adequacy and effectiveness of the organization's activities

- taking prompt, forceful corrective action in response to problems or deficiencies.

The results of this study provides the following information regarding implementation of basic management practices.

3.3.2 Objectives and Philosophy

The objectives and philosophy of the AEC and NRC have been clearly stated and well understood by the industry and AEC/NRC. Objectives have included:

- to protect the public health and safety

- to protect the environment

- to protect and safeguard materials and plants in the interest of national security

- to assure conformity with antitrust laws.

The basic philosophy has been to make the nuclear power plant industry responsible for assuring the safety of its operations and to monitor the industry on a limited sampling basis to verify its fulfilling of this responsibility.

Although the objectives have been clearly stated, the subjective terminology used and vagueness in defining their meaning has resulted in different
perceptions by Congress, the public, the nuclear power plant industry, and the
AEC/NRC of what has been expected in meeting the objectives.

3.3.3 Organization

The organizational structure, functional responsibilities, authorities
and interfaces have been clearly defined and documented in organization charts
and procedures.

3.3.4 Approach

The approach towards accomplishing objectives has been clearly defined in
procedures and other documents which may be easily understood.

3.3.5 Direction

Clear and constant direction has not always been provided to regulatory
personnel and the industry. The AEC and NRC have learned along with the
industry during years of construction and operation of nuclear power plants.
Programs for assurance of quality during design and construction evolved as a
result of the learning process and in reaction to adverse industry events.
There has been insufficient preventive action and planning of programs.

Regulations pertaining to quality assurance and guidance documents for
their implementation have not been sufficiently prescriptive or definitive to
assure their clear understanding by the industry, the AEC and the NRC. Regulations have not adequately defined the applicability of quality program
requirements. Additional discussion of regulations and guidance documents is
included under the Standards Program for Assurance of Quality in this section
of the report.

Licensing activities have not assured that licensees have developed and
implemented adequate quality assurance programs before performing activities
affecting quality. Licensee commitments at the Preliminary Safety Analysis
Report stage have not been made a condition of the Construction Permit.
Additional discussion of licensing is included under the Licensing Program for
Assurance of Quality in this section of the report.

Inspection programs have been intended to serve as guidance to the
inspectors and implementation of the programs has been dependent upon the
engineering judgment of regional management and each individual inspector.
The degree of implementation of inspection programs has varied across the
Regions dependent upon management's approach to regulations and programs and
the capability of personnel.

3.3.6 Decision Making

The decision making process has not always been firm and expeditious.
Licensee and contractor personnel indicated during the NRC Case Studies that:

- the industry needed decisions from the NRC and was guessing for years what to do following Three Mile Island

- the NRC took too long to resolve problems and questions and took one to two years in some instances

- appeal boards resulted in long hearings with few design·changes

- anyone can second guess the NRC and hold up utility programs for years

- the NRC needs to accept the technical views of experts and not hold up work due to unqualified intervenors.

Regional personnel indicated that headquarters was often more of an obstacle than a help. Upon identifying problems, regional personnel would get little assistance from headquarters. Some regulations were viewed as encouraging slow decisions. It was indicated that it may take a year to resolve a 50.55 (e) finding after it is reported. A need for more accountability within the NRC was also expressed.

3.3.7 Communications

Generally, there appears to be good communication within and between NRC headquarters and Regions. Regional personnel did indicate, however, that feedback to Regions on suggestions made by regional personnel was poor, resulting in a reduction of incentive to make suggestions for improvements.

3.3.8 Resources

Adequate resources have not been provided to assure quality in design and construction of nuclear power plants. Budget and manpower restraints have precluded adequate development and implementation of AEC and NRC programs. Programs have tended to be prepared on the basis of available resources instead of defining what must be done to assure quality in design and construction and then obtaining the necessary resources to assure the required activities are uniformly implemented at each facility. Diverting manpower from the inspection program to perform reactionary inspections, investigate allegations and follow-up on special inspection findings, has resulted in missing inspection "windows of opportunity", periods of construction during which an inspection must be performed because it cannot be performed later. Licensee and contractor personnel indicated during the NRC Case Studies that:

- the NRC needs more resident inspectors or roving teams to support all disciplines

- there should be a resident for each discipline

- they questioned the capability of the NRC staff to do adequate technical reviews

- the NRC staff has to be equal or competitive with the utility's and architect engineer's

Regional personnel indicated a need to assure the NRC staff is qualified to perform their jobs and that many auditors from the NRC didn't know enough about the subject being audited to perform meaningful audits. They also indicated that:

- if Regions are to perform all activities for which now responsible, increased resources will be required

- the level of inspector experience has decreased over the years

- the NRC has not remained competitive with the industry regarding salaries

- the NRC needs to hire people with actual experience

- additional training is needed for inspectors, headquarters, and regional personnel.

3.3.9 Assessment of Activities

Although there have been numerous studies of the AEC and NRC, there have not been meaningful, regular internal assessments of the adequacy and effectiveness of AEC/NRC programs for assurance of quality in design and construction of nuclear power plants. No NRC organization has been responsible for auditing all of the activities of the NRC. Review functions of the Quality Assurance Branch and Reactor Construction Branches of the Office of Inspection and Enforcement do not include audits of implementation of NRC programs and the Office of Inspector and Auditor has not fulfilled this function. Findings regarding the accident at Three Mile Island (NRC 1980) included:

- There appears to be no internal technical audit function in NRC. The I&E in Washington, D.C. does review the activities of its inspectors, but there does not appear to be any organization responsible for reviewing and auditing the overall utility overview process. The Office of Inspector and Audit appears to be a legal and administrative audit only, not involved in technical reviews.

- There is no assignment within the NRC organization for overview of critical functions such as problems reporting, failure analysis, and corrective action; systems engineering; and the role of the operator and human factors in plant safety.

- No NRC organization is identified as being responsible for auditing the project management, engineering, and inspection functions of the NRC.

The NRC needs to correct this situation by establishing an audit program that utilizes qualified personnel not having responsibility in the areas audited.

3.3.10 Corrective Action

Forceful action has been taken in response to many problems and deficiencies. However, the promptness of action has been slowed by the organizational structure and procedures of the AEC/NRC and the action taken has tended to be additional requirements resulting from specific events and has not sufficiently included corrective actions regarding the causes of the problems and deficiencies.

3.4 STANDARDS PROGRAM FOR ASSURANCE OF QUALITY

3.4.1 Description

In 1955 and 1956, the AEC issued a set of basic regulations for the civilian nuclear industry. Chairman Strauss emphasized that the regulations were not intended to restrain the industry but to "open the way to all who are interested in engaging in research and development of commercial activities in the atomic energy field" (Langstaff 1982). Providing facilities which did not endanger the health and safety of the employees and the public was to be the industry's responsibility.

In 1967, the AEC published for comment 70 General Design Criteria for Nuclear Power Plant Construction Permits (Appendix A of 10 CFR 50). Criterion 1 specified the quality expected to be incorporated in all aspects of nuclear facilities and required a QA Program "be established and implemented in order to provide adequate assurance that these structures, systems and components will satisfactorily perform their safety functions." Specific criteria for a QA Program were not included.

The need for more definitive QA regulatory criteria was strongly empha-sized at the Atomic Safety Licensing Appeal Board hearing on Zion Nuclear Station in 1968.

The following year, the AEC published Appendix B of 10 CFR 50 for comment, which specifically defined the requirements of the licensee's Quality Assurance program.

Interviews with NRC headquarters personnel revealed that when Appendix B of 10 CFR 50 was published for comment, the criteria were meant to elaborate on the Quality Assurance Program requirements of Appendix A with no intention of separate classes of applicability for items important to safety and safety-related items. Appendix B was to complement Appendix A and apply to all aspects of a reactor, not just seismic category 1. Appendix B was published as an effective rule in 1970 and Appendix A was published as an effective rule in 1971. Since Appendix B was published while Appendix A was still in draft form, references to Appendix A were dropped, including language that indicated Appendix B was to apply to the general design criteria. When Appendix A was published, there was no attempt to revise Appendix B to clarify the intent of applicability. Appendix B was interpreted by AEC staff performing Safety Analysis Reviews to apply to seismic category 1, and was not applicable to any broader class of equipment, systems, or components.

As the AEC reviewed individual nuclear plants, the resolution of issues were negotiated with owners. AEC staff positions gradually emerged in the form of Safety Guides. In 1970, the AEC began to publish Regulatory Guides which clarified the AEC's position and replaced the Safety Guides. A primary purpose of Regulatory Guides (AEC 1972) was to describe and make available to the public methods acceptable to the AEC Regulatory Staff of implementing specific parts of regulations and to provide guidance to applicants concerning information needed by the staff in review of applications for permits and licenses. The Guides were not intended as substitutes for regulations and compliance was not required. Different methods and solutions were acceptable if they provided a basis for findings requisite to the issuance of a permit or license. The AEC delegated the work of devising needed rules to industry committees who would prepare a standard governing a certain aspect of plant design. The AEC would then write a Regulatory Guide that adopted the standard in whole or in part. There are currently 153 Regulatory Guides.

In a report regarding the status and application of ANSI N45.2 Standards (Bernsen and Hellman 1973), the following observations regarding the philosophy of the ANSI Standards were made with assistance from the N45.2 Subcommittee membership:

"Each of the standards issued by the N 45.2 Subcommittee has been subject to an extremely intensive preparation and review process and is believed to contain precise statements of acceptable current practices for commercial nuclear power plants-practices which are practical, currently available and judged necessary to achieve required levels of quality."

"Whereas AEC regulations and the Code are mandatory regulations establishing firm requirements for the areas they cover, and hence, include assignments of responsibilities, the N45.2 series are not written as self-sufficient regulatory documents and are intended to be supplemented by:

a.   a regulatory requirement prescribing its use (i.e., the AEC's codes and standards rules 10 CFR Part 50 or other statements of AEC requirements, such as the AEC Regulatory Guides)

b.   a power plant applicant's license commitments or

c.   an appropriate procurement document.

Another significant difference between the ANSI standards and the regulations is that the ANSI standards are intended to apply to features of the plant which affect operational reliability as well as those which are important to safety. Naturally, the extent to which these standards would be applied to plant features which affect reliability is a matter for determination by the utility and hopefully a mutual agreement between the utility and his principal contractors; but there appears to be a general consensus of opinion that judicious application of quality standards to the total plant will prove beneficial."

In 1971, ANSI N45.2 was published, basically repeating Appendix B of 10 CFR 50 but describing the requirements in more detail. Shortly thereafter the AEC issued Regulatory Guide 1.28 endorsing ANSI N45.2-1971.

In 1973 and 1974, the AEC Regulatory Staff issued "Guidance on Quality Assurance Requirements During Design and Procurement Phase of Nuclear Power Plants" (Gray Book) and "Guidance on Quality Assurance Requirements During the Construction Phase of Nuclear Power Plants" (Green Book) to provide guidance for establishing and implementing Quality Assurance Programs. Most of the guidance was in the form of AEC regulations, Regulatory Guides, and draft standards developed by the American National Standards Institute Subcommittee N45.2.

In 1973, it was recommended to the Director of Regulation (Davis and Brown 1973) that Regulatory host a series of conferences for utilities with participation of the Commissioners to demonstrate the Commission's commitment to QA and to explain the mini-review procedure. During July of 1973, AEC senior staff, including two Commissioners, participated in regional one-day conferences with utilities to explain the role of quality assurance in design, construction and operation of nuclear power plants.

The AEC also announced that it would hold meetings with prospective applicants to discuss in detail the quality assurance criteria in sufficient time for the utility to include the requirements in contracts for design and procurement.

In 1975, the NRC issued a Standard Review Plan to define the scope of review and acceptance criteria for the NRC's approval of Safety Analysis Reports.

The NRC continues to use Appendix B of 10 CFR 50 as the primary requirements for Quality Assurance Programs and supplements Appendix B with Regulatory Guides.

3.4.2 Analysis

3.4.2.1 Evolution of Standards

The role of the AEC, and subsequently the NRC, as a regulator of the commercial nuclear industry has been ill defined since the origin of the program. The primary guidance to the regulators was to protect the "health and safety of the employees and the public". Early AEC interpretation of this mandate minimized specific quality assurance controls, which undoubtedly reflected the then current attitude towards safety in fossil plants or the military nuclear program. The development and application of quality assurance standards have evolved with the growth of the nuclear industry. The role of the AEC and NRC has been a reactive one as both the industry and its regulators have grown to understand the significance of quality assurance.

Although 10 CFR 50 became a regulation in 1954, it was not until 1967 that Appendix A of 10 CFR 50, containing the first mention of a quality assurance program requirement, was published for comment. Until 1967, AEC regulations were intended to encourage research and development of commercial activities and to let the commercial nuclear power industry regulate itself. The AEC recognized the need for defining specific criteria of quality assurance programs as a result of hearings on the Zion Nuclear Station in 1968 and as a result, issued Appendix B of 10 CFR 50 in 1970.

The industry ANSI Standard N45.2 was being prepared about the same time with similar quality assurance requirements. By the use of Regulatory Guides, the NRC has modified the ANSI 45.2 Standard and further defined the Appendix B requirements.

3.4.2.2 Applicability of Standards

The AEC and NRC have failed to fulfill the original intent of Appendix B to 10 CFR 50. As Appendix A and Appendix B of 10 CFR 50 were published, Appendix B was to complement Appendix A and thereby apply to all aspects of a reactor, without separate classes of applicability for items important to safety and safety-related items. The AEC and NRC have failed to clearly define safety-related items and items important to safety and have not adequately defined the applicability of quality assurance program requirements in its regulations. The determination of how and to what extent quality assurance requirements are applied has been left to the discretion of the applicant. Although the applicant must identify safety-related systems in the PSAR, there is no requirement to identify specific safety-related items within the systems and there is no NRC review of classification of such items for completeness or adequacy. Each applicant determines which items it considers safety-related, resulting in lack of uniformity of classification of items as safety-related and lack of uniformity in quality assurance program application.

Several previous studies have suggested that changes be made in the methods used in defining safety related and importance to safety classifications of components and systems.

NUREG 0321 (A Study of the Nuclear Regulatory Commission Quality Assurance Program - 1977) stated:

> "10 CFR 50 Appendix B should be used in the regulation of all areas of power reactor design, construction and operation which are judged to have sufficient importance to safety to fall under NRC regulation. The selective application of QA elements now applied to safety-significant items not interpreted as falling under Appendix B should be replaced by an approach in the degree to which the 18 Criteria of Appendix B are applied would reflect the safety significance of the item."

The Staff Report to the Presidential Commission on the Accident at Three Mile Island - Volume IV - 1979 also addressed the subject as follows:

> "Quality assurance requirements apply only to a narrow portion of the plant defined as safety-related or safety-grade. Many items vital to the safe and reliable operation of the plant are not covered by the quality assurance program because of this definition." And also

> "Safety and reliability requirements and analysis are not required to be applied to many plant systems which may be vital to the safe operation of the plant but are not labeled safety-related."

NUREG/CR-1250, Volume II, Part 1 (Three Mile Island: A Report to the Commission and the Public - 1980) stated:

> "Although the requirements of Appendix B are sufficiently broad to adequately address most aspects of acceptable quality assurance programmatic requirements, one important shortcoming of the regu-

latory program arises from the absence of a definition of "safety-related," a concept central to the entire structure. Although Appendix B contains numerous references and applications of "safety-grade equipment," "safety-related equipment," and "equipment required for safety-related functions," NRC regulations contain no definition of "safety-related" or comparable terms. No other general regulatory guidance for defining or applying these terms is found and NRC staff members have different interpretations of these terms. Failure to define "safety-related" has restricted the scope of the NRC's quality assurance programs. Identification of particular "safety-related" structures, components, and systems is the responsibility of the applicant utility. The absence of definitional guidance supports the applicant's narrow interpretation and, correspondingly, decreases the staff's ability to insist that a particular system or function is "safety-related."

"This lack of clarity has generated staff disagreement concerning the identification of equipment to which Appendix B should be applied and concerning the differences and similarities between Appendix A, which applies to components that are "important to safety" and require a graduated quality standard, and Appendix B, which imposes a higher quality standard on the systems and functions to which it applies. This disagreement has frustrated efforts to formulate a regulatory guide for implementing Appendix B."

Regarding the applicability of quality assurance programs, the EDO stated in SECY-82-352:

"Current rules are not specific on whether or not a licensee or permit holder is required to notify the NRC of changes to the quality assurance program description previously accepted by the NRC in the Safety Analysis Report (SAR). Additionally, current regulations do not explicitly require licensees or permit holders to implement the accepted NRC SAR quality assurance program description. Rulemaking action is currently in progress which will clarify the NRC staff position regarding the types of changes to the licensees' and applicants' quality assurance program descriptions that can be made without informing the NRC and clarify, in the regulations, the requirement to implement the accepted quality assurance program description."

The NRC should more clearly define the applicability of quality assurance program requirements in regulations.

3.4.2.3 Prescriptiveness

Regulations concerning quality assurance have not been sufficiently prescriptive to assure their clear understanding by the nuclear industry, the AEC and the NRC. Because both Appendix A and Appendix B of 10 CFR 50 were vague and contained undefined subjective terminology, there were misunderstandings and differences of opinion in what the requirements were and how to comply with them. As a result, the Gray and Green Books, Safety Guides and later Regulatory Guides were established to clarify the AEC and NRC positions.

In 1973, it was recommended to the Director of Regulation (Davis and Brown 1973) that Regulatory explain precisely what the key QA criteria for design and procurement mean. Davis and Brown reported:

"Some utilities do not know how to implement the 18 QA Criteria. These utilities understand the intent of Appendix B, but, without further guidance from the AEC, they continue at a loss to put them into effect. This has been a problem since the AEC adopted Appendix B, and many of the persons we interviewed emphasized it. One industry representative, for example, stated that both the AEC and the industry have 'all along been fumbling to explain the criteria'."

"Until the 1972 reorgnization, the development of standard to explain the application of Appendix B was not keyed specifically to the practical needs and priorities of Licensing and Regulatory Operations. Substantial efforts have been made since that time to improve this situation by obtaining greater involvement of these Directorates in the development of standards; but, there is still not sufficient interplay among the Directorates in the entire standards-setting process."

In 1976, the Advisory Committee on Reactor Safeguards wrote to the chairman of the NRC (Moeller 1976):

"An increased effort between the NRC and appropriate code or standards groups to develop better criteria and codes or standards comparable to the ASME Nuclear Codes for fire prevention, for electrical systems, and for other safety-related components, is desirable. Current requirements often are ill-defined and amorphous so the "inspector" lacks adequate criteria to determine acceptability. Until these criteria are better defined, there will continue to be confusion concerning acceptable limits as evaluated by the NRC-IE organization."

The difficulty in determining whether a quality assurance requirement is applicable to a particular situation is compounded by the necessary cross references required between the Standard Review Plan, the Safety Analysis Report, the industry codes, the regulations and the Regulatory Guides.

Licensees, in compliance with Appendix B of 10 CFR 50, pass quality requirements on to their contractors and vendors. This typically includes a requirement to implement a quality assurance program that complies with Appendix B of 10 CFR 50 for safety-related items. With approximately 1,000 vendors involved in supplying safety-related items to construction sites, a wide variety of interpretations of requirements has resulted. Normally, quality requirements passed on to vendors have not required or even referenced Regulatory Guides as a source of guidance to the vendors. They have, however, often required compliance with or referenced ANSI Standards.

Utilities, contractors and NRC Regional personnel contacted during this study stated that new regulations are not required, but that better definition of requirements in existing regulations and guidance is necessary. Regulations must be clear and criteria must be well defined. Regional personnel stated that:

- the NRC needs to put some teeth into ANSI standards or Regulatory Guides

- Technical Specifications and the Standard Review Plan need to be upgraded

- existing regulations fail to adequately address timeliness of activities and corrective actions

- requirements are vague enough to permit licensee interpretation to fit their needs at any given time

- regulations have encouraged slow decisions within the NRC (it may take a year to resolve a 50.55(e) finding after it has been reported)

- clear definitions of safety-related items and items important to safety are needed.

The NRC should better define requirements in regulations to assure their clear understanding.

3.4.2.4 Guidance Documents

Guidance documents are not mandatory and have not been sufficiently prescriptive to assure their clear understanding by the nuclear industry, the AEC and the NRC. The Safety Guides and Regulatory Guides adopted industry standards, either in whole or in part. Industry Standards were written by the nuclear industry and tended to reflect the state of the art, not necessarily stringent requirements that might be necessary to assure the health and safety of the public. Guidance documents heavily contain the word "should" and not "shall". Even though utilities commit to using the Gray and Green Books and Regulatory Guides in their Preliminary Safety Analysis Report, confusion has resulted when inspectors tried to verify compliance. Utilities have agreed they committed to use the guidance documents and have then argued it is just guidance and is not even considered mandatory by the AEC or NRC. The need to hold conferences and meetings to define requirements and the need to produce so many industry standards, and guidance documents indicates the regulations themselves have not been in sufficient detail to assure their clear understanding.

3.4.2.5 Changing Standards

Many of the uncompleted nuclear plants have been under construction for a number of years. As a result, current reviews by the NRC may be against standards or regulations that were moderately enforced or non-existent six to eight years ago. With loosely written or reviewed safety analysis report requirements serving as a base, many arguments and discussions between the licensee and the NRC revolve around interpretation of the original commitments and agreements made by the licensee.

The situation is further exerbated by the so called ratcheting or back-fitting requirements. With an increasing number of plants becoming operational, the experience and knowledge level of the NRC has increased. As a result, efforts have continually been made to increase the safety and reliability of the plants under construction and in operation. However, there is a need to establish more stability in the regulatory process. From 1970 through 1979, there were a total of 216 regulatory criteria issued or changed. Design changes and construction modifications made to meet the criteria resulted in longer construction time and more opportunities for errors. Utilities and contractors indicated during the NRC Case Studies that the cost of a nuclear plant had increased significantly in the last 10 to 15 years as a result of AEC/NRC requirements. They questioned whether all the requirements and retrofits were really necessary. They indicated there was too much uncertainty in the regulatory process and there were constantly changing targets. Regional personnel indicated that as construction times increased, there were problems resulting from changes in site personnel and procedures. The quantity of criteria changes indicates insufficient preventive action and planning.

The NRC needs to devote greater attention to preventive action and planning and to establish stability in the regulatory process.

While discussing the effect changing regulations and standards have had on the nuclear industry, it must be kept in mind that two different environments currently exist. The plants now under construction have been built during a period of learning and understanding the beneficial effects of an effective quality assurance program. The requirements have been and will continue to be changing. Therefore, the end result will be somewhat less than had the current requirements been in effect throughout the entire project. We must not judge the entire developing program by the standards we have today.

The next generation of nuclear plants will have the benefit of many man years of construction quality assurance experience. It is vital that the knowledge and understanding gained to date be properly incorporated in the NRC requirements for future nuclear installations.

3.5 LICENSING PROGRAM FOR ASSURANCE OF QUALITY

3.5.1 Description

Prior to 1970, the AEC performed little review of applications before issuance of permits and there was no documented guidance for reviews of Quality Assurance Programs.

Following issuance of Appendix B of 10 CFR 50, the AEC developed a Quality Assurance Program Review Checklist for Nuclear Power Plants and used it in their review of applications. The checklist was based upon Appendix B requirements and provided guidance through defining what was to be included in Quality Assurance Programs. The judgment of the individual reviewers was the determining factor in deciding if the quality assurance information in the application was adequate.

In 1971, 10 CFR 50.34(a)(7) became mandatory requiring applicants to submit a description of their Quality Assurance Program for design, procurement and construction in a Preliminary Safety Analysis Report (PSAR) to the AEC. The program had to satisfy the requirements of Appendix B of 10 CFR 50.

In 1973, it was recommended to the Director of Regulations (Davis and Brown 1973) that:

"Regulatory docket an application only if the utility has a satisfactorily implemented QA Program for existing design and procurement activities and Regulatory upgrade its mini-review of the program."

At that time, the AEC Regulatory Staff initiated the practice of refusing to docket a Construction Permit application until it was determined that it was complete enough to permit substantive review. The reviews performed of the Quality Assurance Program descriptions were primarily a screening for completeness.

In 1973, regulatory procedures were issued which included review by the Directorate of Licensing of the applicant's QA Program description as it applied to design and procurement activities for satisfying requirements of Appendix B of 10 CFR 50. Inspection by Regional Offices of the implementation of the QA Program for these activities was also started.

Regulatory Guide 1.70 was issued covering the preparation of Safety Analysis Reports and included a Standard Format for the Content of Safety Analysis Reports. Chapter 17.0 indicated that the applicant was to provide a description of the Quality Assurance Program which he intended to establish and implement during design and construction. The program was to be started at the earliest practical time consistent with the schedule for accomplishing the activity and the applicant was to provide a schedule for implementation of the portions of the program not yet established at the time the PSAR was prepared. The program was to address each criteria of Appendix B of 10 CFR 50 and could reference appropriate portions of other sections of the PSAR.

In an effort to further define the Quality Assurance Program require-ments, the Rainbow Books were issued in 1974. Each book covered a different area of quality assurance -- gray - design and procurement; orange - operations; and green - construction. The books were intended to provide guidance for establishing and implementing an acceptable Quality Assurance Program. The PSAR was to specifically state which portions of the books were used. The applicant was to indicate any specific alternate methods of accomplishing the Appendix B objectives that were not in conformance with the recommendation of the Rainbow Books.

In 1975, the NRC issued a Standard Review Plan to be used as a reference for evaluating the applicant's PSAR submittal. It also served as a guide which the applicant could use during the preparation of the PSAR. Chapter 17 established the criteria to be used in approving the applicant's Quality Assurance Program. The Plan has been modified several times to reflect the changing conditions in the industry.

In 1979, after the Three Mile Island incident, the NRC added Chapter 13 to the Standard Review Plan. Chapter 13 required the applicant to include information in the PSAR about the organizational structure that was to be used during the construction and operation of the facility. Included was to be a description of the corporate management structure and controls. Further, the responsibilities and duties of any technical staffs was to be stated. There was to be a description of the applicant's past experience in design and construction of nuclear plants or projects of equal magnitude. A program for planning and implementing design and construction activities and responsibil-ities was to be included. The applicant was to identify the general qualification requirements for certain specified positions or classes of positions as well as assigned management and supervisory positions. Required educational backgrounds and experience was to be included for each position.

3.5.2 Analysis

3.5.2.1 Guidance Documents

Prior to 1970, there was no documented guidance for licensing review of Quality Assurance Programs before issuance of permits. Following issuance of Appendix B of 10 CFR 50, early guidance documents for reviews of Quality Assurance Program descriptions indicated that neither the complete program nor a detailed description of how the applicants commitments were to be implemented had to be described. The Quality Assurance Program Review Checklist stated:

"It should also be noted that the applicant is required to submit only "...a description of the quality assurance program..." and not the full program documentation. An appropriate designation for this description of the QA program is the "Quality Assurance Program Plan"; however, the use of this term is not mandatory."

"The QA Program Plan presented in the PSAR should contain sufficient information to enable to reviewer to decide whether an appropriate basis has been established for a detailed QA program which meets the requirements of Appendix B. DRL approves the QA program at an early stage, before it is completely documented, solely on the basis of a description of the program (Quality Assurance Program Plan). Later the detailed QA program (QAP) and its implementation will be under the surveillance of CO. If the QA program or its implementation fails to meet the requirements of Appendix B, this will be duly noted by CO and brought to the attention of DRL. This relieves DRL of a time-consuming review of a detailed program and permits the applicant to set up the program in the course of coordinating the operations of the participating organizations in the project at the appropriate stages."

Guidance for Submittal of Quality Assurance Program Description - Section 17 of PSAR stated:

"To demonstrate the framework for the implementation of 10 CFR 50 Appendix B criteria, a listing of the QA Program procedures which describe the implementation of each of the 10 CFR Part 50 Appendix B criteria, should be provided in the PSAR and identified to the applicable corresponding criterion. In the event that certain required procedures are not established a schedule for their preparation should be provided in the PSAR."

The Standard Review Plan issued in 1975 required evaluation of the entire Quality Assurance Program description included in the PSAR. Section 17.0 of the Standard Review Plan states:

"Prior to docketing a CP application, the NRC performs a substantive review of the applicant's QA program description relative to ongoing design and procurement activities."

"The pre-docketing substantive review places particular emphasis on the areas of organization, QA program, design control, procurement document control, and audit. The application is not docketed unless the established and implemented program in these areas has no substantive deviation from NRC QA guidance applicable to activities conducted prior to docketing."

"Where an NRC-accepted QA topical report is referenced in the application, the referenced QA program is not re-reviewed except for conformance to the applicable Regulatory Guides in effect at the time of tendering the application. For the case of CP applications referencing a standard design that includes an approved QA program directly or by reference, the applicant need not conform to new Regulatory Guides unless they contain regulatory positions determined to be significant to safety."

"The QAB review, after docketing, covers the QA controls to be applied by the applicant and principal contractors to activities that may affect the quality of structures, systems, and components important to safety. These activities include site testing and evaluation (starting with evaluation of exposed excavated surfaces, soil compaction, and testing), designing, purchasing, fabricating, constructing, handling, shipping, storing, cleaning, erecting, installing, inspecting, and testing. This review extends to the determination of how the applicable requirements of the eighteen criteria of Appendix B to 10 CFR 50 are satisfied by the proposed QA program."

"The acceptance criteria include a commitment to comply with the regulatory positions presented in the appropriate issue of the Regulatory Guides including the requirements of ANSI Standard N45.2.12 and Branch Technical Position listed in subsection V. Thus, the commitment constitutes an integral part of the QA program description and requirements. Exceptions and alternatives to these acceptance criteria may be adopted by applicants provided adequate justification is given; the QAB review allows for considerable flexibility in defining methods and controls while still satisfying pertinent regulations. When the QA program description meets the applicable acceptance criteria of this subsection or provides acceptable exceptions or alternatives, the program is considered to be in compliance with pertinent NRC regulations."

The applicant and its contractors were to prepare Quality Assurance Manuals and implementing procedures to fulfill commitments made in the PSAR and to describe the actual program in more detail. The responsibility for reviewing the more detailed Quality Assurance Manuals and implementing procedures to determine if the program to be used complies with the requirements of Appendix B of 10 CFR 50 had been assigned to the Regional Offices.

3.5.2.2 Quality Assurance Program Review

Reviews of Quality Assurance Program descriptions have primarily consisted of determining the completeness of PSAR commitments in meeting the requirements of Appendix B of 10 CFR 50.

A study in 1980 (NUREG/CR-1250) contained the following appraisal of PSAR reviews:

"The review conducted by the Quality Assurance Branch (QAB) in NRC's Division of Project Management is limited to an evaluation of the description of the applicant's QA program in the PSAR and FSAR, and an assessment of whether that proram complies with the 18 criteria of Appendix B. However, no attempt is made by the QAB to determine how or to what extent the QA programmatic requirements are applied. This determination is left to the discretion of the applicant, who is responsible for identifying safety-related items, determining the extent that QA requirements are applied to these items, identifying the activities to which Appendix B applies, and imposing QA requirements on its contractors and vendors. The majority of the applicant's QA programs are found in its implementation procedures, which are not even submitted to the NRC for review or approval. These implementing procedures, which constitute several volumes of documents, are retained by the utility."

"The QAB does not review the applicant's procedures that implement its QA program. Review of implementation is the responsibility of IE. However, IE does not review the substance of the utility's procedures to determine their adequacy or to give NRC approval. The IE review assumes that the utility's procedures for implementing its QA program are adequate, and simply attempt to determine whether they are being followed."

Contrary to the above, regional personnel generally indicated during this study that adequacy of Manuals was reviewed in addition to their compliance with PSAR commitments. However, there were difficulties in determining between commitments and requirements and personnel did not receive training in how to review the Manuals. Procedures were reviewed by project inspectors on a sample basis and the procedures reviewed were documented in inspection reports. There was no formal overview performed to assure all appropriate procedures were reviewed prior to permitting work to proceed.

NRC regulations have not required the QA Program to be included as a condition of the permit and once Licensing approved the PSAR, submittal of changes to the Quality Assurance Program description, for Licensing's review and approval, were not required. A regulation change in 1983 requires that changes to Quality Assurance Program descriptions in PSAR's be submitted to Regional Offices for review. If in the opinion of the licensee, a reduction of commitment to quality occurs, then changes must be submitted prior to their use. Otherwise the licensee has one year in which to make the submittal. Permitting the licensee to make such a determination may result in differences of opinion between the NRC and licensee after the fact. The NRC should consider requiring submittal of all changes for NRC acceptance prior to their use.

Regional personnel stated during this study that there is a great lack of uniformity in what is required during the PSAR review from one reactor to another, especially in the Q-Lists defining safety-related systems, and that too much depends on the whims of the NRC Project Manager and what the licensee is able to negotiate or get by with in the licensing process.

Opinions were expressed that applicants were allowed to do an excessive amount of general design work and purchasing of major components prior to submittal of the PSAR. The applicant's Quality Assurance Program in effect during this period of time has not been reviewed until submittal of the PSAR. Considerable pressure could be placed on Licensing to accept a less than satisfactory applicant Quality Assurance Program if the major components were partially fabricated or areas of design completed. It was suggested that the NRC be involved at the very start of the applicant's work.

3.5.2.3 Management Capability

The NRC has not placed sufficient importance on licensee and contractor attitude and management capability. In 1973, it was reported to the Director of Regulations (Davis and Brown 1973) that:

"The AEC's visible QA efforts date back to the mid-1960's, and there has been some success: a growing number of utilities have responded with improved QA programs. However, it is clear that this success has been gained only through the continuous efforts of the AEC with the utility industry. It is indeed fair to conclude that, throughout this period, status quo considerations have strongly influenced the utilities' attitudes on QA. Today, virtually all utilities are aware that QA is important--but there is still no widespread sense of urgency."

"Some utilities are not philosophically committed--with attitude and resources--to a high level of QA. They do not acknowledge that nuclear technology is in substance different from conventional power technology, and that a new order of management involvement is required. These utilities have successfully constructed and operated fossil fuel plants with unstructured QA programs. They believe that these programs are equally applicable to nuclear reactors."

Following Three Mile Island, it was reported in 1979, (NRC 1979) that: "There is little I&E assessment of the utility's management capabilities."

In an analysis of the experience at problem plants, the EDO stated (NRC 82) that primary problems included:

- "failure of the project management team to provide adequate management controls to prevent a significant breakdown in quality from occurring"

- "failure of the owner's quality assurance program to detect the breakdown in a timely manner and to obtain the necessary corrective action"

He also stated:

"The problem areas are fundamentally derived from a lack of total management commitment to quality at the nuclear projects inception. This lack of commitment has been exacerbated by the lack of understanding of the role of quality assurance in project management and the lack of total understanding of what is required by personnel at all levels of the process."

"Historically, the NRC's licensing and construction inspection programs have not sufficiently examined the project management controls at sites under construction, but have been oriented towards establishing adequacy within major technical and functional areas, e.g., concrete, electrical, etc. The systematic assessment of management performance and evaluation of all other available information have not received the same level of effort as operating sites."

The NRC Case Studies have revealed that the NRC has not sufficiently evaluated whether licensees and their contractors had the experience, knowledge, staffing or ability to effectively manage the design and construction of a commercial nuclear power plant.

Several adverse comments were received about the vague subjective terminology used in Standard Review Plan 13.1.1-Management and Technical Support Organizations. Phrases such as "clear unambiguous management control and communications exist between organizational units" and "substantive breadth and level of experience and availability of manpower to implement the responsibility for the project" used as acceptance criteria makes evaluation on a

uniform basis difficult. Evaluation criteria for management and all elements
of Quality Assurance Programs need to be prescriptive enough to permit a
meaningful review.

3.6 INSPECTION PROGRAM FOR ASSURANCE OF QUALITY

3.6.1 Description

3.6.1.1 General

Prior to 1968, the AEC performed little inspection at nuclear power plants under construction. Few inspection procedures and minimal guidance were available to inspectors. There were 4 or 5 inspectors in each Region (a total of about 20 inspectors) who had nuclear research or nuclear navy experience and were expected to know how to perform adequate inspection.

As a result of many quality related problems at nuclear power plants, including serious problems at Oyster Creek, the AEC recognized a need to look at construction activities and develop more formalized programs. The AEC moved inspectors from operations to construction and later hired personnel with construction background. As the number of inspectors increased, the need arose for more guidance. The AEC began developing a "General Facility Under Construction Inspection Program" and writing inspection procedures. In late 1969, the AEC issued a directive to implement the procedures.

In the early 1970's as Appendix B of 10 CFR 50 became mandatory, there was lack of coordination between the existing inspection procedures and requirements of Appendix B. In 1972, a procedure titled "QA During Design and Construction" was issued addressing Appendix B of 10 CFR 50 and requiring a review of the licensee's Quality Assurance Manual, a meeting with corporate utility management, and an initial inspection subsequent to docketing a Construction Permit application. In 1973, procedures were issued covering pre-docketing and pre-construction permit inspections. The AEC initiated preparation of a more comprehensive inspection program, which was later taken over by the NRC and issued in 1975 as the Inspection and Enforcement Manual.

The NRC used regionally based inspectors to implement the construction inspection program. A generalist inspector, possessing a broad range of technical knowledge and often specific expertise, had overall responsibility for a given plant and assisted in inspecting other plants. Specialist inspectors expert in specific technical areas conducted inspections in their technical specialties at the various plants within their Region. A minimum of six inspections a year were to be performed at construction sites. Until the incident at Three Mile Island, inspections tended to be oriented toward documentation and records review. In 1979, inspection orientation began moving more towards hardware and results.

As a result of numerous problems at construction sites, the NRC regionally based inspection program was criticized for too few inspections, too little of an inspector's time being spent on site, too much onsite time being spent reviewing records instead of observing work in process or conducting independent measurements and tests, and too little evaluation of licensee performance with appropriate NRC response. A General Accounting Office report (GAO. 1978) stated:

> "We believe that NRC's inspection process needs to provide a
> more thorough and independent evaluation of the quality of

powerplant construction work. Without such an evaluation, NRC has to rely to an undue extent on the credibility or validity of evaluations made by utility companies. Thus NRC's inspection program cannot independently assure that nuclear powerplants are constructed adequately. The following simple description of the enormity of nuclear powerplant construction activities and the current NRC inspection level underscores our position."

"Seventy-eight nuclear powerplants are now in various stages of construction. A typical powerplant construction site may involve several thousand construction workers and supervisory personnel--in many cases, working 24 hours a day, 7 days a week. A single powerplant requires making about 25,000 welds, pouring about 360,000 tons of concrete, and using 726 tons of copper and 34,662 tons of iron. Many complex electrical and computerized systems are also involved."

"In answer to our questionnaire to NRC inspectors, the 63 respondents indicated that collectively they each spend only about 22 percent of their official working time, or about 50 days per year, at construction sites. They further indicated that they used only about 34 percent of that time (about 16 days per year) to determine for themselves the quality of construction by performing or observing tests of completed construction work, observing construction work in progress, and talking with construction workers. Therefore, in 1 year, all 76 NRC construction inspectors and supervisors spent about 1,216 staffdays--or about 5-1/2 staffyears effort--in direct inspection work. At each of the 78 powerplants then, NRC's annual direct inspection is about 16 days."

"For most of the past 2 years, however, NRC has been reevaluating its inspection philosophy and approaches. It recognizes many of the shortcomings of the present system, such as the limited amount of direct inspections and verification and the limited time its inspectors spend onsite observing construction work and talking with construction workers. NRC is evaluating the need to perform some type of independent verification of the quality of construction work and is instituting a program to assign resident inspectors to powerplant sites--both under construction and in operation. This, NRC anticipates, will increase an inspector's onsite inspection time from about 22 percent to 75 percent, will permit greater observation and surveillance of construction activities, and will make its inspectors more accessible to construction craftsmen."

"NRC plans to have 20 such inspectors at plant sites by October 1978. Five of these will be assigned to powerplants under construction. Depending on congressional approval, NRC plans to expand the program and provide a resident inspector at every powerplant in operation or under construction by 1981. Currently, a request is before Congress for a supplemental appropriation in fiscal year 1978 to provide 61 people and $2.65 million to get the program started. These people have to be hired now, according to NRC, because it will take a minimum of 2 years of training and experience before they are qualified to take over a resident site. In the meantime, existing NRC inspectors will fill the resident positions."

Due to budgetary restrictions, plans did not envision putting a resident inspector at a construction site until the later stages of construction, when the critical safety-related construction work was being done.

In 1978, the NRC began revising the inspection program. The objectives remained the same but the means of achieving them changed.

Resident inspectors were placed onsite on a full time basis to increase the amount of time spent directly verifying licensee activities and performing independent measurements and to motivate licensees to improve their performance. Resident inspectors were placed at operating plants during 1978-1980 and at construction sites in 1980. The resident is the principal inspector for the site and is supported by specialist inspectors at the Regional offices. Regional offices provide supervisory and administrative support and process noncompliances found by the residents. The current policy is that every construction site have one resident and every operating site have one resident for each operating plant. Residents file a monthly summary inspection report with headquarters and regional inspectors file trip reports. Residents perform both planned and reactive inspections, with planned inspections budgeted for two-thirds of the inspector's time and reactive inspections budgeted for one-third of the time.

In addition to the resident inspectors, the NRC initiated the following inspection activities:

- Performance Appraisal Teams in 1978 to obtain a National perspective of evaluating the effectiveness of the inspection process, assessing licensee performance, and evaluating the objectivity of residents

- Systematic Assessment of Licensee Performance in 1979 to provide an annual review of regulatory performance of licensees

- Construction Assessment Teams in 1980 to provide periodic in-depth inspections of the overall construction project

- Independent Design Verification Program in 1981 to verify design

- Integrated Design Inspection in 1982 to verify the implementation of the licensee's quality assurance program during the design process

A more detailed discussion of the elements of the inspection program follows.

3.6.1.2 Pre-CP Phase

The Light Water Reactor Inspection Program - Pre-CP Phase, issued in May of 1975, is applicable from the time the NRC receives formal notification of a utility's intentions to build a plant, up to issuance of the construction permit. Principal areas covered include inspection of the establishment, execution and administration of the QA Program relating to Preliminary Safety Analysis Report development, design, procurement and construction.

The Light Water Reactor Inspection Program (NRC 1975) provided for examination of objective evidence to determine whether the applicant, consultants, and the constructor have placed into effect:

- Planning and scheduling necessary to assure timely implementation of organizational staffing, procedures and instructions, quality assuring activities and administrative controls consistent with NRC requirements and the description of the quality assurance program provided in the application for a construction permit.

- An implemented quality assurance program consistent with NRC application requirements, which has translated the PSAR commitments into an aggregate collection of procedures and instructions (QA Manual), and is being executed as required for each organization performing and/or verifying the attainment of quality objectives established for the design, procurement and construction of safety-related structures, systems and components of the nuclear facility.

- The means to ascertain and document the adequacy and utilization of procedures and instructions necessary to achieve quality objectives.

- The means to evaluate and document the effectiveness of the implemented quality assurance program for each organizational element assigned responsibility for attainment or verification of safety-related quality objectives.

Until docketing of the application, the inspector was to use the "Guidance on Quality Assurance Requirements During Design and Procurement Phase of Nuclear Power Plants" (Gray Book) as guidance in evaluating activities. After docketing he was to use the PSAR commitments. QA Manual inspection was to be performed at the Regional Office prior to conducting implemention inspection.

Inspection Procedure 35100B (Review of QA Manual), issued in March of 1975, had the objective of ascertaining whether quality assurance plans, instructions and procedures have been established in the QA Manual and conform to PSAR commitments for organizational structure and QA personnel, audits, quality requirements, work and quality inspection procedures, control of material, control of processes, corrective action, document control, test control and control of test equipment and quality records.

Inspection procedure 35003B (QA Manual Review), issued in May of 1975, had the objective of providing for uniform application of IE inspection requirements when reviewing and examining procedures and instructions of the implemented QA Program. The inspector was to complete review requirements of the procedure only when another procedure of the LWR Inspection Program Pre-CP Phase or other MC 2500 program referenced the procedure as a requirement for that inspection activity. The procedure referenced three attachments to be considered in reviewing the QA Manual of the applicant where major elements of the applicant organization perform a significant part of design, procurement and construction but identified all three as being under development. The same procedure is currently in the IE Manual without the specified attachments. Enclosure 1 to the procedure was identified as partially completed and still exists in that same form.

Inspection Procedure 35016B (Initial Pre-CP QA Inspection), issued in May of 1975, had the objective of determining if the establishment and execution of the quality assurance program for activities of design, procurement and planning for construction was consistent with the status of the project and the program described in the application. A Quality Assurance manual review was to be performed during the fourth month after docketing and an inspection of program implementation was to be performed following the manual review.

Inspection Procedure 35004B (Initial Predocketing QA Inspection), issued in October of 1976, had the objective of determining if the establishment and execution of the quality assurance program relating to criteria I-VII and XVI-XVIII of Appendix B of 10 CFR 50 was being implemented consistent with the status of activities of PSAR development, design and procurement without substantive deviations from NRC QA guidance for design and procurement. A Quality Assurance Manual review was to be performed with deficient findings forwarded to IE headquarters for submittal to NRR before the application was tendered. If serious deficiencies did not exist, an inspection of the program implementation was to be performed and results were to be forwarded to IE headquarters for submittal to NRR.

Inspection Procedure 35012B (Second Predocketing QA Inspection), issued in July of 1975, had the objective of repeating initial predocketing activities for areas determined deficient after applicant corrective action.

Implementation reviews were to include availability of instructions, understanding of their content and purpose by personnel using them, establishment of in-process and permanent files for records, acceptable implementation of the program, and consistency of the planning and scheduling of program implementation with engineering schedules.

Inspection Procedure 35100 (Review of QA Manual) issued in 1983 has the objective to determine whether quality assurance plans, instructions, and procedures for specific safety-related activities have been established in the QA Manual and implementing procedures and whether these documents conform to the QA Program as described in Chapter 17 of the facility Safety Analysis Report (SAR). The review is to be performed by the inspector, who is to refer deficient items to the Region for resolution.

3.6.1.3 Construction Phase

The Light Water Reactor Inspection Program - Construction Program, issued in March of 1975 and effective in October of 1975, is applicable from the time a Construction Permit or Limited Work Authorization is issued until issuance of an Operating License. Final activities of the program overlap with the preoperational testing and operational preparedness phase activities, which are covered by another program.

Upon notification that a utility intends to seek a license for construction of a nuclear power plant, the NRC meets with the utility to describe the NRC inspection program and procedures and gives the utility a copy of the NRC Standard Review Plan to be used in the review of the utilities Preliminary Safety Analysis Report (PSAR).

Upon receipt of an application for a Construction Permit and the PSAR, the NRC reviews PSAR commitments for compliance with regulations and accepts or rejects the application. The Division of Licensing of the Office of Nuclear Reactor Regulation directs a program for safety and environmental review and evaluation of applications, including a review of organizational structure of the utility, qualification of management and acceptability of the Quality Assurance Program description.

Following acceptance of an application, Inspection and Enforcement performs a review of implementation of the organizational structure and management controls over design, procurement and project management of the utility, Architect Engineer (AE) and Nuclear Steam System Supplier (NSSS) to ensure programmatic controls are in place prior to their use. If the same AE or NSSS was recently reviewed on another project, it has not been necessary to review them again.

If a Limited Work Authorization is requested, the implementation review includes verification of capability to perform the work identified in the request.

During design, procurement and construction activities, Inspection and Enforcement's efforts have been about equally divided between reviewing programmatic controls, observing work in process and reviewing records. As a result of criticisms of looking too much at paper and not enough at hardware, the emphasis has changed to expending 60% of the effort observing work and 20% of the effort in reviewing programmatic controls and 20% in reviewing records.

The Inspection Manual consists of inspection modules prepared by Inspection and Enforcement at headquarters and provides a framework for inspection. Inspections are performed to the commitments of the licensee, which can be a different vintage for different plants. As regulatory standards are upgraded, backfit is often required of older plants. Old plants then have mixtures of old and new standards to comply with. Implementation of changes in the inspection program is determined by the Regions based upon the construction status of the plant.

About two years prior to fuel loading, the NRC begins inspection of startup operational procedures. The project passes from construction to operations within the NRC at the time of hydrostatic tests and an additional resident inspector is assigned for preoperational and startup activities. The Division of Licensing reviews the Final Safety Analysis Report (FSAR) commitments for compliance with regulations. The preoperational inspection program consists of verifying implementation of FSAR commitments covering preoperational tests and startup, reviewing test procedures, witnessing tests, evaluating test results, and reviewing management control systems for operations. Normally the resident inspector assigned for preoperational and startup activities becomes the resident inspector for operations.

At the end of preoperational activities, the Regional Administrator sends a report to the Office of Nuclear Reactor Regulation indicating the status of construction and preoperational and startup inspection programs, identifying a list of open items, recommending any conditions for the Operating License and stating the reactor can startup.

Following issuance of an Operating License, NRC holdpoints are established for fuel load, low power testing, power ascention testing, and full power testing. The resident inspector for operations performs operations inspections.

The objective of construction inspections is to ascertain whether construction and installation of safety-related components, structures and systems meet applicable requirements. Since inspection activities must be coordinated with construction activities, inspectors must be cognizant of construction status and must plan their inspections in the proper sequence of activities. Inspection procedures identify frequency of inspections and time frames for completion based upon milestones relating to the status of work activities. Inspectors are to conduct inspections outside the scope of the program and are to annually determine if QA Manual changes have been made and if such changes are appropriate and adequate. For multi-unit sites, inspection is required for each unit under construction. Records for material or items are to be reviewed prior to use or installation.

Limitations on construction inspection resources has precluded completion of all procedures at all sites. To provide guidance to inspectors concerning which procedures and portions of procedures should be completed, a Construction Inspection Program Priority Plan was established which varied emphasis on different facets during the construction period. The use of the plan was optional but preferred.

On the basis that the amount of inspection required to assure the same degree of confidence that construction was adequate would vary from site to site and that different types of activity at the same site may require varying levels of inspection to provide the same degree of assurance, Regional management has been permitted to modify the priority plan. Reductions or additions in inspections could be initiated by an inspector with concurrence of his supervisor.

Regional Section Chiefs have been responsible for the inspection program implementation and inspection status for their assigned plants.

The construction inspection program has been intended to provide the framework for managing resources without being totally prescriptive. Inspectors have been expected to apply judgment regarding the need to complete each line item of inspection procedures.

3.6.1.4 Performance Appraisal Teams

Performance Appraisal Teams (PAT) were established in 1978 to obtain a national perspective of evaluating the effectiveness of the inspection process, assessing licensee performance and evaluating the objectivity of residents.

Inspections were designed to determine how well all levels of licensee management and operational personnel understood and performed their duties.

Inspections were conducted through a series of interviews with both corporate and operations personnel, and review of licensee-generated records, logs, and other documents. For each functional area examined, the PAT determined whether:

- the licensee had written policies, procedures, or instructions to provide management controls

- the policies, procedures, and instructions were adequate to assure compliance with NRC regulatory requirements

- the personnel with responsibility in any given area were adequately qualified, trained, and retrained to perform their duties

- the individuals assigned responsibilities in a given area understood their responsibilities

- the requirements for a given area were implemented to achieve full compliance and appropriately documented

As part of PAT, the NRC established the "Module Sample Performance Inspection" as the means of assessing the adequacy of the NRC's modular inspection program and to determine if the NRC's current sampling rates were adequate for detecting noncompliance. Inspections were performed for procedures previously performed by regional inspectors. The time period reviewed and the procedures used were identical to those used by the regional inspectors in every aspect except the sampling rate, which was much higher.

The NRC planned to have four teams of five or six personnel each, performing about six inspections a year and appraising each operating plant every three to four years. PAT activities were interrupted as a result of Three Mile Island and resumed in 1980. In 1981, there were two teams performing 10 to 12 inspections a year. In 1981-1982, the Institute of Nuclear Power Operations (INPO) initiated programs for inspecting operating plants every 15 months. The NRC entered into agreements with INPO, reviewed their program, participated in their inspections as observers and backed off the PAT program. Currently, there is one team performing 3 or 4 PAT inspections a year with the objectives being to assess Regional performance and the inspection program and to determine the effectiveness of INPO activities. PAT monitors INPO by reviewing their reports, talking to INPO personnel and performing inspections after INPO has performed their inspection at a site. PAT findings are followed up by Regional personnel.

The NRC requires corrective action for PAT findings within the NRC inspection program and presents findings outside the NRC inspection program as weaknesses without requiring corrective action.

3.6.1.5 Systematic Assessment of Licensee Performance

Following problems at Three Mile Island, the NRC initiated a program for Systematic Assessment of Licensee Performance (SALP), consisting of annual reviews of regulatory performance of licensees by a team of inspectors and regional supervisors involved at the site and headquarters personnel.

Chapter 0516 of the NRC Manual addresses the Systematic Assessment of Licensee Performance (SALP) program and identifies the objectives to be:

- to improve the NRC Regulatory Program with emphasis on resource allocation

- to improve licensee performance

- to collect available observations on an annual basis and evaluate licensee performance based on those observations.

The SALP assessment is intended to be sufficiently diagnostic to provide a rational basis for allocating NRC resources and to provide meaningful guidance to licensee management.

For construction activities, the functional areas reviewed are:

- soils and foundation

- containment and other safety-related structures

- piping systems and supports--including welding, NDE and preservice inspection

- safety related components--includes vessel, internals, pumps

- support systems--includes HVAC, radwaste, fire protection

- electrical power supply and distribution

- instrumentation and control systems

- licensing activities

- others (as needed).

For reactors in the preoperational phase, functional areas from the listing for either Operating Reactors or Reactors under Construction are selected as appropriate for evaluation.

The evaluation criteria are as follows:

- management involvement in assuring quality

- approach to resolution of technical issues from safety standpoint

- responsiveness to NRC initiatives

- enforcement history

- reporting and analysis of reportable events

- staffing (including management)

- training effectiveness and qualification.

The evaluation process is comprised of a SALP Board assessment, a meeting with licensee management to discuss the assessment, and issuance of the

B.53

report. To provide a consistent evaluation, attributes associated with each criterion are listed in the procedure to describe characteristics applicable to the three categories. The attributes are intended only as guidance.

Each functional area evaluated is assigned a Category. Not all functional areas need be covered in a given review. If a functional area appropriate to a licensee is not covered, the reasons are to be given in the report. The functional area being evaluated may have some attributes that would place the evaluation in Category 1 and others that would place it in either Category 2 or 3. The final rating for each functional area is a composite of the attributes tempered with judgment as to significance of individual items.

Performance Categories

Category 1. Reduced NRC attention may be appropriate. Licensee management attention and involvement are aggressive and oriented toward nuclear safety; licensee resources are ample and effectively used such that a high level of performance with respect to operational safety or construction is being achieved.

Category 2. NRC attention should be maintained at normal levels. Licensee management attention and involvement are evident and are concerned with nuclear safety; licensee resources are adequate and are reasonably effective such that satisfactory performance with respect to operational safety or construction is being achieved.

Category 3. Both NRC and licensee attention should be increased. Licensee management attention or involvement is acceptable and considers nuclear safety, but weaknesses are evident; licensee resources appear to be strained or not effectively used such that minimally satisfactory performance with respect to operational safety or construction is being achieved.

Regional Administrators are responsible for implementing the SALP Board assessment including the following activities:

- obtaining assessment data from NRR, AEOD and NMSS applicable to the appraisal period

- tabulating and analyzing the data obtained, including summary of numbers and types of inspections performed and findings, number of LER's submitted under each cause category, number of Construction Deficiency Reports and Part 21 reports submitted, abnormal occurrences, and number and nature of unplanned trips

- developing the performance analysis for each functional area

- conducting the SALP Board meeting with senior regional management, the NRR Project Manager, resident inspectors and others as determined by the Regional Administrator to review the analysis and supporting data and to develop the report

- conducting meetings with licensees to provide assessment findings to utility management

- after considering the licensee's oral and written comments, transmitting the report by letter to the licensee with the letter including a characterization of overall safety performance.

3.6.1.6 Construction Assessment Teams

In 1980, the NRC initiated Construction Assessment Team inspections to provide periodic in-depth inspections of the overall construction project by concentrating on the examination of safety-related hardware after installation and the licensee's inspection is completed. Objectives of the CAT program (NRC 1982. SECY-82-150A) are:

- to evaluate the effectiveness of design controls and construction practices used to ensure that as-built conditions are in accordance with the design basis

- to provide a means of monitoring the progress of INPO activities related to construction reviews performed by INPO and the INPO-sponsored utility self-evaluation program

- to assess the effectiveness of regional implementation of the IE inspection program at reactors under construction.

During 1980-1981, eight trial CAT inspections of two weeks onsite were performed by five man teams from Regional offices. The inspections obtained useful results but strained Regional resources and reduced normal inspection efforts. In 1982-1983, the CAT program was revised and CAT inspections are now performed out of headquarters by teams of Inspection and Enforcement personnel and consultants with Regional participation. A team consists of one leader and 10 engineers and consultants and spends two weeks at the site, one week at headquarters and two more weeks at the site. The NRC plans to perform four CAT inspections a year and is monitoring the INPO Construction Project Evaluation Program.

3.6.1.7 Independent Design Verification Program

The Independent Design Verification Program (IDVP) was created in 1981 as a method of design verification after a serious design error was discovered in a nearly completed nuclear plant.

Although originally intended to be a program to review the one error, the verification process revealed that other design errors had occurred at the same plant. As a result, the NRC concluded that other plants nearing completion should be considered for a design review. The NRC examines several factors about a licensee in deciding if a review is necessary. Included is previous plant construction experience of the licensee, architect engineer and constructor, the complexity of the design interfaces, the general plant construction record and the length of time since the Construction Permit was issued.

The qualifications of the Independent Design Verification team is carefully reviewed by the NRC. Technical competence and complete objectivity are of major importance.

The program, to date, has revealed some design inadequacies but nothing of major proportions. The NRC is currently using the IDVP approach as an interim one and intends to utilize the continuing IDI/CAT program for long term license review. The NRC does suggest that the licensees institute their own ongoing IDVP program for their own benefit.

3.6.1.8 Integrated Design Inspection

The Integrated Design Inspection (IDI) program was started in 1982 as a means to verify the implementation of the licensee's quality assurance program during the design process. IDI teams consisting of personnel from the Office of Inspection and Enforcement, Office of Nuclear Reactor Regulation, the applicable Region, and consultants spend about four weeks examining procedures records, and training of design personnel and inspecting a system as installed in the plant. Emphasis is placed upon reviewing the adequacy of design details as a means of measuring how well the design process functioned for the selected system. Sample systems are chosen from five major disciplines. Common areas with in these disciplines are examined for adequacy and consistency of design details.

The results of the IDI program are being used with similar information from the CAT program to evaluate the licensee's compliance to commitments. Three Integrated Design Inspections are currently planned per year and are to be performed midway through the plant construction period.

3.6.1.9 AEC and NRC Philosophy

The basic philosophy of the AEC and NRC has not changed significantly over the years. IE Office Procedure 0300 presents the current philosophy and policy upon which the IE program is based. The philosophy can best be stated that the industry is responsible for the safety of its activities and safeguarding of nuclear facilities and materials used in its operation and the NRC ensures the industry adequately discharges this responsibility.

Inspection is on a planned sampling basis with the focus on areas of greatest safety significance in order to evaluate the overall adequacy and effectiveness of licensee performance. Objectives of inspection are: to provide a basis for recommending issuance, denial, continuation, modification, or revocation of an NRC permit or license; to identify conditions within areas inspected that may adversely affect public safety; and to ascertain the status of compliance with NRC regulations, licenses and orders.

Enforcement actions are to ensure licensees comply with Commission requirements with a goal of making noncompliance more expensive than compliance. Objectives of enforcement are to assure maximum compliance practicable with Commission requirements through consistent application of reasonable enforcement actions in accordance with established and well understood procedures and to ensure that licensees who do not comply with regulatory requirements will promptly implement corrective action to do so.

Regional Administrators have the authority to modify the inspection program at individual facilities based upon licensee's performance during the SALP process. The scope, depth and emphasis of inspection is affected by the program requirements in NRC rules and regulations, the relative safety signif-icance of licensee functions and aspects of operations being inspected, and the budgeted inspector resources to perform the program.

Inspection requirements and guidance is expressed in the form of perfor-mance objectives and evaluation criteria.

3.6.1.10 Inspector Qualification

Inspectors receive regional, formal classroom and on-the-job training. Inspectors must attend required training classes or successfully complete a written equivalency examination. The passing grade for examinations is 70%. Training activities encompass regulatory, administrative and technical prac-tices pertinent to each area of inspection. Self-study is required in the subjects of the Code of Federal Regulations, NRC and IE Manual Chapters, technical areas of inspection, methods and knowledge.

If regional management evaluates the background and performance of an individual inspector and concludes the inspector has demonstrated an ability to perform inspections in specific areas, it can authorize the individual to perform inspections in those areas while completing training.

Training at the site or regional offices consists of:

- Regional and/or Site Orientation

- Code of Federal Regulations

- Final Safety Analysis Report

- Regulatory Guides

- NRC/IE Manual

- Industry Codes and Standards

- Onsite Training

- Construction Inspection Accompaniments

Training at the NRC Technical Training Center consists of:

- BWR Technology Course

- PWR Technology Course

- Concrete Technology and Codes Course

- Welding Technology and Codes Course

- NDE Technology and Codes Course

- Electrical Technology and Codes Course

- Instrumentation Technology and Codes Course

- Fundamentals of Inspection Course

Optional courses, dependent upon the inspector's previous work experience and planned inspection activities, are required for performing inspections in specific areas. For resident inspectors in construction, optional courses consist of:

- Inservice Inspection Course

- Radiation Contamination Protection Course or equivalent plant training

- Quality Assurance Construction Course

- Quality Assurance Modifications Course

Personnel assigned as resident inspectors after January 1, 1984, must complete the training/qualification requirements for self-study, on-the-job training, and required training identified in the Regional Training and Qualification Journal.

Inspectors who have been trained/qualified under existing Regional Inspector Journals do not have to requalify under the Regional Training and Qualification Journal.

At the discretion of regional management, inspectors currently working to complete their training/qualification under existing Regional Inspector Journals may transfer appropriate self-study, on-the-job training, and required training courses to the Regional Training and Qualification Journal.

All newly hired personnel and new assignees are required to complete the required regional training activities or take and pass equivalency examination(s) within the first 24 months after being assigned.

Refresher training is required in concrete technology and welding technology every 48 to 60 months after completion of the concrete and welding courses and in NDE 36 to 48 months after completion of the NDE course.

3.6.2 Analysis

3.6.2.1 General

Inspection Programs of the AEC and NRC have evolved as a result of the learning process and in reaction to industry events.

- Prior to 1968, the AEC performed little inspection at construction sites.

- Following quality problems at construction sites, including Oyster Creek, the AEC formalized inspection programs.

- As a result of the Zion hearings in 1968, the AEC issued Appendix B of 10 CFR 50 in 1970, specifying quality assurance criteria.

- Two years after issuance of Appendix B of 10 CFR 50, the AEC issued an inspection procedure addressing its requirements.

- Three years after issuance of Appendix B of 10 CFR 50, the AEC issued procedures for pre-docketing and pre-Construction Permit inspections and developed a more comprehensive inspection program.

- Following the Browns Ferry Fire in 1975, the NRC developed additional programs for fire protection.

- Following problems at construction sites and criticism of the regionally based inspection program, the NRC began revising the inspection program in 1978 and initiated Performance Appraisal Teams.

- Following problems at construction sites and at Three Mile Island in 1979, the NRC initiated Systematic Assessment of Licensee Performance, Construction Assessment Team, Independent Design Verification, and Integrated Design Inspection programs.

Budget and manpower restraints have generally precluded completion of construction inspection programs. In recognition of this problem, the NRC established a Construction Program Priority Plan varying the emphasis on different facets of activity during the construction period.

The regionally based inspection program was changed in 1978 as a result of too few inspections, too much of an inspector's time being spent reviewing records instead of observing work in process or conducting independent measurements and tests and too little evaluation of licensee performance.

The current inspection program is being rewritten in recognition of budget and manpower restraints. A goal in rewriting the inspection program is to reduce it by 40% to bring it in line with available resources.

Regional personnel stated during this study that:

- the level of resources is inadequate for the inspection required

- completion of the inspection program has ranged from 60-70% to 90-100%

- inability to complete the program has resulted from diverting personnel from the program to perform reactionary inspections, investigate allegations and follow-up findings of team inspections and program evaluations (10 man-years of effort at one site and 14 man-years of effort at another site)

- diverting of personnel has resulted in missing inspection "windows of opportunity", periods of construction during which an inspection must be performed because it cannot be performed later (i.e., placement of rebar before pouring concrete)

- about 50% of the inspection program was being implemented at most sites in one Region as a result of diverting personnel to other sites

- a Construction Assessment Team inspection at one site took three times as many man-hours as was budgeted for program inspections for a year and it was expected follow-up activities would take as long.

The impact of investigating allegations of poor construction work on normal inspection work was reported in an earlier study (GAO. 1978).

"Commission inspectors are spending more of their time investigating allegations of improper construction activities, often at the expense of their normal inspection activities. A new regulation requires utility companies to post notices informing workers that they may report suspected defective work to the Commission. This new publicity will increase the number of allegations received by the Commission. However, the Commission should review organizational elements and seek additional staff to investigate these allegations without disrupting the normal inspection work."

3.6.2.2 Pre-CP Phase

The AEC and NRC have done too little too late in the pre-Construction Permit stage. Pre-Construction Permit inspection activities are designed to verify the establishment, execution and administration of the quality assurance program relating to PSAR development, design, procurement and construction activities before issuance of a Construction Permit. Inadequate attention has been paid to design and the inspection program for design has not been changed two years after the problems at Diablo Canyon. General design work may be performed and major components may be purchased 18 months ahead of issuance of a Construction Permit, without prior review of the applicable Quality Assurance Program. Reviews of Preliminary Safety Analysis Report quality assurance commitments have been to assure completeness in addressing requirements of Appendix B of 10 CFR 50. Project Inspector reviews of licensee quality assurance manuals and procedures have been for ongoing work only and have tended to be cursory in nature in determining the compliance of management controls with PSAR commitments. Evaluations of licensee and contractor management have not been adequate to assure management had the ability to assure quality in design and construction activities. It was previously recommended to the Director of Regulation (NRC 1974) that inspection effort be increased in the management of QA inspection programs.

During this study, regional personnel stated that quality assurance manuals are so general that procedures can be changed by deleting requirements and yet still comply with the manual. There are also difficulties in differentiating between commitments and requirements and manual and procedures reviewers have received little training in how to perform their tasks. The review of manuals and procedures by a number of inadequately trained inspectors has produced inconsistent results.

The AEC and NRC have not made adherence to Preliminary Safety Analysis Report commitments a condition of the Construction Permit and until 1983 did not require submittal of PSAR changes for acceptance. Once the PSAR was accepted by the NRC, licensees could change their commitments.

3.6.2.3 Construction Phase

Inspection programs have been viewed more as guidance than mandatory requirements. The AEC and NRC have relied on qualified engineers to use their best judgment in determining which inspections are to be performed and the degree of inspection necessary. During the 1960's and early 1970's, inspections were performed by inspectors with years of broad experience in research or Navy reactors using little procedural guidance. 'These inspectors were expected to have good engineering judgment and to know what to do and how to do it. As the number of plants to be inspected increased and more inspectors were required, additional guidance was provided through written programs. The inspection program permits Regional management to adjust the priority plan of inspection to meet the specific needs required at each site. This may cause the level of inspection activities to vary from site to site and different types of activities at the same site may receive varying amounts of inspection. Ultimately each inspector is responsible for determining the total inspection effort he feels is necessary.

The Reactor Inspection Program states:

"The credibility of the inspection program is based upon completion of inspection procedures and the conduct of each procedure in a technically adequate manner."

"Line items in inspection procedures reflect the collective judgment and experience of personnel responsible for program development and personnel responsible for program implementation. Line items are to be placed in the perspective of the objective of the inspection and considered in the inspector's evaluation of whether activities are safe and in compliance with requirements."

"Failure to complete the inspection program is inferred that less than the desired level of assurance is obtained and the Division Director's decision to relax inspection program requirements is to be governed by whether the resulting level of safety assurance remains adequate to allow issuance of a license."

Implementation of inspection programs has varied among the Regions as a result of management's attitude toward regulations and programs and the capability of personnel. In some Regions, all problems are documented and reported to regional management, while in other Regions some problems are handled more informally at an inspector/craft level. Meaningful NRC data on inspection program implementation was difficult to obtain during this study. Regions which have tracked the status of the completion of inspection modules can produce computer printouts listing the modules implemented. However, the degree of implementation of the modules cannot be easily determined. As the inspection program has evolved over the years, it has been possible to inspect an area once, close out the inspection module as complete, and never go back and inspect that area again. Since many construction activities may extend for a period of years, personnel performing the activity and procedures may have changed. Initial acceptance of the adequacy of the activity does not ensure continued adequacy. When plants drag out in time, this situation becomes more acute. Instructions to Regional personnel in implementing the inspection programs have varied. Some Regions have relied more heavily upon engineering judgment than paperwork while other Regions have placed more emphasis on paperwork. Portions of inspection modules may be worked over a long period of time by as many as four different inspectors resulting in a need for good recordkeeping so that each inspector is aware of the effort previously expended . Inspections of continuous activities generally need to be performed throughout the duration of the activities. Licensee and contractor personnel indicated during the NRC Case Studies that legality was often a matter of geography and compliance was a matter of where you are.

There is a general feeling within Regions that the inspection program has been too fragmented and more attention should be paid to meshing inspection requirements more closely with the construction schedule.

3.6.2.4 Performance Appraisal Teams

To supplement the resident inspection program, and to obtain a national perspective of inspection activities, Performance Appraisal Team inspections were initiated. The Performance Appraisal Team Program has been an effective method of measuring one aspect of operating plant performance. The program does not apply to plants under construction. Most of the subjects covered during the review relate to the licensee management and are therefore not covered during the normal inspection program.

Due to budget and manpower restraints, the program has not been implemented as intended and has been modified to utilize INPO efforts. The use of INPO teams with NRC observers and later spot follow-up of an NRC team has been successful.

3.6.2.5 Systematic Assessment of Licensee Performance

The Systematic Assessment of Licensee Performance is an annual review of licensee performance by inspectors and supervisors involved at the site and by headquarters personnel. Available observations on licensee performance is collected and evaluated to provide a rational basis for allocating NRC resources and to provide meaningful guidance to licensee management. While being a trend analysis of licensee performance, SALP is limited in effectiveness to the available observations. If the observations are inadequate or misleading, the SALP results will also be inadequate or misleading.

3.6.2.6 Construction Assessment Teams

To provide periodic in-depth inspections of overall construction, the NRC initiated Construction Assessment Team inspections in 1980. These inspections concentrate on examination of safety-related hardware after installation and license inspection is completed. The inspections have obtained useful results but have been resource intensive. Initial CAT inspections were performed by Regions and reduced the normal inspection efforts by diverting personnel to the CAT inspections. Now personnel for the inspections is furnished by headquarters. Performing follow-up activities resulting from CAT findings has been found by regional personnel to take as much time as was spent in performing the original CAT inspection. Current inspections are taking five weeks to perform. At one site, the CAT inspection took three times as many man-hours as was budgeted for routine inspections for one year. The performance of four CAT inspections a year will result in all plants under construction not having a CAT before construction is completed. It appears that budget and manpower restraints will prohibit the CAT program from being effective at all plants under construction. Sites selected for CAT's may feel singled out for unwarranted extra NRC inspection.

3.6.2.7   Independent Design Verification Program

The Independent Design Verification Program (IDVP) is another positive step in NRC review of the design process. It is to be applied on a selective basis at the near term operating license period. All plants under construction will not have an IDVP inspection before being granted an operating licence. It appears that budget and manpower restraints will prohibit the IDVP from being effective at all plants and that sites selected for IDVP's may feel singled out for unwarranted extra NRC inspections.

3.6.2.8   Integrated Design Inspection

The Integrated Design Inspection program is a positive step in NRC review of the design process and inspections performed to date have produced meaningful results. Since the inspections are of a limited portion of work, problems detected are an indicator of potential problems on a more widespread basis. The NRC needs to assure that licensee's response to adverse findings include a review of similar activities in other areas or systems and root causes of the problems are identified and corrected.

The IDI is to be performed midway through the plant construction period. Since much of the design work is completed before or early into construction and extensive design changes tend to complicate attaining assurance of quality during design and construction, the NRC should supplement the IDI with a program performed earlier in the design process. Thorough review of the design process at the Pre-Construction Permit stage or before would result in early detection of design process deficiencies and permit their correction before the start of construction.

Three IDI's are to be performed a year. Such a limited number of inspections will result in all plants under construction not having an IDI before their construction is completed. It appears that budget and manpower restraints will prohibit the IDI program from being effective at all plants under construction. Sites selected for the IDI's may feel singled out for unwarranted extra NRC inspection.

3.6.2.9   AEC/NRC Philosophy

Regulatory agencies in other industries are generally perceived to be on the side of the general public. Because the original AEC mandate was to both promote and regulate nuclear power the NRC has struggled with the image that they are more favorably inclined towards the nuclear industry than the general public. Even though the NRC has an adversarial role and, through its enforcement actions, has levied large fines, any attempt made by the NRC to work with the industry is taken as showing favoritism.

Changing such incorrect perceptions is a lengthy but worthwhile process. It can best be accomplished by maintaining a vigorous enforcement program and implementing it in the design and construction areas.

The AEC and NRC have made nuclear utilities responsible for assuring that the health and safety of the public is not adversely affected by the operation of their nuclear plants. The role of the NRC as a regulator has been to see that the industry discharges that responsibility. The NRC performs its function through inspections and reviews during design, construction and operation activities. The extent of the overview is governed by engineering juagment and available resources.

It was stated by headquarters and regional personnel that about one percent of the licensee design and construction activities are currently reviewed by the NRC. Budgetary limitations may cause this level of inspection to remain about the same in the future.

In order to achieve maximum benefit from the current program it becomes imperative that ways be found to:

- allocate additional resources.

- Upgrade the quality of inspectors.

- Provide the inspector with a workscope which will best utilize his time and knowledge.

- Require the licensee to perform more effective internal audits and utilize more outside organizations to review their operations. The scope of such audits and reviews should be controlled by the NRC.

- . Upgrade the status and earning potential of the resident inspector.

- Provide all inspectors and other employees involved in this area an opportunity to contribute to the identification and solution of problems.

The AEC and NRC attitude toward construction deficiencies and inadequacies has been that there is no threat to the health and safety of the public until a plant becomes operational. If construction deficiencies were found and rework was required, even on a repetitive basis, it was not an area of great concern. Plants would not be licensed for operation unless ready for operations, which would be determined by prerequisite, preoperational and start up testing. As a result, the threshold for enforcement action in some Regions was too high.

This study found that there was still resistance to recognizing the importance of quality assurance in both the NRC and the licensee organization. The NRC must continue to work with all employees of the Commission by having lectures, workshops and training sessions on the subject. Meetings should be held between NRC and licensee management to verify that the proper quality attitude is present on the licensee organization.

The AEC and NRC have often been unable to identify specific problems as a symptom of a larger system problem. Hardware problems have been easier to isolate and identify. It has been necessary to build a history and volume of hardware problems before recognition of a system problem. The NRC must recognize that problems found during inspections on limited sampling of work activities and records is an indicator of more widespread problems and must require licensees to determine the extent of the problem and to take effective action to correct the cause of the problem.

One of the recommendations included in the Staff Report to the Presidential Commission on the Accident at Three Mile Island - Volume IV 1979 stated:

"Region's on-site inspections appear to miss signals and symptoms that indicate potential plant operating problems and weak utility management."

NRC management didn't recognize the significance, magnitude or complexity of problems. Licensee, AEC and NRC management has tended not to listen unless there has been a major problem or a "smoking gun." Management has tended to think quality control instead of quality assurance. The AEC and NRC have not forced expeditious handling of corrective action.

The AEC and NRC has had a lack of understanding of quality assurance. The Compliance Manual didn't address quality assurance. Appendix B of 10 CFR 50 was not initially used as the basis of inspections. Quality Assurance for operations has only been required since 1977-1978. It was stated during this study that the practice was to look at quality assurance up front and then not look at it again.

3.6.2.10 Inspector Qualifications

A part of the training program requires self study in the Code of Federal Regulations, NRC and IE Manual chapters, and various other technical areas. Self study has been recognized as a means of obtaining basic information and knowledge, however, it does not provide adequate training in how to apply the basic information.

AEC and NRC training programs have not kept pace with the increasing needs of the organization. It was stated during this study that the level of experience of inspectors has declined over the last 10 years.

This has been partially attributed to:

- expansion of the nuclear industry in the early 1970's and the resulting need for more inspectors

- implementation of the resident inspection program, in which experienced inspectors were initially placed at operating sites and replaced with less experienced inspectors

- promotion of good inspectors as part of a career path and replacement with less experienced inspectors

- NRC inability to remain competitive with the industry in salaries.

An NRC Office of Personnel and Management study indicated NRC inspection salaries to be 21 percent below an industry average. Frequently, inspectors have left the NRC for higher salaried positions in the utility industry. This is particularly disturbing if it occurs right after they have completed the initial training period and are just becoming a major part of the NRC program.

Early training to Appendix B of 10 CFR 50 was through on the job training with experienced personnel. In 1975, training in Appendix B consisted of self-reading. In 1976, one hour of a fragmented course whose schedule was diverted by the class, was allocated to Appendix B. A longer formalized course on Appendix B was not developed until 1983. During this study, it was stated there is a great need for more training in quality assurance, standards and Appendix B of 10 CFR 50. It was also stated that there was practically no training in how to apply modules or how to do inspections. These skills come mainly from on-the-job training. More training is needed to improve the caliber and qualifications of inspectors.

Regional personnel stated during this study that inspectors in one discipline have been assigned duties in disciplines for which they have not been trained and that they would like more guidance from headquarters to better understand their responsibilities.

3.7 LICENSEE CONTRACTOR AND VENDOR INSPECTION PROGRAM

3.7.1 Description

Prior to 1969, AEC philosophy regarding vendor activities was that qualification and monitoring of vendors was the licensee's responsibility. If problems with vendor equipment existed, they would be identified during start up testing.

Major quality problems in the reactor pressure vessel, piping systems and installation of second hand, non-pedigree valves at Oyster Creek and subsequent problems having safety significance at other facilities, made the need to re-evaluate the NRC policy evident.

AEC recognized that new standards needed to be written, old standards needed to be upgraded, and all standards needed to be enforced. They also recognized that inspection of work and enforcement of standards cannot always wait until final assembly at the site and that it was frequently impossible to make a repair at the site without compromising the final quality of the product.

In 1970, 10 CFR 50 Appendix B introduced the quality assurance concept and made the licensee responsible for the evaluation and selection of procurement sources. Regional site construction inspectors were directed to evaluate licensee vendor inspection programs as part of evaluating the licensee's QA Program and to periodically accompany selected licensees on their inspections of selected vendors. This "Host - Concept" didn't work well and was discriminatory in that the selected licensees were expected to follow through on corrective action of generic type problems for all licensees. Inspections were difficult to coordinate and administrate and were ineffective. The presence of the NRC inspector as an observer inhibited the detection of deficiencies.

In 1973, the AEC initiated a trial vendor inspection program covering fuel fabricators and discovered that greater conformance to quality standards and a subsequent reduction in major quality problems could be achieved through an effective direct vendor inspection program.

Analysis of Licensee Event Reports indicated that about 63% of construction and operation problems were traceable to vendor errors in design or fabrication performed off-site during the design and construction stages and indicated a need for improved vendor performance.

In 1974, the NRC initiated the Licensee Contractor and Vendor Inspection Program (LCVIP) as a 2-year trial program covering all types of vendors. The program was administered by Region IV in Arlington, Texas. In about the same time frame, a special Regulatory Task Force study (Study of Quality Verification and Budget Impact) recommended expansion of the trial vendor inspection program as a result of increases in the number of reported problems and difficulties experienced in performing inspections of vendors.

With the large number of vendors and suppliers worldwide involved in the U.S. Nuclear industry and with budget and manpower restraints, a priority for inspection of vendors was established. Emphasis was placed on vendors supplying important safety-related products or services, such as the 5 Nuclear Steam System Suppliers (NSSS), fifteen Architect Engineers (AE) firms and approximately 120 suppliers of ASME class 1 and other safety-related parts or components.

Vendors of NSSS and AE services were inspected to assure their Topical Reports, previously approved by the Office of Nuclear Reactor Regulation (NRR), were transferred into procedures and the procedures were implemented. Vendors without Topical Reports were inspected to PSAR commitments.

Vendors of mechanical components having ASME Certification were inspected to their ASME program and vendors without ASME Certification were inspected for the same type of detail required by the ASME. Vendors to be inspected were selected based upon their doing a large volume of business on a continuing basis.

In 1977, the importance of inspecting electrical equipment vendors was recognized and two inspectors with electrical experience began a limited program of reviewing 4 to 5 Quality Assurance Program areas of vendors every 2 to 3 years. Inspections were often performed to draft procedures and some procedures were never formally issued.

In 1978, the LCVIP began looking at the effectiveness of vendor design programs, including verification of design inputs and checking design calculations at suppliers of NSSS and AE services.

Until 1978, the LCVIP functioned under an edict of not identifying the project or site to which the vendor being inspected was supplying equipment or services, resulting in the inability of Regions, headquarters and resident inspectors to correlate problems to the sites under their responsibility. In 1978, the policy was changed to identify such sites.

In 1979, following the problems of Three Mile Island, the LCVIP began getting requests for performing reactive inspections and follow-ups at vendors. There was no guidance for these inspections so Region IV prepared a program which was issued through headquarters. Vendors are chosen for reactive inspections based on the number of requests for inspections and the significance of problems. As a result of more sensitivity within the NRC, there has been an upward trend in requests for reactive inspections, increasing to about 200 requests in 1981 and about 350 requests in 1982.

The NRC issues a Letter of Acceptance to NSSS, AE, and Fuel Fabricator organizations verifying the capability of their program to meet PSAR Commitments and uses withdrawal of the letter to obtain corrective action. Licensees may accept the NRC Letter of Acceptance as evidence of qualification of the vendors but must retain the final responsibility for acceptability of the product or service provided.

If a utility performs its own engineering function, the Region in which it is located has the responsibility for inspection activities, including reactive inspections.

Inspections of vendors are performed 1 to 4 times a year on a 3-year repetitive cycle with a detailed review of the QA Program and its implementation in the first year and sampling the quality of work to determine QA Program effectiveness in the second and third years. There is no scheduled interface of the LCVIP with the licensing process.

The NRC is currently re-evaluating the LCVIP and is studying the licensing of vendors as well as utilization of third party inspection.

3.7.2 Analysis

The Licensee Contractor and Vendor Inspection Program (LCVIP) evolved as a result of the learning process and of licensee inability to assure the quality of items and services supplied by their vendors.

- Prior to 1969, vendor qualification and monitoring was viewed as the licensee's responsibility.

- In 1970, regional inspectors evaluated licensee vendor inspection programs.

- In 1973, a trial vendor inspection program was initiated for fuel fabricators.

- In 1974, a trial LCVIP was initiated because 63% of construction and operation problems were traceable to vendor errors in design or fabrication.

- In 1974, a task force recommended expansion of the LCVIP as a result of increases in vendor-related problems.

- In 1977, electrical equipment was added to the program.

- In 1978, the effectiveness of vendor design programs began being evaluated.

- In 1979, inspections became reactionary as a result of Three Mile Island.

The legal base for direct NRC inspections of vendors and enforcement is not clearly addressed in Section 206 of the Energy Reorganization Act of 1974 or in 10 CFR 50 Part 21, which results in difficulty in taking enforcement action with vendors. It is not easy to determine, for example, if an executive willingly and knowingly fails to report a deficiency. The NRC may conduct reasonable inspections to insure compliance with part 21. However, corrective action must occur through the licensee. There has only been one civil penalty issued as a result of 10 CFR 50 Part 21. That penalty was issued to Babcock & Wilcox for failure to notify the NRC of precursor events to Three Mile Island.

The NRC has been slow to respond to findings and recommendations of previous studies of the LCVIP.

In 1977, it was recommended (F. Muller and others. 1977) that:

- NRC take steps to assure each vendor inspected under the LCVIP is aware of the responsibility and authority of the licensee

- vendors to be inspected under the LCVIP be selected on a basis which ensures every vendor has a likelihood of being inspected

- IE inspection of material produced under the ASME Code be eliminated provided ASME requirements are expanded to include operation.

In a report to Congress in 1978 (Controller General 1978) it was stated that:

- the LCVIP has had a positive effect but improvements were needed in inspector's reporting practices, attention to inspection details, documentation of inspections and in investigations

- there was no systematic method of selecting vendors for inspection and all vendors of safety related equipment were not identified

- vendors manufacturing electrical components and instruments controlling critical operations were neglected

- more inspectors be assigned to vendor inspector activity. (There were 11 inspectors reviewing over 200 suppliers at the time)

In a report to the NRC in 1978 (TRW 1978) it was reported that over 50% of a plant by dollar value was designed and/or fabricated off site, that a review of Licensee Event Reports between January 1975 and September 1977 indicated that 60.8% of problems were related to component failures and design errors, (51.2% component and 9.6% design) and that on-site inspection was roughly four times off-site efforts. The report conclusions and recommendations included:

- NRC should perform independent inspections of nuclear contractors and vendors

- third party inspection would supplement and extend vendor inspection effort

- the NRC program should be functionally integrated with programs of licensees

- formalized procedures were necessary for selecting vendors for inspection based on the operating record of the product, previous inspection findings and the safety significance of the product

- emphasis of inspections should be changed from systems administration and management to evaluation of procedures used, implementation of procedures, and quality of resulting product

- reporting include a mechanism through the White Books for licensee acquisition of inspection reports, data relating to vendor performance, and statements pertaining to program compliance with Appendix B and implementation of the program

- documentation of the LCVIP in a Topical Report

- there was under representation of several skill areas among inspectors

- inspection bases other than Appendix B were used prior to 1977

- sampling was based on coverage of prior inspections and areas of suspected weakness

- the statistical adequacy of the sampling process and sample size were not determined

- the LCVIP was not implemented in accordance with MC-2700 of the IE Manual in that suggested schedules weren't followed, no explicit verifications of program content or implementation were issued for competent vendors, and little product sampling was performed.

The LCVIP was being implemented by 21 personnel who in 1977 conducted 236 inspections with about 25% being reactive inspections. The TRW report identified the following issues as needing to be addressed by the NRC:

- NRC must decide who is to perform certification of vendor's Quality Assurance Manuals for conformance with Appendix B

- NRC must determine whether some group should certify that a vendor is implementing its Quality Assurance Program

In addition, the report included an analysis of several alternative approaches in certifying and monitoring vendors.

The current NRC re-evaluation of the LCVIP includes consideration of findings and recommendations of previous reports, but is being performed five years after the last report. Coordination of vendor qualification activities with a third party, the American Society of Mechanical Engineers (ASME) has been ongoing since 1972.

The NRC has assumed licensee responsibility for evaluation of vendors through implementation of the LCVIP. Appendix B of 10 CFR 50 places the responsibility of assuring the quality of vendor supplied items and services, including evaluation and selection of procurement sources, on licensees. The NRC has concentrated its efforts in resolving quality problems with vendor supplied items and services by conducting evaluations and inspections of vendors and has not taken enforcement action against licensees to force them to fulfill their responsibility. Whenever the NRC performs a function that falls within the licensees responsibility, the NRC assumes at least a partial responsibility for the success or failure of that function.

The perception that because the NRC has a "Vendor Inspection Program" it is inspecting all vendors leads to greater expectations by the general public than can be realized. The failure of any vendor therefore, becomes a reflection of the perceived NRC inadequacies to do its job and, hence, the public's health and safety are endangered. As a regulator, the NRC can only monitor that the licensee is performing its functions in a proper and correct manner and take enforcement action when deemed necessary.

It would appear that the requirements of Appendix B of 10 CFR 50 pertaining to the evaluation of procurement sources warrants revision. Multiple evaluations of vendors by licensee and contractors has resulted in ineffective redundancy.

A solution to this problem could be in the establishment of a more intensive vendor evaluation and monitoring program using CASE (Coordinating Agency for Supplier Evaluation) or INPO, or by a Certification program administered by the NRC or a third party.

Standard evaluations could be conducted for different levels of contractors and suppliers, incorporating a graded inspection of "Important to Safety" items as well as the full inspection of safety related products. If a vendor licensing program was installed, the NRC I&E office could certify the licensees, AE's and NSSS vendors with a third party certifying the balance.

The subject of licensing vendors met with a mixed reaction from licensees, contractors and regional personnel. The general attitude seemed to be to try licensing all AE and NSSS vendors but restrict it to that level. Licensing of vendors at lower levels would tend to force vendors out of the industry.

3.8 ENFORCEMENT PROGRAM FOR ASSURANCE OF QUALITY

3.8.1 Description

3.8.1.1 General

Initial AEC enforcement actions consisted of providing written notification of nonconformances to licensees and requesting corrective action. Licensees responded with action to be taken and correspondence continued between the licensee and AEC until the nonconformance was resolved.

In 1970, the AEC issued the Enforcement Procedure For Reactors Under Construction (0700/3), which provided general guidance for the Regions on enforcement actions. The criteria used to determine enforcement action and categories of noncompliance were first published in 1972 (37 FR 21962).

In 1973, the AEC issued Chapter 0800 -- Enforcement Actions to describe the policy and guidelines for the enforcement Program implementation.

In 1975, the criteria used to determine enforcement action and categories of noncompliance was revised (40 FR 820) and the NRC reissued Chapter 0800 -- Enforcement Actions as part of the IE Manual.

In December of 1979, following Three Mile Island, the NRC again revised the criteria used to determine enforcement action and categories of noncompliance (44 FR 77135).

The approval of Public Law 96-295 in June of 1980 amended section 234 of the Atomic Energy Act and raised the maximum civil penalty from $5,000 to $100,000 and eliminated the provision limiting the total civil penalties payable in any 30-day period to $25,000.

In October of 1980, the NRC issued the Proposed General Statement of Policy and Procedure for Enforcement Actions (45 FR 66754) for implementation and public comment. In March of 1982, a revised policy statement, based upon experience gained in implementing the proposed policy statement and comments received during and following public meetings on the policy, was adopted and codified as Appendix C to Part 1 of Title 10 of the Code of Federal Regulations. The fundamental basis of the revised policy remained the same as the proposed policy with changes made in how the steps are accomplished and in clarifying the language.

A more detailed discussion of enforcement programs follows.

3.8.1.2 Chapter 0800

The NRC defined a noncompliance as a failure to comply with a regulatory requirement and categorized noncompliances by severity levels into violations, infractions and deficiencies.

Fabrication, construction, or testing of a Seismic Category I system or structure in such a manner that the safety function or integrity was lost was a violation. Fabrication, construction or testing of a Seismic Category I system or structure in such a manner that the safety function of integrity was impaired and inadequate management or procedural controls in the QA implementation was an infraction. A deficiency was an item of noncompliance in which the threat to the health, safety, or interest of the public or the common defense and security was remote and no undue expenditure of time or resources to implement corrective action was required. When a licensee failed to conform to commitments which were not licensee requirements, it was referred to as a deviation.

Enforcement actions consisted of notices of violations, civil penalties, and orders.

A Notice of Violation was a written notice to a licensee of a nonconformance. Deviations were identified in the cover letter transmitting a Notice of Violation, on a separate page forwarded with a Notice of Violation or by separate correspondence.

If an acceptable response was not received from the licensee or if items were uncorrected, repeated, or chronic, an enforcement conference was held and/or a strong Notice of Violation from headquarters bearing the signature of the Director of Field Operations or higher authority was issued. An enforcement conference was a meeting arranged by supervision or management of an IE Regional office to discuss with representatives of a licensee's management the status of its compliance with regulatory requirements, the licensee's proposed corrective measures and schedules for implementing corrective action, and the enforcement options available to the Commission. Enforcement conferences could be held at the licensee's facility, in the Regional Office, at IE Headquarters or in any mutually designated place.

If the licensee's program was not brought into compliance with regulatory requirements, a civil penalty could be issued. Civil penalties were monetary penalties to be issued for chronic, deliberate, or repetitive items of noncompliance where a Notice of Violation was not effective and for first of a kind violations if considered serious. Failure to meet licensee commitments was not a basis of a civil penalty but could aggravate items of noncompliance.

The NRC had authority to issue orders to "cease and desist," and orders to suspend, modify, or revoke licenses. Such orders were to be ordinarily preceded by a written Notice of Violation to the licensee providing him with an opportunity to respond as to the corrective measures being taken. In the event the licensee failed to respond to the notice or to demonstrate that satisfactory corrective action was being taken, an order to show cause could be issued requiring the licensee to show why the order should not be made effective. In some instances where the health, safety, or interest of employees or the public so require or deliberate noncompliance with the Commission's regulations was involved, the notice provision could be dispensed with and the particular order could be made immediately effective pending further order.

The signatory of enforcement correspondence was to be escalated as the importance of the enforcement action was escalated. Forms signed by the inspector who performed the inspection and routine notices of violation from the Regional Offices were signed by the appropriate Branch Chief. The Branch Chief escalated the enforcement correspondence with the signature by the Regional Director if difficulties concerning enforcement matters were encountered with the licensee or when a reply was required to significant items of noncompliance of safety items. Notices of violation escalated to the Headquarters level were to be signed by the Director of Field Operations or by higher authority. Notices of intent to impose civil penalties and orders to invoke civil penalties, to cease and desist, or to suspend, modify or revoke a license were to be signed by the Executive Director for Operations, the Director of Inspection and Enforcement, the Director of Nuclear Reactor Regulation of the Director of Nuclear Material Safety and Safeguards as appropriate.

Inspection and Enforcement Bulletins could be issued for a group of licensees to inspect, report and make commitments to implement certain controls or remedial actions as a result of safety, safeguards, or security related conditions resulting if inadequacies or failures that have occurred at the same or a similar facility, or in similar operations. If a licensee did not make commitments for remedial action as specified in a Bulletin, the NRC could issue an order to require the proposed action.

Inspection and Enforcement Immediate Action Letters could be issued by the Regional Director (with Headquarters' concurrence) for a licensee to inspect, report and make commitments to implement certain controls or remedial actions as a result of safety, safeguards, or security related conditions resulting from inadequacies or equipment failures at the licensee's facility. If a licensee did not respond to an Immediate Action Letter, the NRC could issue an order to make the proposed action a requirement of the license. The Immediate Action Letter was also used to confirm verbal commitments by licensees to take immediate action.

Chapter 0800 also contained guidance to elaborate upon the proper application of the enforcement criteria. Each item of noncompliance was to be categorized as a violation, infraction or deficiency. A review of the licensee's history of noncompliances was to be performed to determine if items identified involved the same basic requirement as items identified during other inspections or investigations based on the last several inspections and generally covering a period of one to three years. Each item of noncompliance was assigned action points. A violation was assigned 100 points, an infraction 10 points and a deficiency 2 points. For a repeated or uncorrected item of noncompliance with the same basic requirement, action points were to be successively increased by a factor of two each time it occurred. When a total of 100 action points or more resulted from an inspection or investigation and items of noncompliance included one or more violations or repetitive infractions or deficiencies, the regional office staff was to review the case to determine whether a civil penalty or show cause was warranted. As a general rule, a civil penalty was to be imposed for noncompliances which did not represent an immediate threat to the health and safety of the public and orders were to be issued for noncompliance that did. Where civil penalties or orders were not issued for violations or cases having 100 or more action points, the mitigating conditions or circumstances were to be documented. A civil penalty or Notice of Violation from Headquarters was to be issued when one letter identified several items of noncompliance in the infraction and deficiency categories with a total of 100 points or more.

An order to suspend, modify or revoke a license was appropriate when there was an apparent breakdown in the licensee's Quality Assurance program, based on the significant nature and number of items of noncompliance resulting in construction of discrepant Seismic Category I structures, systems, and/or components. The items of noncompliance were generally in the infraction category. It was to be considered a breakdown if there were several significant items of noncompliance with several of the Appendix B of 10 CFR 50 criteria. Procedural matters in themselves were not generally considered to be of prime significance. Failure of a system or failure to implement a program due to failure to develop, review and approve procedures was considered a manifestation of QA breakdown. If several items of noncompliance constituted a QA breakdown, the sanction was to be selected as follows:

"If the licensee cannot demonstrate that the quality of Seismic Category I systems, components or structures under construction or undergoing maintenance meet the stated requirements, an order may be issued to suspend operations or activities which have resulted in doubtful quality. The activities in question will not be resumed until the licensee has properly demonstrated that quality meets the requirements for Seismic Category I structures, systems or components."

"An order to suspend or modify a license may also be issued for a breakdown in quality assurance program implementation which results in a threat to the health, safety or interest of the public or the common defense and security."

"If inspection or investigation findings demonstrate that the quality assurance breakdown has not placed the quality of Seismic Category I systems or components in doubt and that there is no immediate threat to the health, safety, or interest of the public, or the common defense and security, a civil penalty may be the appropriate sanction."

"A civil penalty is the appropriate sanction in those cases where a licensee's history is one of chronic and numerous violations which do not involve an immediate threat to the health, safety, or interest of the public or the common defense and security, and provided that (as a general rule) the licensee's management has been properly apprised of the items of noncompliance. Normally this is done through enforcement conferences."

"The progression of the enforcement conferences resulting from inspections of such cases will normally include, in addition to the inspector's review of his findings with management, a meeting of the appropriate Branch Chief with an appropriate representative of the licensee's management at the site and a telephone discussion or a meeting at the Regional Office, or other designated place, between the Regional Director and the president or a corporate vice president who has authority to implement corrective measures. The Director or Deputy Director of the Office of Inspection and Enforcement may attend enforcement conferences with corporate management in appropriate situations."

"Since one of the basic parameters for civil penalty is items of noncompliance which represent a significant threat (but not immediate) to the health and safety of people or the common defense and security, the basis for this sanction is those items of noncompliance with regulatory requirements in the violation and infraction categories. However, the additive effect of deficiencies in the third category is one of the parameters considered in selecting this sanction. Each item of noncompliance with a regulatory requirement may carry a monetary penalty. Deviations from the provisions of commitments, codes, guides and standards will be listed separately and will carry no monetary penalty. Civil penalties based exclusively on deficiencies would be difficult to justify and their use for such items of noncompliance, while not excluded, is highly unlikely. Civil penalty or a "Notice of Violation" from Headquarters is the appropriate sanction when one enforcement letter identifies several items of noncompliance in the infractions and deficiencies categories with a total of 100 action points or more. The determination as to which sanction will be used is based on whether the licensee has been duly notified of the probability of such sanctions in previous correspondence and enforcement conferences, and on such judgment factors as the severity of the items of noncompliance, the nature and number of such items, the licensee's past performance, the frequency of noncompliance, and length of time the items of noncompliance have existed, the steps taken to correct them and the licensee's stated intentions of performance in correcting them promptly."

"A Notice of Violation will be issued from the Regional Office for all other items of noncompliance or combinations of items of noncompliance (a Form AEC-591 will be issued in the field by the inspector as appropriate for cases involving materials). The total sanction points for items of noncompliance in such notices from the Regional Offices may, on occasions, be greater than 100."

The above considerations were guidelines and Regional Directors could recommend any enforcement action available if the rationale was provided to support the recommendation.

3.8.1.3 General Statement of Policy and Procedure for Enforcement Actions

Appendix C to Part 1 of Title 10 of the Code of Federal Regulations describes the purpose of the enforcement program as:

"The purpose of the NRC enforcement program is to promote and protect the radiological health and safety of the public, including employees' health and safety, the common defense and security, and the environment by:

. Ensuring compliance with NRC regulations and license conditions;

. Obtaining prompt correction of noncompliance;

. Deterring future noncompliance;

. Encouraging improvement of licensee performance, and by example, that of industry, including the prompt identification and reporting of potential safety problems.

Consistent with the purpose of this program, prompt and vigorous enforcement action will be taken when dealing with licensees who do not achieve the necessary meticulous attention to detail and the high standard of compliance which the NRC expects of its licensees. It is the Commission's intent that noncompliance should be more expensive than compliance. Each enforcement action is dependent on the circumstances of the case and requires the exercise of discretion after consideration of these policies and procedures. In no case, however, will licensees who cannot achieve and maintain adequate levels of protection be permitted to conduct licensed activities."

The first step in the enforcement process is to identify the relative importance of each violation. Violations are categorized in five levels of severity as described in Appendix C. Severity Level I has been assigned to violations that are the most significant and Severity Level V violations are the least significant. Severity Level I and II violations are of very significant regulatory concern. In general, violations that are included in these severity categories involve actual or high potential impact on the public. Severity Level III violations are cause for significant concern. Severity Level IV violations are less serious but are of more than minor concern, i.e., if left uncorrected, they could lead to a more serious concern. Severity Level V violations are of minor safety or environmental concern.

The severity level of a violation may be increased for careless disregard of requirements, deception, or other indications of willfulness. The severity level of a violation involving failure to make a required report to the NRC is based on the significance of and circumstances surrounding the matter.

A Notice of Violation is the standard method for formalizing the existence of a violation and is to normally require the licensee to provide a written statement describing corrective action taken and results achieved, steps taken to prevent recurrence, and the date full compliance will be attained. The NRC does not generally issue a Notice of Violation for a violation identified as severity level IV or V if it was reported, if required, it was or will be corrected within a reasonable time, including measures to prevent recurrence, and if it was reasonably not expected to have been preventable by action to a previous violation.

A Civil Penalty is a monetary penalty generally imposed for Severity Level I and II violations, considered and usually imposed for Severity Level III violations, and may be imposed for Severity Level IV violations that are similar to violations discussed in a previous enforcement conference.

Enforcement conferences are normally to be conducted for all Severity Level I, II and III violations and for Severity Level IV violations considered symptomatic of program deficiencies.

The NRC imposes different levels of civil penalties for different severity level violations, taking into account the gravity of the violation as a primary consideration and ability to pay as a secondary consideration. Civil penalties are not intended to put a licensee out of business or to adversely affect his ability to safely conduct licensed activities. Orders are used when the intent is to terminate licensed activities. The NRC considers increases or decreases to base civil penalties on a case-by-case basis. Civil penalties for continuing violations may be issued on a per day basis up to $100,000 per violation per day. Civil penalties may be increased by as much as 25% based upon enforcement history, prior notice of similar events, multiple occurrences and if initiation of corrective action is not prompt or the action is minimally acceptable. Civil penalties may be decreased by as much as 50% based upon prompt identification and reporting and prompt and extensive correction action.

An order is a written NRC directive to modify, suspend, or revoke a license; to cease and desist from a given practice or activity; or to take such other action as may be proper. Orders are effective immediately without a hearing when determined the public health, interest or safety so requires or for violations involving willfulness. Otherwise, a hearing is held for the licensee to show cause why the order should not be issued in the proposed manner. Where necessary, the NRC is to issue orders in conjunction with civil penalties. Enforcement actions are to escalate for recurring similar violations.

In addition to Notice of Violation, civil penalties and orders, the NRC uses enforcement conferences, bulletins, circulars, information notices, generic letters, notices of deviation and confirmatory action letters. The NRC expects licensees to adhere to any obligations and commitments resulting from these processes and may issue orders to make sure such commitments are met.

Alleged or suspected criminal violations of the Atomic Energy Act and other relevant Federal laws are referred to the Department of Justice for investigation.

The Director, Office of Inspection and Enforcement is the principal enforcement officer of the NRC and has been delegated the authority to issue Notices of Violation, civil penalties and orders.

The Severity Categories for facility construction as shown in Appendix C are:

A. Severity I -- Very significant violations involving a structure of system that is completed in such a manner that it would not have satisfied its intended safety related purpose.

B. Severity II -- Very significant violations involving:

    1. A breakdown in the quality assurance program as exemplified by deficiencies in construction QA related to more than one work activity (e.g., structural, piping, electrical, foundations). Such deficiencies normally involve the licensee's failure to conduct adequate audits or to take prompt corrective action on the basis of such audits and normally involve multiple examples of deficient construction or construction of unknown quality due to inadequate program implementation; or

    2. A structure or system that is completed in such a manner that it could have an adverse effect on the safety of operations.

C. Severity III -- Significant violations involving:

    1. A deficiency in a licensee quality assurance program for construction related to a single work activity (e.g., structural, piping, electrical or foundations). Such significant deficiency normally involves the licensee's failure to conduct adequate audits or to take prompt corrective action on the basis of such audits, and normally involves multiple examples of deficient construction or construction of unknown quality due to inadequate program implementation.

2. Failure to confirm the design safety requirements of a structure or system as a result of inadequate preoperational test program implementation; or

3. Failure to make a required 10 CFR 50.55(e) report.

D. Severity IV -- Violations involving failure to meet regulatory requirements including one or more Quality Assurance Criteria not amounting to Severity Level I, II or III violations that have more than minor safety or environmental significance.

E. Severity V -- Violations that have minor safety or environmental significance.

3.8.2 Analysis

3.8.2.1 General

Early AEC enforcement action consisted of correspondence between the AEC and the licensee.

In 1973, in a report to the Director of Regulation (Davis and Brown) it was stated:

> "The AEC has neither imposed civil penalties nor taken significant enforcement or procedural actions against utilities which fail to implement the requirements of Appendix B. Regulatory's efforts to upgrade utility QA programs have relied on "jawboning" sessions with utility executives and routine enforcement letters, while the utilities have been permitted to continue construction or operation of their facilities notwithstanding QA deficiencies."

In the 1970's, guidance was provided for enforcement action which permitted issuance of Notices of Violation, civil penalties and orders and provided for escalation of enforcement action if the licensee was nonresponsive or if responses were not acceptable. However, the AEC and NRC did not aggressively implement enforcement action and the emphasis of enforcement action was in the area of operating plants.

3.8.2.2 Chapter 0800

The categorizing of each noncompliance required judgment of each inspector and was more difficult in design and construction than in operations. To categorize a nonconformance as a violation required determining that the safety function or integrity of a Seismic Class I system or structure was lost as a result of the noncompliance. To categorize a nonconformance as an infraction required determining that the safety function or integrity of a Seismic Class I system or structure was impaired as a result of the noncompliance. To categorize a nonconformance as a deficiency required determining that the threat to the health, safety, or interest of the public on the common defense or security was remote and no undue expenditure of time or resources to implement corrective action was necessary. Since the plant was under construction and not being operated, there was no immediate threat to the health and safety of the public or interest of the public on the common defense or security. In most cases, the safety function or integrity could not be clearly shown to be impaired or lost. For these reasons, nonconformances in design and construction tended to be categorized as deficiencies or deviations with some infractions and few violations.

Inspections at construction sites during the 1970's were performed by regional inspectors on a periodic basis of about six inspections a year at each site. Inspections were planned and scheduled with the licensee in accordance with construction schedules. It was not uncommon for an activity to be inspected during one site visit and not to be inspected again for a year or longer if at all. A relatively long period of time could elapse before a history on noncompliances to the same basic requirement developed. The AEC and NRC have tended to accept a fix to specific problems without requiring a review for identifying the magnitude of the problem to other areas of activity or action to prevent the problem from recurring.

The categorizing of noncompliances to lower action levels and the infrequency of inspections contributed to action point totals that were below the level for issuance of civil penalties or orders. If the action point totals did reach the levels for civil penalties or orders, AEC and NRC management tended to hold enforcement conferences instead of issuing the civil penalty or order.

The enforcement program did not encourage licensee conformance to commitments. Failure to conform to commitments such as the PSAR and provisions of applicable guides, codes and standards, when such lack of conformance did not constitute an item of noncompliance, was considered a deviation, the lowest level of enforcement action.

The AEC and NRC had difficulty in recognizing breakdowns in quality assurance programs and were hesitant to take permitted enforcement action. A breakdown was to be determined based upon the significance and number of items of noncompliance resulting in construction of discrepant Seismic Category I structures, systems, and/or components and several significant items of noncompliance with several criteria of Appendix B of 10 CFR 50 were required. A civil penalty or an order could be issued for a quality assurance program breakdown but neither were mandatory. The periodic nature of inspections, categorizing of noncompliances, and low level of attention afforded nonconformances with licensee commitments resulted in difficulty in recognizing quality assurance program breakdowns. The attitude that since the plant was not operating there was no immediate danger to the health and safety of the public and it was the licensee's responsibility to correct problems before an operating license would be issued resulted in hesitancy of NRC management to take permitted enforcement action. The fact that the Atomic Energy Act specified a maximum civil penalty of $5,000 and limited the total civil penalties payable in any 30 day period to $25,000 may have further influenced management reluctance to issue a civil penalty. Further, investigation into civil penalties and orders issued during the 1970's may be warranted.

Inadequate significance was attached to procedural matters. Procedural matters were not generally considered to be of prime significance. Failure of a system or failure to implement a program due to failure to develop, review and approve procedures was considered a manifestation of a QA breakdown. The failure to follow prescribed procedures is an indicator of potential problems.

B.81

Although primary concern is the adequacy of the end item, adherence to good procedures enhances the attainment of the desired adequacy.

3.8.2.3 General Statement of Policy and Procedure for Enforcement Actions

Following Three Mile Island, Public Law 96-295 was issued raising the maximum civil penalty from $5,000 to $100,000 and eliminating the limiting provision of total civil penalties payable in any 30 day period. The raising of civil penalties that could be issued put more strength into the enforcement program.

The 1980 and 1982 NRC General Statements of Policy and Procedure for Enforcement Actions also put more strength into the enforcement program. The policies more clearly defined severity categories, eliminated the action point system, and recognized quality assurance as an important aspect of construction activities by mentioning it in three of five severity categories. The intent of the new policies was that noncompliance should be more expensive than compliance. Severity levels of noncompliances could be increased for careless disregard or requirements, deception, or other indications of willfulness.

Notices of Violation are to require the licensee to provide a written statement describing corrective action taken, results achieved, steps taken to prevent recurrence and the date full compliance will be attained. During this study, regional personnel indicated that the AEC and NRC should have been more aggressive in requiring licensee determination of the extent of problems and correction of the cause of the problems and in following up of licensee open action items. The Notice of Violation, if properly used, and prompt follow up on all open action items can be strong points of the enforcement program.

Licensees are encouraged to report safety-related problems and the NRC may decrease civil penalties by as much as 50% for prompt identification and reporting of problems and for prompt and extensive corrective action. During the NRC Case Studies, licensees indicated there was little incentive to identify problems. The licensee would identify a problem, take corrective action to eliminate the problem, promptly report the problem to the NRC and then receive a fine from the NRC and publicity in the public media inferring poor quality of construction. Licensees also indicated there was lack of uniformity in application of civil penalties and orders.

Past enforcement programs of the AEC and NRC were not as effective as they could have been as a result of inconsistency in requiring licensees to determine the extent of problems and to correct the causes of the problems, inability to recognize that the problems detected were but symptoms of larger problems, and inability to raise problems to the threshold of action. New enforcement programs tend to correct these deficiencies. However, the new programs do not appear to encourage licensee conformance to commitments or attach greater significance to procedural matters. The NRC should consider making commitments a condition of a permit and placing greater emphasis on procedural matters.

3.9 NRC Inability To Prevent Problems And Slowness To Identify And Act On Problems

An analysis of the root causes of the inability of the NRC to prevent problems and slowness to identify and act on problems at Diablo Canyon, Marble Hill, Midland, South Texas and Zimmer nuclear plants follows.

3.9.1 Diablo Canyon

The major problem was identified as ineffective design control.

The licensee received its Construction permit in 1968. The NRC issued an Operating License in 1981 and then revoked the license two months later following identification by the licensee of an error in seismic response spectra for some piping and equipment restraints. NRC investigation determined that the cause of the problem was informality in the procedures used for design document control and lack of independent review of data by the licensee prior to submittal to its seismic consultant. Prior to reinstatement of its operating license, the licensee is required to complete an extensive design verification program.

Appendix B of 10 CFR 50 was issued in 1970, some 2 years after the construction permit date and as a result there were no quality assurance program requirements at the time much of the design work was performed.

The inspection program concentrated on construction activities and did not focus attention in the area of design. Since PG&E was their own AE, the Licensee Contractor and Vendor Inspection Program started in 1975 did not apply to them.

Although the licensee had no commitment to implement an Appendix B type of quality assurance program on Unit 1, he agreed to implement such a program, as applicable.

Since the work had progressed beyond design and emphasis was on inspection of construction activities, design control activities were not reviewed when Appendix B became applicable.

The root cause of NRC inability to prevent the problem and slowness to identify and act on the problem is:

. Insufficient attention in the design area.

3.9.2 Marble Hill

The major problems were identified as concrete consolidation, improper repair of the imperfections, inadequate or nonexistent records traceable to the repairs, inadequate training and supervision of personnel responsible for the repairs, welding, and insufficient awareness of the problems and control by the licensee. The NRC Case Study identified licensee inexperience to be a root cause of the problems. The licensee had not built a nuclear power plant.

The licensee received its Construction Permit in 1978. In 1979, the NRC shut down all safety-related construction activities.

The NRC had detected nonconformances in concrete work from the outset of the project. About one year after CP issuance, the NRC requested the licensee to upgrade its quality assurance program. The licensee agreed to upgrade the program and determine if previously poured concrete was adequate.

About a month later, a former employee of the Civil Construction contractor alleged that surface defects in concrete had been improperly patched.

Concurrently, the National Board of Boiler and Pressure Vessel Inspectors confirmed problems with piping installation identified by a mechanical subcontractor.

These events led to an NRC team inspection which confirmed concrete consolidation problems and improper repair of the imperfections and resulted in the shut down of safety-related construction activities. Work was not permitted to resume until the licensee upgraded its QA program and that of its contractors and the adequacy of completed construction work was verified.

The root causes of NRC inability to prevent problems and slowness to identify and act on the problems are:

- Inadequate review of the licensee and contractor experience and ability to manage construction of a nuclear power plant.

- Irregular non-constant presence.

- Inability to recognize the significance and magnitude of the problems.

3.9.3 Midland

The major problem was identified as settlement of the diesel generating building in 1978 as a result of inadequate and poorly compacted soil. Licensee investigation revealed that other safety-related systems and structures were affected. NRC investigation determined specifications had not been followed for soil fill activities and there was insufficient control and supervision of the activities by the utility and its contractors.

The licensee received its construction permit in 1972. The NRC issued a Show Cause Order regarding the soils problem in 1978. Rework is in progress and the application for an operating License is in litigation before a hearing board.

NRC personnel were aware of problems at Midland. Between 1973 and 1978, problems were reported with cadwelds, omitted rebar, tendon installation and bulgeing of the containment liner. Problems were identified on multiple occasions separated by about one year. There were meetings at the Region to determine if action should be taken and meetings were held with Midland management. Regional requests to stop work at Midland were not supported by NRC headquarters until 1978. In response to the Show Cause Order, Midland requested a hearing. The hearing process is still going on and Midland has been permitted to continue soils work.

Since 1978, additional problems have been identified in HVAC welding, reactor vessel anchor bolts, pipe supports and hangers, electrical cable separation and in the diesel generator building inspection performed by the licensee. Mechanical equipment, piping and electrical systems were poorly installed and supervisors had ordered QC inspectors to suspend inspections if they found too many deficiencies.. A civil penalty of $38,000 was issued in 1979 for the HVAC problem and a $120,000 civil penalty was issued in 1982 for breakdown of the Quality Assurance Program. Reinspection and finishing of the plant is to be performed in accordance with an NRC approved plan under the oversight of an independent contractor.

The root causes of NRC inability to prevent problems and slowness to identify and act on the problems are:

- Irregular non-constant presence until 1980.

- Reluctance to take enforcement action.

- Mindset that it is the licensee's responsibility to properly construct the plant and it would not be licensed until ready for operation as determined by pre-operational and startup tests.

3.9.4 South Texas

The major problems were identified as concrete placement, welding activities, procedural violations, records falsification and personnel qualification. Additional problems were identified as harassment and intimidation of inspectors and insufficient design work to support construction. The NRC Case Study identified licensee and AE/Constructor inexperience to be a root cause of the problems. Neither the licensee nor the AE/Constructor had built a nuclear power plant.

The licensee received its Construction permit in 1975. The NRC issued the licensee a $100,000 fine and show cause order in 1980. Work was allowed to restart only after upgrading of QA for that area and verification by the NRC.

In 1977, the NRC received reports of intimidation of QC inspectors at the construction site. Between July of 1977 and November of 1979, the NRC performed 10 investigations of allegations. In 1978, the NRC held a meeting with licensee management to discuss morale problems. An FBI probe into allegations of forged documentation in 1979 reported widespread problems. A NRC special investigation was performed which determined shortcomings in management and implementation of the QA/QC Program.

A summary report (Gower 1981), prepared after reviewing headquarters files of inspections performed from 1974 through 1979, stated there was good inspections procedure coverage of the major problem areas but the degree to which the procedures would have turned up similar problems is strongly influenced by the experience, practical knowledge ad technical depth of the inspectors. In an analysis of 72 allegation relating to problems at the site, the report indicated 34 were substantiated, 28 were refuted and 10 were neither substantiated nor refuted. NRC inspections had detected problems concerning procedures, records, personnel qualifications, audits, and concrete and welding activities.

The root causes of NRC inability to prevent problems and slowness to identify and act on the problems are:

- Inadequate review of the licensee and contractor ability to manage construction of a nuclear power plant.

- Irregular non-constant presence.

- Inability of the NRC to recognize the significance and magnitude of the problems.

- Mindset that it was the licensee's responsibility to properly construct the plant and it would not be licensed until ready for operation as determined by pre-operational and startup tests.

3.9.5 Zimmer

The major problems were identified as Q.C. documentation, procedure violations, inadequate nonconformance reporting system, deficiencies in drawings, specifications, instructions and procedures, material control and licensee audits and corrective action.

The licensee received its Construction Permit in late 1972. In November of 1981, the NRC issued a $200,000 fine for Quality Assurance breakdowns following investigation of allegations of shoddy construction practices. In November of 1982, the NRC suspended all safety-related work in response to concerns about the quality of construction and management controls. The licensee has been required to complete a Quality Confirmation Program of the as-built condition and to correct any problems before additional consideration for an Operating License.

A summary report (Gower 1981), prepared as a result of reviewing headquarters inspection files and the draft report on investigation of Zimmer, revealed inspection coverage appeared to be extensive and comprehensive, inspections up through 1976 appeared to have been in line with the inspection program, and that during 1977, 1978 and 1979, inspection-hours per year (600 to 1200) exceeded planned hours (400-500) and 12 to 16 different inspectors contributed to the inspection effort during one year. Three to six different inspectors were thought to be sufficient for adequate coverage. There were signs of problems with the licensee/constructor audit programs in 1973, 1975, 1977 and 1979. Up to 1981, there were 13 investigations performed addressing allegations in depth and dealing primarily with QA/QC problems. There were numerous instances of enforcement citations in QC documentation, procedure violations, materials control, and deficiencies in instructions, procedures and specifications. The notices of violation were limited to the item of noncompliance with little, if any, inference that the concern may be indicative of a larger more pervasive problem that should be looked into and corrective action taken.

Another study (Torrey Pines 1983), reported inadequate management controls of the project citing GC&E and H.V. Kaiser inexperience in building nuclear plants as a cause of problems at Zimmer. The study reported inadequate staffing, procedures, and control systems and an ineffective audit program. Problems remained uncorrected partly through a lack of attention and follow through on corrective action by the NRC. CG&E was allowed to continue construction while being lulled into a false sense of satisfactory performance until the late 1970's.

An NRC report (NET 1983, NUREG-1969) indicated inspections and investigations revealed inadequacies in administration of the Quality Assurance Program and that the quality of plant systems, structures and components was indeterminate.

During this study, Region III personnel indicated lots of the problems were noted but they didn't reach a threshold of action. Reviews of the Action Item Tracking List revealed every criteria of Appendix B of 10 CFR 50 was cited in the first and last years of construction. They indicated that the NRC failed to follow-up on open action items (approximately 12,000) and failed to require reviews for determining the extent of problems and determination and correction action regarding the cause of the problem. They also indicated that until 1974, teams of four or five cross-discipline inspectors performed quarterly inspections. In 1974 and 1975, six experienced inspectors with accumulated experience of about 120 years left the Region for other assignments and were replaced by less experienced, more specialized inspectors.

The root causes of the NRC inability to prevent problems and slowness to identify and act on problems were:

.   Inadequate review of the licensee and its contractor's ability to manage construction of a nuclear power plant.

.   Failure of the NRC to require licensee reviews of problems to determine their extent and to take corrective action regarding the cause of the problem.

.   Inability to recognize the significance and magnitude of the problems.

.   Loss of inspection experience in the Region.

.   Problems didn't reach the threshold for enforcement action.

.   Mindset that it was the licensee's responsibility to properly construct the plant and it would not be licensed until ready for operation as determined by pre-operational and startup tests.

4.0 STUDY GROUP

4.1 GROUP MEMBERS

John L. Heidenreich - N. C. Kist & Associates

Robert W. Hubbard - N. C. Kist & Associates

Richard M. Kleckner - N. C. Kist & Associates

Willard D. Altman - Nuclear Regulatory Commission

E. William Brach - Nuclear Regulatory Commission

5.0 LIST OF ACRONYMS

AE	Architect Engineer
AEC	Atomic Energy Commission
ANSI	American National Standards Institute
ASME	American Society of Mechanical Engineers
CAT	Construction Assessment Team
EDO	Executive Director of Operations
ERDA	Energy Research and Development Administration
HVAC	Heating, Ventilating, and Air Conditioning
I&E	Office of Inspection and Enforcement
IEEE	Institute of Electrical and Electronic Engineers
IDI	Integrated Design Inspection
IDVP	Independent Design Verification Program
INPO	Institute of Nuclear Power Operations
LCVIP	Licensee Contractor and Vendor Inspection Program
NMSS	Office of Nuclear Material Safety and Safeguards
NRC	Nuclear Regulatory Commission
NRR	Office of Nuclear Reactor Regulation
NSSS	Nuclear Steam System Supplier
PAT	Performance Appraisal Team
PSAR	Preliminary Safety Analysis Report
RES	Office of Nuclear Regulatory Research
SALP	Systematic Assessment of Licensee Performance
SAR	Safety Analysis Report

6.0 REFERENCES

Bernsen, S. and Hellman, S. 1973. The Status and Application of ANSI N45.2 Standards, Revision 1. American National Standards Institute.

Blaha, J. Sept. 27, 1983. Management Information Report (MIR) Base Report Through August 31, 1983. Memorandum. U.S. Nuclear Regulatory Commission.

Bland, W. and Reilly, D. 1979. Staff Reports to the President's Commission on the Accident at Three Mile Island, Vol. IV. U.S. Government Printing Office.

Davis, J. and Brown, H. 1973. Quality Assurance and the Utilities: Is Regulatory Doing Enough? U.S. Atomic Energy Commission.

DeYoung. 1982. Organizational Changes In The Office of Inspection and Enforcement. Memorandum. U.S. Nuclear Regulatory Commission.

Langstaff. October 25, 1982 and November 1, 1982. A Reporter At Large – The Cult of The Atom. New Yorker.

Moeller, D. 1976. Report on Nuclear Reactor Inspections. U.S. Nuclear Regulatory Commission.

Muller, F. et. al. 1977. A Study of the Nuclear Regulatory Commission Quality Assurance Program. NUREG-0321. Sandia Laboratories.

Olds, F. June 1974. The AEC Bears Down on Nuclear Quality Assurance. Power Engineering. 41-47.

Teknetron Research, Inc. 1981. Evaluation of NRC's Revised Inspection Program. NUREG/CR-1904, Vol. 1. U.S. Nuclear Regulatory Commission.

Torrey Pines Technology. 1983. Independent Review of Zimmer Project Management. GA-C17173. Torrey Pines Technology.

TRW. 1978. Report of a Study of the Licensee Contractor and Vendor Inspection Program (LCVIP). NUREG/CR-0217. U.S. Regulatory Commission.

U.S. AEC. 1972. Preamble to Regulatory Guide Series. U.S. Atomic Energy Commission.

U.S. AEC. 1974. Guidance on Quality Assurance Requirements During Design and Procurement Phase of Nuclear Power Plants. WASH 1283. U.S. Atomic Energy Commission.

U.S. AEC. 1974. Guidance on Quality Assurance Requirements During the Construction Phase of Nuclear Power Plants. WASH 1309. U.S. Atomic Energy Commission.

U.S. AEC. 1974. Study of Quality Verification and Budget Impact. WASH 1292. U.S. Atomic Energy Commission.

U.S. Congress, House and Senate. 1974. Energy Reorganization Act of 1974. Public Law 93-438, 94-77, 95-39, 95-209, 95-238 and 95-601. U.S. Government Printing Office.

U.S. Congress, House and Senate. 1983. Authorization of Appropriations - Fiscal Years 1982-1983. Public Law 97-415. U.S. Government Printing Office.

U.S. General Accounting Office. 1978. The Nuclear Regulatory Commission Needs to Aggressively Monitor and Independently Evaluate Nuclear Power Plant Construction. U.S. General Accounting Office.

U.S. NRC. 1975. Inspection and Enforcement Manual. U.S. Nuclear Regulatory Commission.

U.S. NRC. 1979. Review of the Continuing Implementation of NRC's Resident Inspection Program. U.S. Nuclear Regulatory Commission.

U.S. NRC. 1980. Three Mile Island: A Report to the Commissioners and the Public. NUREG/CR-1250, Volume II, Part 1. U.S. Nuclear Regulatory Commission.

U.S. NRC. 1981. Standard Review Plan, Sections 13.11 and 17.1. NUREG-0800 (Formerly NUREG-75/087). U.S. Nuclear Regulatory Commission.

U.S. NRC. 1982. Title 10, Chapter 1, Code of Federal Regulations - Energy, Part 2, Appendix C. U.S. Nuclear Regulatory Commission.

U.S. NRC. 1982. Assurance of Quality. SECY-82-352. U.S. Nuclear Regulatory Commission.

U.S. NRC. 1983. U.S. Nuclear Regulatory Commission Functional Organization Charts. NUREG-0325. U.S. Nuclear Regulatory Commission.

U.S. NRC. 1983. Title 10, Chapter 1, Code of Federal Regulations - Energy, Part 50 - Domestic Licensing of Production and Utilization Facilities. U.S. Nuclear Regulatory Commission.

U.S. NRC. March 1983. Licensee Contractor and Vendor Inspection Status Report. NUREG-0040, Vol. 7, No. 1. U.S. Nuclear Regulatory Commission.

U.S. NRC. 1983. Report of the NRC Evaluation Team on the Quality of Construction at the Zimmer Nuclear Power Station. NUREG-0969. U.S. Nuclear Regulatory Commission.

U.S. NRC. 1988. Proposed General Statement of Policy and Procedure for Enforcement Action. 45 FR 66754. U.S. Nuclear Regulatory Commission.

APPENDIX C

ASSURANCE OF QUALITY IN NUCLEAR CONSTRUCTION PROJECTS:
AN EXAMINATION OF SELECTED CONTRACTUAL, ORGANIZATIONAL,
AND INSTITUTIONAL ISSUES

PRINCIPAL CONTRIBUTORS

M. E. Walsh (a)
M. V. McGuire (a)
B. L. Hansen (a)

January 1984

Prepared for
Division of Quality Assurance, Safeguards,
  and Inspection Programs
Office of Inspection and Enforcement
U.S. Nuclear Regulatory Commission
Washington, D.C. 20555

Pacific Northwest Laboratory
Richland, Washington 99352

(a) Battelle Human Affairs Research Centers

EXECUTIVE SUMMARY

This appendix presents preliminary analyses, findings, and conclusions of a study on the contracting and procurement process used for the construction of nuclear power plants. The objectives of the study were:

- to characterize the aspects of contracts and procurement that appear to affect the quality during construction of a nuclear power plant

- to determine the types of contract and procurement provisions and arrangements that could contribute most to enhanced quality

- to develop guidelines for construction contracts and procurement that could assist in achieving overall quality objectives

- to examine the contributions of selected organizational and institutional arrangements to nuclear construction projects.

To accomplish these objectives, a series of site visits to utilities constructing nuclear power plants, architectural-engineering firms, constructors, and subtier contractors was planned and partially implemented. Specific contractual, organizational, and institutional factors were investigated at each site. Findings and conclusions contained in this appendix are based upon four such site visits (three to nuclear construction projects and one to an architectural-engineering firm) made to date. Information used in the analyses was obtained in the field, from secondary source materials, and from telephone and personal contacts with informed sources.

The following, preliminary findings and conclusions were reached:

- Bid evaluation and selection processes should be based upon functional criteria related to the work to be performed.

- Because designs are usually not complete before construction is begun, fixed-price contracting for most aspects of nuclear power plant construction projects is not appropriate. Instead, cost-reimbursable contracts with fixed fees are recommended most frequently by those involved in nuclear construction projects, particularly for assuring quality performance.

- In achieving quality objectives, the focus should be on the implementation of QA and QC programs rather than on the level of detail in contract and procurement documents.

- Previous nuclear experience appears to provide a significant advantage in a nuclear construction effort. Utilities that do not possess such experience internally should consider hiring either a project staff or contractors who can provide such expertise.

- Together with the NRC, state public utility commissions provide a major source of regulatory oversight for nuclear construction projects. Historically, state pubic utility commissions have not frequently disallowed construction costs that may have resulted from lapses in quality assurance or project management. Recent developments suggest that this position is changing.

- A nuclear construction project appears to benefit when its procurement agent is large enough and experienced enough to exert "marketplace presence." A large procurement entity offers the advantages of market familiarity and commercial power (based upon frequency and continuity of purchasing) as well as the expertise needed to secure satisfactory performance on procurements.

Possible recommendations resulting from these preliminary findings include the following:

- As part of their management review, the NRC should require applicants for construction permits to stipulate their proposed contracting methods, bid evaluation and selection procedures, and their reason(s) for choosing them.

- The NRC should examine methods to focus more attention on how a licensee proposes to insure that quality work is being performed rather than on the documents that describe general QA and QC programs.

- The NRC should examine the implications for its own mission of state public utility commission scrutiny of and policies toward nuclear construction project costs and management.

ACKNOWLEDGEMENTS

The authors wish to express their sincere appreciation to the following individuals and organizations for their help in this effort:

- Dr. W. D. Altman, NRC Project Manager, for his tireless devotion to, interest in, and support of this project.

- Mr. G. T. Ankrum, NRC Branch Chief, for his interest in raising and exploring new issues.

- E. Baker, USNRC, and J. Heidenreich, N. C. Kist and Associates, for their collaboration in conducting site visits.

- Members of the Review Group for their invaluable comments and suggestions as well as their encouragement of our work.

- Ms. Jean Botcky, Ms. Christine Norell, and Mrs. Charleen Sager for their skills, assistance, and support of the project.

- The management and staffs of the several utilities, construction projects, and A-E firm visited, who gave so freely of their time and shared so generously their insights and experiences. Without the interest and support of these people, this work could not have been done.

CONTENTS

APPENDIX C

ASSURANCE OF QUALITY IN NUCLEAR CONSTRUCTION PROJECTS: AN EXAMINATION
OF SELECTED CONTRACTUAL, ORGANIZATIONAL, AND INSTITUTIONAL ISSUES

C.1 INTRODUCTION

This appendix summarizes and discusses the findings to date of a case
study project examining the contract and procurement processes at nuclear
power plants under construction. Section C.1 presents introductory material on the study's purpose and objectives, background and scope, technical approach and study limitations. Section C.2 summarizes project
findings, conclusions, and recommendations. Section C.3 examines QA contractual issues in nuclear power plant construction, focusing primarily on
insights gained at the sites visited. Section C.4 discusses organizational issues affecting quality assurance in nuclear power plant construction. Section C.5 examines institutional issues and their implications
for the course and success of nuclear construction projects. Section C.6
lists references, and a bibliography is provided in Section C.7.

C.1.1 Purpose and Objectives

This appendix is intended to serve several purposes: 1) to constitute
an extended progress report on project activities; 2) to communicate preliminary findings and suggest future directions; and 3) to form the basis
for developing NRC staff recommendations regarding a course of action to
improve the assurance of quality in nuclear power plant construction projects in response to Congressional directives in the FY 1982-83
Authorization Act. Because of this last purpose, project recommendations
are offered, although these are based upon preliminary findings and
conclusions.

C.1.2 Background and Scope

The complexity and extent of problems that have been identified in
recent years at some nuclear power plants under construction have raised
questions regarding the quality assurance (QA) programs required by the
NRC and implemented by NRC licensees. As part of an effort to better
understand and address these problems, the NRC initiated a study of the
contract and procurement process employed by licensees at nuclear power
plants under construction and of the organizational and institutional
environments in which such projects are initiated and financed.

The purpose of the study is to examine how "quality" responsibilities
are delegated, managed and controlled by the licensee in the contract and
procurement process. This study is also to determine what improvements
may or should be made to the QA programs required by the NRC, based on the
review and study of the contract and procurement process and of organizational and institutional arrangements in the nuclear power industry. The

NRC will use the study's results in their analyses of QA programs and in the preparation of a report to Congress required by the FY 1982-83 Authorization Act.

Specific objectives of the study are to 1) characterize the aspects of contract and procurement that appear to affect quality during construction of a nuclear power plant; 2) determine the types of contract and procurement provisions and arrangements that could contribute most to enhance quality; 3) develop, to the extent possible, guidelines for construction contracts and procurement that could assist in achieving overall quality objectives; and 4) examine the contributions of selected organizational and institutional arrangements to the quality and success of nuclear construction projects.

This project is one of several initiatives undertaken by the NRC to improve the assurance of quality in the design and construction of nuclear power plants. The full range of initiatives that have been undertaken involve the following issues or topic areas:
- Measures at Near-Term Operating License Facilities
- Industry Initiatives
- Construction Inspection Programs
- Qualification and Designation of QA/QC Personnel
- Management
- Long-Term Review.

This project is included in "Long-Term Review." Three projects, also part of the Long-Term Review, that are closely linked to this project are described in Appendix A, "Quality in the Design and Construction of Nuclear Power Plants: Case Studies of Successes and Failures," Appendix B, "Management Review of the NRC QA Program," and Appendix D, "Outside Programs for Assurance of Quality in Design and Fabrication."

C.1.3 Technical Approach

The findings and recommendations in this appendix are based upon insights gained through site visits, review and analyses of selected secondary source materials, and other project activities including telephone contacts with state regulatory personnel. These activities were combined to allow the project team to examine both actual contracting and procurement practices used by firms involved in nuclear construction projects and the organizational and institutional environments in which these projects are initiated, guided, and completed. Approaches used to examine these topics are described separately below.

Examination of Contracting and Procurement Practices

To examine contracting and procurement practices, the perspectives, experiences, and practices of key groups involved in constructing nuclear power plants were collected through site visits. The following criteria were used to select the sites: geographic location, site reputation/ success, experience of site personnel, and structural, contractual, and organizational arrangements of the site. Initially, four nuclear construction projects and the utilities constructing them, two architect-engineering (A-E) firms, one constructor, and two subtier suppliers were to be examined. However, because of time constraints, only a subset of the planned site visits have been completed to date (three construction projects and one A-E). During actual site visits, representatives of the sites provided useful insights into the contracting and procurement process and its relationship to the assurance of quality.

Site Visit Protocol. Each visit was conducted according to a site visit protocol, which consisted of personal interviews with designated individuals and examination of relevant documents and materials. Personnel interviewed included legal, contracting, and procurement specialists; key managers involved in bid evaluation and selection; construction and project managers; representatives of the utility's QA organization; inspection and audit specialists; and contractor on-site managers.

The following documents were identified as important to obtain and review (if possible): bid evaluation procedures/guidelines, standard procurement forms/guidelines, standard contracts for major project contractors (at construction sites visited), and special conditions for all major contracts (if separate from standard contract document).

Contractual, Organizational and Institutional Factors Examined at the Sites. The site visit protocol was tested and revised on the basis of a pilot site visit to a nuclear construction project. During the pilot visit, a series of contractual, organizational, and institutional factors thought to have potential significance for the assurance of quality was examined.

Contractual factors investigated included the following:

- types of contracts executed

- the use of incentive provisions in contracts

- assignments of responsibilities and risk sharing relative to quality between the utility and its contractors

- requirements for demonstration, review, and/or approval of QA programs

- procurement practices and procedures

- approaches used to communicate QA/QC requirements to subtier contractors and to monitor compliance with these requirements

- provisions for source and on-site inspections.

The following organizational factors were examined:

- prior nuclear experience

- project structure and participation in the engineering, construction management, procurement and constructor roles

- owner involvement in the engineering, construction management, procurement and constructor roles

- labor arrangements.

Two institutional factors were examined:

- project ownership arrangements

- the effects of state public utility commission policies on nuclear construction projects.

Profiles of Sites Visited. The site findings contained in this appendix are based upon visits to four sites: three nuclear construction projects and one A-E firm. Site 1 is a nuclear project being built by a relatively small investor-owned utility, and a small rural electric cooperative, neither of which has previous nuclear construction experience. The utility is being assisted by an experienced A-E and contractors. Quality-related problems identified at the site several years ago led the utility to stop construction and change its contracting and project management style. Site 1 served as the pilot site for this project.

Site 2 is a joint venture nuclear project being built by a group of utilities (four investor-owned and one public utility). The investor-owned utility serving as project manager for the owners had no previous nuclear construction experience. This utility is being assisted by a large, experienced prime contractor that serves as A-E and constructor and acts as the utility's agent for procurement and the management of project subcontractors.

Site 3 is another joint venture nuclear project being constructed by three small public utilities and one large investor-owned utility. The investor-owned utility, the subsidiary of a large holding company, has previous nuclear experience, owns just over 50% of the project and serves as the construction manager and agent for the owners' group. The utility is assisted by an experienced A-E and many contractors. The utility controls all project procurement. The project, which was delayed initially for financial reasons, has a solid reputation within the industry.

Site 4 is a large and experienced A-E firm that has been a major force in the nuclear construction industry. This firm has played all project roles either alone or in combination with others for many nuclear construction efforts. The firm has worked for owners and with contractors and suppliers possessing a wide range of expertise and experience in nuclear construction.

Examination of Organizational and Institutional Environments

Investigation of organizational and institutional factors at the sites visited, combined with findings from other NRC quality assurance studies contained in the Long-Term Review, stimulated the NRC Project Monitor to determine that the following two issues needed to be examined further: 1) effects of state public utility commission (PUC) policies on nuclear construction efforts, and 2) effects of ownership arrangements on project success. Subsequently, a separate set of project activities was initiated to provide additional information on these issues.

First, the project team consulted a wide range of relevant secondary sources pertaining to PUC actions and policies; and to the legal, organizational, and institutional parameters of the nuclear industry. All sources examined by the project team are contained in the Bibliography.

To examine the actions of PUCs, secondary source materials were supported by telephone contacts with state commissions having operating plants in their jurisdictions. In several cases, members of commission staffs provided materials related to special commission actions. In other cases commission staffs identified documents relevant to PUC decisions, and the project team obtained them. These telephone contacts and materials provided additional insights into the attitude of state PUCs toward nuclear construction projects.

Finally, the project's investigation of ownership arrangements in the nuclear industry was assisted by the NRC Review Group, a group of experts advising the NRC on its entire program of QA initiatives.[a] Group members provided insights from their own experiences and suggested sources of further information.

C.1.4 Study Limitations

Several factors of timing and approach necessarily limit the breadth of findings, conclusions, and recommendations discussed in this appendix. To provide the NRC staff with preliminary findings and to assist in preparing the Congressional report, the results had to be summarized based upon only a subset of the planned field work. In addition, because examination of the public utility commission and project ownership issues was initiated after completing some of the site visit activities, they have not been thoroughly studied. Finally, this project is based primarily on a series of case studies, which are intensive examinations of the experience of individual firms and/or construction projects. Therefore, while the experience of these firms may not be unique, it also may not be representative of, or generalizable to, other nuclear construction projects.

C.2 CONCLUSIONS, FINDINGS AND RECOMMENDATIONS

Based upon insights drawn from site studies, secondary sources and other project activities, the several general conclusions found below were reached. Second, findings specifically related to the site studies are discussed. The

---

(a) J. Christensen. Draft. Pacific Northwest Laboratory, Richland, Washington.

study team visited three nuclear construction sites and a large architect-engineering firm. At each site, contractual, organizational and institutional factors were examined to determine their relationship to and influence on the assurance of quality in nuclear construction programs. For each of the factors investigated, findings from the site visited are cited. Third, several observations growing out of project activities, but requiring further examination, are made. Finally, recommendations are offered, based upon the project's preliminary findings and conclusions.

C.2.1 General Conclusions

Based upon the findings at the sites visited, examination of relevant secondary sources, and other project activities, the project team has drawn several general conclusions.

First, no substitute appears to exist for an objective bid evaluation and selection process based upon relevant technical criteria. Where such criteria are rejected in favor of "people we're familiar with" or "country club cousins" problems can result. This is particularly significant for first-time owners because the "people we're familiar with" are not likely to have nuclear experience. (See Section C.3.1.)

Second, without substantially more complete designs before construction is begun, fixed-price contracting for most aspects of nuclear power plant construction projects does not appear to be justified. Instead, utilities involved in nuclear projects most frequently recommend cost-reimbursable contracts with fixed fees, particularly for assuring quality performance. Although such contracting de-emphasizes cost, it may be most cost-effective in that it is more likely to result in getting the job done correctly the first time. (See Section C.3.2.)

Third, the level of detail of the QA and QC requirements in contract and procurement documents is less important than the degree to which QA and QC programs are actually implemented. Actual checks of work done, source and on-site inspections, the implementation of worker and supervisor training programs, and required demonstrations of contractor expertise and commitments in both pre- and post-bid award periods are all examples of actions that demonstrate more about QA/QC programs than do written QA/QC requirements. (See Section C.3.3.)

Fourth, previous nuclear experience appears to provide a significant advantage in a nuclear construction effort. Utilities that do not have such experience internally should hire either a project staff or contractors who can provide such expertise. (See Section C.4.3.)

Fifth, with the NRC, state public utility commissions provide a major source of regulatory oversight for nuclear construction projects. Regulatory influence in this case is exercised through the rate base treatment of such projects. Historically, state PUCs do not appear to have been active in disallowing construction costs that may have resulted from lapses in quality assurance or project management. This position results in shifting the risks of quality lapses from the utility to its ratepayers. Recent developments suggest that this position is changing. (See Section C.5.1.)

Sixth, a nuclear construction project appears to benefit when its procurement agent is large enough and experienced enough to exert "marketplace presence." A large procurement entity offers the advantages of market familiarity and commercial power (based upon frequency and continuity of purchasing) as well as the expertise needed to secure satisfactory performance on procurements. (See Section C.5.2.)

C.2.2 Factors Affecting the Assurance of Quality in Nuclear Construction Projects: Site Study Findings

Contractual Factors

Site examination of contractual factors yielded the following findings:

- Kinds of contracts executed. The type of contract universally preferred for most aspects of nuclear power plant construction is the cost-reimbursable contract. This contract type offers several advantages for assurance of quality: 1) it permits extensive monitoring of contractor performance; 2) it encourages the taking of corrective action; 3) it flexibly accommodates scope and design changes; and 4) it allows construction to begin before design work is complete. (See Sections C.3.2 and C.3.4.)

- Use of incentive provisions in contracts. Incentive contracting is used at only one of the sites visited. In general, those interviewed argued that incentive provisions tended to place too much emphasis on cost and schedule, to the detriment of quality objectives. (See Sections C.3.2 and C.3.4.)

- Responsibilities and risk sharing between the utility and its contractors relative to quality. By using cost-reimbursement contracts, which limit or remove contractor liability for rework and errors, utilities assume virtually all the risks of completing nuclear construction projects successfully and on time. Generally, the owner's assumption of risk is reflected in relatively small fees earned by contractors. (See Sections C.3.2 and C.3.4.)

- Requirements for demonstration, review, and/or approval of contractors' QA programs. In addition to the necessary review and approval of contractors' QA programs, utilities visited felt that requiring contractors to demonstrate their approach to assurance of quality was important. The argument was that while some contractors might be able to describe an acceptable QA program on paper, the only way to evaluate their real understanding was to ask them to demonstrate how they planned to implement such a program. At some sites, pre- and post-award meetings with contractors were used for this purpose. (See Section C.3.3.)

- Procurement practices and procedures. At all sites, procuring materials, supplies, and equipment was helped by pre-screened and/or
  evaluated suppliers' lists. Such lists are typically updated as project experience warrants. Where adequate resources are available,
  surveillance programs may provide additional feedback on vendor performance. (See Section C.3.1.)

- Approaches used to communicate quality requirements to subtier contractors and to monitor their compliance. Communicating quality
  requirements appears to be divided between two approaches. One
  approach is to have detailed contract and procurement documents,
  incorporating directly all applicable QA/QC requirements, codes, and
  standards. The other approach relies on more general statements of
  quality expectations in procurement documents. With this latter
  approach, suppliers would be required to verify their compliance, for
  example, through mutually agreed upon audit procedures or through the
  submission of acceptable test and inspection data. In either case,
  all utilities visited agreed that communicating requirements would
  not assure quality unless compliance was actually monitored in some
  way--quality assurance requires follow-through. (See Section C.3.3.)

- Provision for source and on-site inspections. Procurement documents
  at all sites required suppliers to make their facilities available
  for inspection. The provision relating to this issue was nearly
  identical in documents reviewed at each site. Similarly, inspections
  on receipt were standard practice at all sites. Sites differed, however, in the scope of receipt inspection activities performed and in
  the resources available for inspections. In general, larger, more
  commercially active purchasers possessed both the economic incentive
  and resources to monitor systematically contractors' shops. (See
  Sections C.3.1 and C.3.3.)

Organizational Factors

Site examination of organizational factors yielded the following findings:

- Various combinations of engineering, management, procurement, and
  construction roles. The wide array of construction project role
  arrangements in the U.S. nuclear industry suggests that no one
  arrangement will insure success. External A-E firms performed engineering at all sites studied. Either the utilities or the A-E handled construction management. Procurements were managed by the A-E
  with utility supervision or by the utility alone. Construction
  arrangements at the sites studied varied greatly, ranging from one
  prime contractor to more than 30. (See Section C.4.)

- Various combinations of engineering, management, procurement, and
  construction roles for owners. Owners generally assumed the role of
  project manager, combining this role with that of construction manager at two sites. In-house engineering tended to be used only for
  A-E oversight or non-safety-related design work. Procurement was

generally a function shared by the owner with the A-E and contractors. Construction at all sites studied was performed by external
firms. (See Sections C.4.1 and C.4.2.)

- Prior nuclear experience. All sites recognized the value of previous
  nuclear experience although only two of the sites visited had such
  experience. In particular, previous experience increased familiarity
  with quality requirements and expectations; improved the selection of
  contractors; and permitted the utility, A-E, or contractor to anticipate the inexperience of others and to take steps to compensate for
  it. (See Section C.4.3.)

- Different Labor arrangements. All of the construction projects
  visited had negotiated broad labor agreements with major craft unions
  and locals before the project began. Given the size, duration, and
  complexity of nuclear projects, unions had sufficient incentive to
  enter into such agreements. Generally, these agreements benefited
  both sides: labor was guaranteed work for a long period of time, and
  the utility/contractors won concessions on job rules, work interrup-
  tions, and walk-outs. (See Section C.4.4.)

Institutional Factors

Site examination of institutional factors yielded the following findings:

- State public utility commission (PUC) policies. At all sites
  visited, state PUC policies were not reported to have been pivotal
  either in the original decision to build a nuclear plant or in later
  project and contracting decisions. (See Section C.5.1.)

- Project ownership arrangements. The sites visited exhibited a range
  of ownership arrangements. No one arrangement appeared superior to
  the others in producing project success. Joint projects did appear
  to offer some advantage for financial stability. (See Section
  C.5.2.)

C.2.3 Project Observations

In the course of project activities, the study team made a number of
observations deserving of mention here as well as of further investigation and
analysis. These observations are as follows:

The nuclear construction industry does not appear to make extensive use of
incentive contracting of either a reward or punitive nature. Therefore,
because most of the contracts are cost reimbursable, virtually all the risk
lies with the utilities. Further examination of incentive contracting might
reveal some particular advantages that would have implications for both the
quality and cost of nuclear power plants. (See Section C.3.2.)

State PUCs, while not particularly active in scrutinizing nuclear power
plant construction costs, appear quite aggressive in their examination of

operating and maintenance costs. It was suggested that this is because better-accepted methodologies exist for evaluating costs associated with operations and maintenance than are available to assess construction expenditures. If this is the case, it would appear worthwhile to explore the development of such a methodology for assessing construction costs. (See Section C.5.1.)

Several commentators have suggested the value of greater consolidation and coordination of the nation's nuclear generating capacity. While initially there do not appear to be legal barriers to such consolidation, neither do there appear to be particular incentives to coordination. Should enhanced coordination be deemed desirable, antitrust and other potential legal issues would require more extensive examination. (See Section C.5.2.)

Finally, while some project ownership arrangements appear to have advantages over others, careful empirical examination of utility and project ownership arrangements and their relationship to construction project outcomes is lacking. Further study could begin to identify some of the relative strengths and weaknesses of different types of arrangements. Through such additional study, it might also be possible to determine the appropriate vehicle for advocating increased coordination, assuming that further investigation offered evidence of its merits. (See Section C.5.2.)

C.2.4 Recommendations

The focus of this project was on case studies of individual sites and their nuclear construction experience. This approach necessarily limits both the ability to generalize the project's findings and the development of recommended actions. With these limitations in mind, the following recommendations are offered:

- As part of their management review, the NRC should consider requiring applicants for construction permits to explain their proposed contracting methods, bid evaluation and selection procedures, and their reasons for choosing them.

  Given the overwhelming consensus about contractor selection processes and cost-reimbursement contracting, this item clearly seems to warrant NRC attention. Utilities are advised to require bidders to demonstrate their approach and commitment to a project, the NRC could demand the same of licensees. This would force the potential licensee to think through the contracting process with all its implications for risk sharing, cost control, and quality performance requirements.

- The NRC should examine methods to focus more attention on how a licensee proposes to insure that quality work is being performed rather than on the documents that describe general QA and QC programs.

An overemphasis on what is written about quality assurance and
quality control appears to contribute little to the actual assurance
of quality and may be detrimental. This is particularly true if such
an emphasis diverts attention from how the elements of QA and QC programs will be implemented. The issue here is the difference between
examining a utility's QA manual and examining the number and qualifications of the staff it assigns to QA functions. The former audits
writing ability; the latter contributes to an assessment of the capacity to carry out a QA objective.

- The NRC should examine the implications for its own mission of state
  public utility commission scrutiny of and policies toward nuclear
  construction project costs and management.

  State PUCs appear to be taking more action in their examination and
  disallowance of unnecessary and unwarranted expenses. How this new
  posture affects execution of the NRC's safety mission, PUCs' expectations of the NRC, and the assurance of quality in nuclear construction projects is not yet clear. This shift represents what may be a
  major change in the institutional environment of nuclear power plant
  construction; thus, the NRC should examine carefully its implications.

C.3 CONTRACTUAL ISSUES

A key aspect of any major construction effort is the contracting and procurement process used. This process defines the scope and level of involvement
for all project participants, establishes their relationships with each other
as well as with the owner, and secures all of the materials, supplies, tools,
and equipment for building the plant.

At each of the study sites, several features of the contracting and procurement process were examined to determine their contribution to assurance of
quality in a nuclear construction project. Three aspects of the contracting
and procurement process were of particular interest: 1) the procedures used to
evaluate and select contractors and vendors; 2) the terms and conditions
(including incentives, if any) of contracts and procurement documents; and
3) the nature and scope of quality-related requirements incorporated in contracts and procurement documents. The project team's findings in each of these
areas are discussed below. A concluding section summarizes the findings.

C.3.1 Bid Evaluation and Selection Procedures

Selecting qualified contractors to perform construction tasks requires
more than careful drafting of contract documents. For this reason, contracting
and procurement guides typically view an objective bid evaluation and selection
process as fundamental to successful contracting (Cibinic and Nash 1981; Federal Procurement Regulations). Such a process has several characteristics. It
is independent of any particular procurement that is undertaken and involves

procedures that are clear and can be readily communicated. Finally, for contractor selection, it uses criteria that are rationally related to the product or service being procured.

Recommended features of contractor and vendor bid evaluation and selection procedures suggested by those interviewed are described in the following discussion. To clarify the discussion, the term "contracting" is used to denote the process of selecting on-site suppliers of labor, expertise, and services. The term "procurement," on the other hand, is used to denote the selection of off-site suppliers/vendors of materials, supplies, and equipment.

Bid Evaluation and Selection Procedures for Project Contractors

The use of bid evaluation and selection procedures to select project contractors at the sites studied did not appear to affect the number of contracts and vendor agreements executed. Two of the sites studied involved utilities that had executed many contracts directly with construction contractors. One of these sites (Site 3) uses a detailed bid selection procedure for all contracting. The other site (Site 1) indicated that it had selected its original contractors because they were "people we were familiar with" from previous, non-nuclear construction efforts. In the last case, lack of contractor experience appeared to play a role in quality-related problems that resulted in a self-imposed work stoppage. Of the other two sites, Site 2 involved a utility that executed one prime contract and relied on the prime contractor to use its own bid evaluation and selection process to subcontract with others performing services for the construction project. However, the utility's selection of the prime contractor was highly formalized. At Site 4, the major A-E firm visited, the staff interviewed gave several examples of unsatisfactory contractor performance that resulted from inappropriate contractor selection procedures. One person noted that "hiring country club cousins" over technically or functionally superior bidders is not conducive to quality construction efforts.

Formalized bid evaluation and selection procedures were more likely to be used if they were expected to improve bid outcomes. Those who adopted such procedures stressed several important characteristics of an effective sound bidder evaluation and selection.

First, criteria must be established for developing a bidders' list that would not only restrict the number of proposals to be reviewed, but also pre-screen prospective bidders. Having such criteria appeared more important than the content of the criteria used. People interviewed at all sites visited gave examples of problems that resulted from unrestricted bidders' lists or improper additions to such lists. One example involved a supplier added inappropriately to a bidder's list. Although the firm was not selected for the procurement, it was shortly revealed that the firm's president had been arrested for drug smuggling.

Those interviewed also felt it was important that bidders demonstrate their expertise to undertake procurement tasks. This expertise could be demonstrated partly by prior experience. Staff at each site also stressed the value

C.12

of indicating the level of commitment of staff and other corporate resources that prospective bidders were willing to devote to the buyer's project.

Another feature of bid evaluation and selection procedures found useful at several of the sites was pre-award meetings with potential contractors. Such meetings typically occur after initial selections have been made from all the bidders. The purpose of these meetings is to provide the owner with a way to judge technically acceptable bidders on the basis of formal presentations of corporate capabilities and commitment to the project and staff to be committed to the project. Because contracting for nuclear construction projects is usually initiated before much of the design work is complete, contractors cannot demonstrate how they would accomplish specific tasks. However, those who used pre-award meetings as part of a bid selection process felt that the meetings made a significant contribution to successful contracting.

A final recommended component of an effective bid evaluation and selection process is establishing post-award meetings with the chosen contractor. The purpose of these meetings is to work out the process to develop job-related procedures, to communicate site-specific work rules, and to develop the details of QA and QC plans. These meetings also provide a way for utility and contractor personnel to build a project team philosophy and approach since they will be interacting regularly. Those who had used this process found both pre- and post-award meetings to be very helpful in establishing positive contracting and project relationships. (Specific benefits of the project team approach are discussed further in Section C.4.2.)

Bid Evaluation and Selection Procedures for Vendors/Suppliers

As noted earlier, the procurement process in this chapter refers to acquiring supplies and materials as opposed to acquiring labor and expertise. All the sites visited use some degree of formalized vendor selection process. In many cases this process is necessary because of the many potential vendors, particularly for non-safety-related or non-Q-class items.

Each site had some type of evaluated supplier listing or supplier review process. The formality and complexity of the process depended in part on the size of the reviewing body. A larger and therefore more commercially active reviewing body normally resulted in a more elaborate supplier selection and evaluation process. At Site 3, for example, a large utility involved in several construction efforts and servicing several operating plants performed the procurement function; there was also a formal vendor selection and review process. The A-E firm visited reflected a similar situation. Because of its size, the firm has developed an extensive vendor evaluation process, combined with quality surveillance procedures (discussed further in Section 5.3.3). While the extent of the process varied, each site attempted to maintain information on vendors and suppliers to make informed commercial and technical decisions in purchasing materials, supplies, and equipment.

The process for preparing purchase orders with suppliers was the same at all sites studied. Technical specifications are usually developed by design and engineering, reviewed by quality assurance, reviewed for commercial aspects by procurement, and then sent out to a pre-established list of acceptable bidders. The same actors or functional groups are involved in the review of bids received. If a supplier takes exception to any of the bid specifications, those points must also be reviewed and resolved before a procurement can take place. Subsequent changes generated by either the vendor or the purchaser are also subjected to technical and administrative review procedures.

The purpose of procedures to review and make changes in procurements is to ensure that necessary changes are made and reviewed in a timely manner so that everyone is working with the most recent and accurate specifications. Site visits indicated that success of this process depended upon the extent to which commercial aspects of a procurement were allowed to affect technical aspects. For example, at Site 1 purchase orders were written in such a way that suppliers understood that they were to respond to the latest technical direction given, with the commercial paperwork to follow. This was to avoid complete renegotiation of a procurement every time a technical change occurred. Interviewees stressed how critical it was for suppliers to be confident that good faith performance, as requested, would be fully compensated. The dictum, "You never want your suppliers to be losing money," was noted time and again.

Depending on the grade of materials being procured, purchase orders generally prescribed the appropriate shop inspection and monitoring provisions, as required by specific codes or standards. Typically, such provisions consisted of notifying vendors that their premises were to be available for inspection by the purchaser or his agent. Sites differed in the extent to which these inspection or monitoring provisions were followed. This is discussed further in Section C.3.3. However, all sites tended to include the same quality-related specifications and provide for the same level of monitoring. They differed in the resources made available to perform shop inspections, audits, and/or other monitoring activities. Again, the larger the entity executing the purchase order, the more likely adequate resources will be available for vendor surveillance activities.

C.3.2 Terms and Conditions of Contractual and Procurement Documents

Contractual documents themselves are generally not as important as the expertise, experience, and attitude of the contracting parties. Nevertheless, such documents often represent the only formal statement of the intended relationships among project participants. The specific terms and conditions of contractual documents may reflect not only the contracting parties' preferred style of interaction, but also the contract writer's preferences and experiences. In the absence of other formalized statements, the contracts set the parameters for project relationships.

Three features of contractual documents were reviewed at the construction sites visited: the primary type of contract being used to undertake construction tasks; some of the contract's general terms and provisions guiding project relationships; and the use of incentive provisions to effect particular project relationships or to achieve specific project goals.

Types of Contracts Used at the Sites Visited

The preferred form of contracting for nuclear power plant construction appears to have gone through several cycles. Johnson et al. (1976) describe contracting changes occurring in the nuclear power industry to date, indicating that initially (early 1950s) power plant construction was executed by cost-type contracts. By the late 1950s to mid-1960s, fixed-price, turn-key projects became more typical. With more stringent licensing requirements imposed and construction costs rising, modified fixed-price contracts with escalation clauses came into prominence by the late 1960s (Johnson et al. 1976).

The projects visited all began construction in the early to mid-1970s. By this time, fixed-price contracting was used only under special circumstances, and cost-reimbursement contracts were the most frequently used type. This preference remains strong today. For example, construction at Site 1 began in the early 1970s, with most contractors working under fixed-price contracts. During the project, several construction deficiencies were discovered, construction was halted, and the type of contracting used was changed completely. Resulting modifications to the original contracts transformed them from fixed-price to cost-reimbursement (Johnson et al. 1976).

Changes in contract preference reflect as much a response to external conditions as to any changes in relationships among utility owners, constructors, or design and engineering firms. Thus, fixed-price contracts are most appropriate under the following conditions: scope and specifications are known in advance; few changes are expected; and/or costs are not expected to fluctuate widely or increase substantially (Business Roundtable 1982). Cost-reimbursement contracts, on the other hand, are more appropriate in the following situations: full project scope is uncertain; changes and modifications are expected during the project; and little exists on which to base a firm fixed-price bid (Cibinic and Nash 1981; Business Roundtable 1982).

Since the mid-70s, cost-reimbursement has clearly been predominant in nuclear power plant construction. The use of the cost-reimbursement contract, then, reflects the industry's response to the situation in which power plant construction begins before the design is complete, inflation results in the expectation of widely fluctuating costs of materials and labor, and regulatory and economic uncertainties make architect-engineers and contractors reluctant to "lock in" fixed-price contracts.

Different forms of contracts also provide different levels of owner involvement with the contractor. The form of contract selected reflects the need of the parties, particularly the owner, to monitor contractor performance (Cibinic and Nash 1981). A fixed-price contract does not typically permit the owner extensive surveillance and monitoring of contractor records and activities. The owner has no basis for monitoring because performance is up to the contractor, within the parameters of the contract. In cost-reimbursement contracts, however, the owner must be involved in monitoring both the schedule and expenditures because the contractor's payment is based on demonstrated expenses of the work performed.

If quality of construction is the owner's primary goal, requiring owner acceptance of all work is one useful approach. Where the owner requires construction according to exact specifications, a contractual arrangement that permits the owner to closely monitor the costs and quality of the construction, i.e., the cost-reimbursement type contract, may be most appropriate (Cavanagh 1974).

The main difference between fixed-price and cost-reimbursement contracting for assurance of quality is not the method used to compensate the contractor, but the risks that each party assumes because of the method of compensation. In the fixed-price contract, the contractor assumes virtually all of the risk. He has made a firm bid and runs the risk of failing to perform as expected within a given budget. In a cost-reimbursement contract, on the other hand, the owner assumes the risk that the cost may exceed the estimate. Either way, the contractor will expect to be reimbursed for any changes the owner chooses to make. But with a fixed-price, lump-sum contract, additional compensation must be negotiated with the owner, or failing this, the contractor must seek legal redress. With cost-reimbursement contracts, the mechanism for providing additional compensation is built in. Responsibilities for risk associated with timing, costs, and quality of performance, therefore, shift back to the owner.

Many persons interviewed at the sites felt that utilities should assume the risks associated with nuclear power plant construction because, as owners, they clearly have the responsibility for the safety and quality of their plants. Therefore, the use of cost-reimbursement contracts seemed consistent with the owner's duties and obligations. The interviewees also preferred the cost-reimbursement contract because of the number of design changes typically involved in a nuclear project. Many of these changes were the result of the evolutionary nature of most projects, with design substantially incomplete at project initiation. Other changes were often the result of new regulatory requirements or guidelines. Because of the many changes, it was felt that these were risks that the owner, as licensee, should assume, rather than passing them on to contractors. Thus, it was viewed as unrealistic to expect contractors to anticipate the risks of unspecified changes by making firm price bids. Instead, the course recommended by interviewees was for licensees to recognize the likelihood of many regulatory and design changes and to plan to reimburse contractors for incorporating those changes as they were executed. It was agreed that this position not only serves time and budget needs but quality objectives as well.

Convertible contracts have been suggested as a way to combine the advantages of both fixed-price and cost-reimbursement contracts (Business Roundtable 1982). This type of contract can be changed once a specified level of project completion is reached. For example, in a nuclear project, such a contract might begin as a cost-reimbursement type when scope and design are still not completely defined. As the project continues and the design and scope of each contractor's area of performance is more clearly defined, cost-reimbursement contracts might be converted to fixed-price, target-price, or unit-price contracts to complete the job. Therefore, the owner would assume the risks initially, but over time, as circumstances change, risks would shift to the contractors in a well-defined and predetermined manner.

Convertible contracts permit some of the advantages of both major methods of contracting. In addition, responsibilities and risks are assumed more evenly by both parties. Ultimately, however, from NRC's point of view, the responsibility for cons ruction and construction quality lies with the licensee. Most licensees have found themselves recently in a situation of considerable uncertainty in nuclear construction projects, emanating not only from the NRC but from the financial markets as well. In this environment, the cost-reimbursement form of contract has proven to be the most flexible.

In general, while a cost-reimbursement contract may be written in several ways, the preferred style at the sites visited was cost-reimbursement with a fixed-fee. This type of cost-reimbursement contract represents an attempt to control costs because the contractor's fee, or profit, is not tied to the size of the underlying contract, but is fixed. With this contract the contractor does not have the incentive to enlarge his contract scope or to engage in extensive rework to increase the amount of the contract and thereby increase his profit. Even where a fixed percentage fee has been negotiated, such fees are not typically large. To some extent, then, the contractor's guaranteed fee or profit is reduced to reflect the owner's assumption of most of the risks associated with project completion. Because of the duration of most nuclear projects, some argue that large fees might be preferable, to avoid loss of contractor interest or commitment over time. At Site 3, this situation has been addressed through periodic renegotiation of contractor fees to reflect current project and external conditions.

Not all contracting at the sites studied is the cost-reimbursement type. Fixed-price contracts continue for specific jobs associated with nuclear power plants. Generally, for these jobs, the scope was known early enough for detailed specifications to be written so that a realistic and firm price bid could be solicited. Such jobs also reflect a situation where it is reasonable for the contractor to assume the risks associated with executing the contract. Fixed-price contracts are frequently used for elements of heating, ventilating, and air conditioning work; and at one project, the cooling tower contracts were fixed price as well.

The General Provisions of Contracts

Several provisions of the cost-reimbursement contracts negotiated at the sites studied relate to quality assurance. Each of the sites visited dealt with some of these provisions in similar ways, although the exact language used differed. For example, each site visited had provisions on the assignment of contractor's key personnel to the project. This type of provision reflects the owner's interest in the assignment of individuals from the contractor's staff (particularly for key management and supervisory positions). The reason for inserting a key personnel clause appeared to be related to unpleasant experiences in prior nuclear projects or construction efforts where contractor personnel had proven difficult to work with or had demonstrated a level of commitment inconsistent with the owner's. In a nuclear construction project, trying to work with problematic contractor personnel was not considered satisfactory. Each project visited therefore included provisions in their contract reserving the right to approve and pass judgment on contractor personnel involved in their projects. At Site 1, the owner not only reserves the right to approve such personnel but also requires the personnel to participate in a series of team-building workshops to build project commitment.

Contractor liability for error, accident, or negligence by their personnel is another provision affecting quality assurance. In general, such liability was limited and was restricted to gross negligence or deliberate actions, not to typical errors or accidents since errors invariably occur. The utility at Site 2, for example, views the contractor's personnel as an extension of its own staff and expects the same level of errors among contractor personnel as among its own staff.

Support for this position is found in a recent court case in Virginia. In the case, a utility tried to avoid corporate liability by shifting responsibility to its employees and their "human error." The court rejected that argument, stating that "human error" was to be expected (Virginia Electric v. Division of Consumer Counsel 1980). Many interviewed at the sites argue further that if contractors are liable for simple errors, there might be a tendency to cover up problems, which could have serious quality-related implications. Therefore, most owners choose to control primarily for gross or deliberate negligence of contractor personnel and to accept simple error or accidental behavior as a normal part of the project.

Other contract provisions relating to quality assurance concerned the scope of contractor costs deemed reimbursable under a cost-reimbursement contract, for example, costs related to training. Site 3 placed particular emphasis on on-site training of craft and supervisory personnel and invested much time and energy in training programs directly related to the project work. At this site all contractor training costs were reimbursable. The utility felt that incurring those costs, up front, increased the likelihood that the work would be done properly. Thus, reimbursement for training costs was argued to yield substantial savings because costly rework would be avoided. In general, the scope of costs the owner is willing to reimburse reflects the importance that owner places on various aspects of the project. By reimbursing training costs, the owner is stressing that the quality of work is more important than cost or the schedule.

Incentive Provisions in Contracts

Site 1, when renegotiating contracts, developed some fairly detailed and profitable incentive provisions concerning productivity and completing certain key milestones on schedule. Of the sites visited, this was the only one, however, in which incentive provisions were extensively used.

Site 2 used negative incentives for contractor performance. The first x-millions of dollars of rework became the contractor's responsibility, after which the owner would begin to assume liability. Thus, there was a limited penalty for mistakes or rework, but only up to a certain amount. These incentives operated somewhat like a deductible in a standard insurance policy. Rework penalties constitute a financial check on unlimited contractor rework and shift some risks back to the contractor. For that reason, they are often suggested as important cost control leverage in cost-reimbursement contracts, where the contractor may not otherwise have incentive to perform efficiently.

An alternative to rework penalties is the use of positive incentives, i.e., rewarding the contractor for avoiding rework. Such provisions might be based on rework avoided over time or on performance of specific tasks. The efficiency of such provisions, however, depends on how well QA and QC programs detect unsatisfactory performance.

To meet an incentive's objective, a contractor must be able to control performance in that area of activity. This may require careful management of other contractors and their relationships with the incentive work. If the contractor cannot control the performance of others, the construction manager must adjust the incentive program. In a construction project with many contractors and a great deal of interdependence, incentive contracts may not be practical.

Another problem with incentive contracting is establishing equitable goals. Often cost-related targets are set because they seem easy to agree to, but negotiating cost-related incentives may send an inappropriate message that could negatively affect performance or other kinds of objectives. Thus, incentive contracting, despite its potential for cost efficiency, may actually result in increased costs because of the additional project management and project administration required to monitor contractor performance.

The performance incentive, a type used at Site 1, may have some value as the power plant nears completion. Such incentives encourage a contractor to surpass particular performance targets by providing a reward for exceeding them. Often such targets relate to overall contractor performance or completion of a particular installation or set of tasks. These incentives can aid in the timely completion of a construction project. This is particularly important if the cost of money is significant and is expected to increase over time and therefore could result in considerable savings.

Incentives typically are not developed for objectives that are essential or required under the general contract provisions. Therefore, quality-linked incentives in nuclear construction are inappropriate because quality is nonnegotiable. Construction is required to be completed according to quality requirements and the design's technical specifications. Other aspects of contractor performance are more appropriate for incentive contracting than quality objectives. Quality-related incentives might be considered if an owner wants a level of performance exceeding the minimum contract standards.

C.3.3 Quality Assurance Provisions in Contracts and Procurement Documents

At the studied sites, several approaches were used to incorporate quality assurance provisions in contracting and procurement documents. The approaches varied mainly in the level of detail that quality requirements were stated in the documents or in materials attached to those documents. In one case, the contract's quality assurance provision stated that all work related to the project was to be accomplished in accordance with applicable NRC and utility QA/QC standards. In another case, actual quality assurance manuals and quality control procedures were incorporated directly as appendices to a contracting document.

At sites where only general statements of quality assurance were contained in contract documents, specific job-related quality requirements and procedures were developed as work progressed. In this way, such requirements developed in the context of the project itself before a set of tasks was begun. This also permitted QA and QC procedures to be tied to the particular technical specifications guiding those tasks.

The logic behind handling quality requirements as work progressed is that incorporating detail into contract documents is too rigid. In detailed contract documents, a QA or QC change would accompany every specification change and requires a contract modification. To avoid these problems, at some sites a contract document containing a statement of basic quality requirements was developed, incorporating by reference whatever document or set of documents (e.g., 10 CFR 50, Appendix B, ANSI standards, or ASME codes) are appropriate for a contractor to consider. The contract therefore gives a general framework of quality expectations, with specific guidance on quality requirements coming from job-related experience that would develop with the project.

The alternative point of view is that quality requirements should be made very clear and explicit in contract documents. This argument led to a basic contract document that refers to appendix materials, including intended quality procedures and quality assurance manuals. In negotiating such contracts, the appendix materials, as well as the body of the contract, have to be negotiated and agreed upon by the parties.

In general, documents for procuring equipment and supplies incorporated the appropriate standard or code provisions, stipulating that, if appropriate, a supplier would be expected to have his premises open for inspection at the discretion of the owner. The specific quality requirements are typically contained in the technical specifications rather than in the general provisions of the procurement documents.

Those interviewed generally felt that the written detail of the quality requirements or their method and level of communication to a contractor or supplier is not as significant as the extent of QA/QC follow-through. The strong feeling was that it was possible to produce documents that reflect detailed quality control and quality assurance programs that might not ever be implemented. Real assurance of quality, it was argued, comes first from effective people performing the work and then from actual checks being made, materials actually being inspected and audits occurring at a supplier's facility or on a job site. The fact, as well as the expectation, were the most important aspects of a QA program.

The capacity to provide such effective QA/QC programs varies greatly, depending on the parties' resources. An owner may require very specific quality standards of his contractors but not have the capacity to review contractor performance. At the sites visited, most generally understood the requirements needed; for example, nearly every person could recite 10 CFR 50 Appendix B requirements. However, the sites had varying amounts of resources to ensure that such requirements were being met. If an owner doesn't have the resources to follow through on QA/QC requirements, some other mechanism is needed to assure that the follow-through occurs. For example, a project structure and organization that can assure that a QA/QC program actually "lives" may be required. Issues such as project structure and organizational arrangements are discussed in Section C.4.

C.3.4   Contractual Issues--A Summary

Three aspects of the assurance of quality in the contracting and procurement process were of particular interest during the site visits: 1) the procedures used to evaluate and select contractors and vendors; 2) the terms and conditions (including incentives, if any) of contracting and procurement documents; and 3) the nature and scope of the quality-related requirements incorporated in contracts and procurement documents.

An effective bid evaluation and selection process is fundamental to successful contracting. Those interviewed stressed the importance of developing a bidders' list based on established technical criteria. The exact criteria were regarded as less important than their implementation. It was felt that bidders should be required to demonstrate their expertise, both by prior experience and by a representation of the bidders' staff and other resources to be devoted to the work. Several of the sites found pre-award meetings with the finalists useful in assuring that the successful bidder would devote the required effort. Post-award meetings were also felt to be very helpful for establishing contracting and project relationships.

A formalized vendor selection process was also thought helpful for procuring equipment and supplies. A large owner/utility or project manager is generally better able to conduct a thorough and formal selection process than a small utility or "one-time" plant builder. This process can be successful whether it is performed by a large utility or by a large or experienced A-E. Interviewees recommended that purchase orders be written to allow suppliers to make rapid technical changes before the commercial paperwork is completed.

The cost-reimbursement type of construction contract was favored by interviewees over the fixed-price, lump-sum contract, because of the large number of changes in work during construction of nuclear generating facilities. Cost-reimbursement contracts typically include a fixed fee, set either as a lump sum or as a percentage of base costs. Some suggested that the advantages of both fixed-price and cost-reimbursement contracting can be realized by "convertible" contracts, which change compensation from cost reimbursement to a fixed price after the design is completed or at some other logical point. Where cost-reimbursement with fixed-fee contracts are used, provisions for renegotiating the "fixed fee" have been useful to avoid loss of contractor interest or commitment and to accommodate changed circumstances on the part of the owner or contractor.

In general, reimbursing contractor expenses like those associated with training indicates that how the work is performed is most important, not its cost. Also, interviewees suggested that limiting the contractor's liability to the owner for errors of key contractor personnel removes much of the incentive that contractor staff may have to cover up discovered error, avoiding even more costly rework and contributing to achieving quality objectives.

Positive incentives for exceeding performance expectations can be useful nd may be most effective when the project is nearly completed and many uncer-ainties have been removed. However, incentive provisions are generally inappropriate when developed around essential or required objectives (such as quality requirements).

The detail of QA provisions in contract and procurement documents has varied from site to site. Generally, however, detail was not as important as the implementation of QA/QC requirements by actual inspections, checks, and tests, which require an adequate commitment of resources.

C.4 ORGANIZATIONAL ISSUES

One aspect of a major construction project that will influence the nature, course, and outcome of that effort is its organizational framework. Therefore, the relationships between the assurance of quality and four organizational issues associated with contracting and procurement were examined: 1) the structure of construction projects; 2) the owner's role in the construction effort; 3) the owner's or utility's prior nuclear experience; and 4) arrangements for labor and labor relations. Each of these is discussed separately below.

C.4.1 Structure of Nuclear Power Plant Construction Efforts

Nuclear construction projects involve four major roles or functions: engineering, procurement, construction, and management. These roles may be performed singly or in combination and by individual or multiple firms. The U.S. nuclear industry demonstrates a wide array of project arrangements, suggesting that no one arrangement will insure success. However, some standard practices appear to reflect the industry's collective experience.

The engineering role is generally filled by a single A-E firm. However, the A-E may contract out parts of the engineering to other firms and maintain overall responsibility. Also, the A-E may or may not be responsible for on-site inspection of work during construction. If the utility has engineering expertise, it may use that expertise to support the outside A-E or may use it to design non-safety-related buildings and facilities. At each project visited, external A-Es had been hired for the design and engineering functions. One project, Site 3, used "in-house" engineering capabilities for non-power block, non-safety-related work.

The engineering role is frequently combined with other project functions. For example, at Site 2, the A-E was also performing project procurement functions and serving as construction manager and prime contractor for most of the construction work (assisted by subcontractors). The A-E firm visited by the project team, Site 4, had served in every conceivable role in nuclear construction projects, sometimes playing a single role and at other times undertaking responsibility for all project functions.

The only project function A-E personnel felt should be closely tied with engineering was procurement. Here, there was a strong feeling that the A-E should either undertake procurement as well or at least have direct input into the procurement process. The reason given for this important linkage was that the A-E knows to what extent the safety margin built into a design might be eroded by suppliers. If the design cannot compensate for deviations in materials, the A-E can advise the procurement functionary of an unacceptable shipment and/or an unsuitable supplier.

As noted above, the A-E at Site 2 was also serving in the project procurement role. This was not true at the other sites, although the A-E had input into procurement in each case. At Site 1, for example, each contractor provided his own materials and supplies, with major equipment purchases handled by the utility with A-E consultation. At Site 3, all procurement was handled by the utility itself, with the A-E providing the specifications and handling the bid evaluation and selection process for many items.

The utility at Site 3 had strong feelings about the procurement role. In the words of one top manager of the nuclear construction effort: "[Suppliers] are notorious for not delivering on schedule. This can bring a construction project to its knees. That's why we knew we wanted to control procurement ourselves." This view is supported by a 1981 study of construction productivity where the unavailability of needed materials, tools, and supplies was found consistently to be the most frequent source of delay on the construction projects studied (Borcherding and Garner 1981).

The number of firms performing the wide range of construction tasks involved in a nuclear project varies greatly. Sites 1 and 3 reflect some of the complexity that may characterize the constructor role. At both these sites, about 30 contractors have been on the project site at the same time, performing a variety of construction functions. At Site 2, on the other hand, a single prime contractor serves as constructor, assisted by its subcontractors and a few limited scope contractors.

Having many contractors on-site not only increases the number of participants in the construction effort but also increases the need for close coordination of contractor efforts. The coordination function is the responsibility of the construction manager, who must assure that the construction effort progresses within time and budget constraints. Where a prime construction contractor has been hired, the construction management role may also be assumed by this firm since most of the project personnel already report to him as sub-contractors. Construction management may also be combined with the engineering function although some argue that in this arrangement no one oversees the A-E's work.

At Sites 1 and 3, the utility is serving actively as construction and pro-ject manager and employs approximately 900 on-site staff to carry out the func-tion. The largest and most experienced utility visited (at Site 3) views the construction management function as one that the owner must play. As one of the utility's key managers noted: "If you intend to build one of these plants, you have to develop as well the expertise to build, license, and manage it." In this view, a utility that can not manage the construction effort probably would not be able to run the completed plant.

An entirely different approach exists at Site 2 where an experienced A-E serves as construction manager and is overseen by the utility. There, an inex-perienced utility, rather than incurring the substantial staff investment for carrying out the construction management role, decided to take advantage of the experience and expertise of its A-E and prime contractor.

Whether performed by the utility itself, an independent construction man-agement firm, or another project participant, the best interests of a construc-tion effort in terms of quality, safety, schedule, and cost are served when the project has clear objectives and the resources to achieve them. The project manager should exert his authority within the project itself, in his absolute power to stop work, to order rework, and to mediate disputes. Personnel at all sites visited expressed frustration at the often lackluster performance of ven-dors and contractors alike unless continued surveillance was maintained. Those most successful in exacting contractor and vendor performance were the largest and most experienced firms. As one key manager at Site 3 noted: "This is all about tomorrow. They [suppliers and contractors] have to satisfy us because we're part of their future."

Where the utility assumes the project management functions, it must commit sufficient staff and resources to effectively direct, coordinate, and support the contractors' work. In addition, because the construction manager must assume a strong role, a weak owner (in terms of staff, economic backing, expertise, or experience) may not be in a position to be an effective construc-tion manager. If the construction manager is not strong, the contractors may not be motivated enough to perform fully on the project, regardless of their contractual obligations. The advantages and disadvantages of the various roles of the owner are discussed in the next subsection.

C.4.2 Owner's Role in the Construction Effort

The extent of utility involvement in each of the job functions just dis-cussed varies markedly. For example, the utility may handle all procurement for the construction project, as did Site 3, or procurement might be managed by the prime contractor with utility concurrence (Site 2). Similarly, the utility may assume little or no responsibility for construction management (Site 1, initially), or the utility may take a very forceful and active role in managing the construction project and its contractors (and subcontractors), as Site 1 does now.

The nature and level of utility involvement in construction projects typically falls into one of five categories that can be thought of as a continuum (Theodore Barry and Associates 1979). First, the "project management" arrangement is one in which the utility hires a project manager to select and coordinate project contractors and activities. In this case, the utility delegates not only the project management function but much of the responsibility for the project to its independent project management firm. This arrangement has been used rarely in nuclear power plant construction.

A second level of owner involvement is the "design-build" arrangement, in which the utility has a minimum of involvement in the project and contracts with one firm to design and build the power plant. This firm ordinarily handles the design, construction, and procurement work for project scheduling or auditing. Site 2 conforms most closely to this model.

A third category of owner involvement is the "general contractor" arrangement in which the utility enters into separate contracts for project engineering and construction. This arrangement gives the utility greater control over the project and requires greater staff involvement and commitment by the utility than either of the first two arrangements.

A fourth type of owner relationship is the "prime specialty contractor" arrangement, characterized by a utility that serves as its own construction manager and general contractor, hiring all its prime contractors for each of the major divisions of work. Sites 1 and 3 are variations of this organizational arrangement.

Finally, in the fifth category, the "in-house construction" arrangement, the utility handles virtually all aspects of the project. In this case, utility staff actually construct, if not design, the project. This arrangement requires the development of a tremendous level of expertise within the utility and a large utility staff.

As this discussion suggests, the level or degree of complexity involved in construction project organization can vary dramatically. For example, a utility that hires one large engineering and construction firm to design and build a nuclear plant ("design-build" arrangement) can be expected to evolve fairly simple contractual and organizational relationships with its contractor, and between the contractor and the subcontractors, vendors, and suppliers with which it works. On the other hand, a project for which the utility contracts directly with each contractor and subcontractor (a "prime speciality contractor" arrangement such as Site 1) is likely to involve a much more complicated project structure.

Regardless of what role(s) the utility assumes, an extremely effective strategy for coordinating the construction effort and for assuring a successful construction project involves establishing a project team. A utility may attempt to interact with its construction project through its traditional departments and sections. However, a strong project team appears to facilitate project relationships and to enhance performance.

The project team concept draws all participants in the construction project (utility, engineers, constructors, etc.) into a single unit focused on the project itself, regardless of each actor's organizational or disciplinary background. Furthermore, the project team approach can support matrix management, which allows team members to benefit from technical supervision and support without diffusing responsibility or authority, which has caused some matrix management arrangements to fail.

All of the sites visited had adopted a project team orientation, and utility staff were able to cite examples of how the team approach facilitated relationships among contractors, the utility, and construction managers. For example, at Site 3, the assignment of procurement staff to the nuclear project appeared to streamline the cost and administrative procedures involved in purchasing. The procurement staff were able to obtain guidance from corporate procurement and simultaneously work closely with the construction project group to assure that the project's procurement needs were being satisfied. The project team can create and maintain cooperation and clear lines of authority, two ingredients that the Electrical Power Research Institute found in a recent study to be important in assuring a successful construction project (Bauman, Morris and Rice 1983).

C..3 Prior Nuclear Experience

The value of experience in the construction of nuclear power plants was frequently mentioned as important by individuals interviewed at all four sites visited. Obviously, experience increases the ability to perform effectively in most any activity.

Individuals at Site 4 suggested that experience of all project participants is particularly important because nuclear power plant construction is unique in several respects. First, it is extremely complex. Construction typically begins well before design work is completed. Also, design specifications change throughout the course of construction--not only in response to the construction work itself, site characteristics, and contractor input, but also because of changing regulatory requirements. A second reason nuclear construction differs from other construction efforts is that the individuals and firms involved in nuclear power plant design, construction, and operation comprise a rather small community characterized by fairly effective channels of communication. Experience in nuclear construction establishes the utility, architect-engineer, construction firm, or other contractors in the nuclear community; facilitates the sharing of expertise and experience; and encourages working effectively with that community and marketplace.

A third aspect that makes nuclear power plant construction unique is the stringency of standards. While other types of construction projects are built to exacting standards, the stringency of their standards may vary from discipline to discipline. In other construction, material substitutions may be liberally permitted and documentation of work may be performed infrequently. None of this is typical of a nuclear plant construction effort, and experience with nuclear projects makes it possible to anticipate and deal with the potential inexperience of other project participants. Utility staff at Site 3, for example, suggested that their prior nuclear experience made it possible for them to effectively use contractors with strong disciplinary capabilities who lacked specific nuclear experience.

Depending on the role the utility assumes in project management, particularly in procurement, the utility's nuclear experience becomes crucial. Some argue that a utility considering construction of its first nuclear plant should rethink its decision (Bauman, Morris and Rice 1983). However, experience indicates that a utility can have a successful first nuclear construction project by hiring experienced, effective staff to serve on a project team and experienced contractors to act on its behalf. For example, combining the project team concept, several strategic hiring decisions (extremely capable and highly experienced individuals), and an experienced A-E/prime contractor has led to a successful first nuclear construction project at Site 2. Although the utility had no prior nuclear experience, project costs and schedules, as well as construction quality, benefited from the experience (and expertise) of the utility's project team and its prime contractor.

C.4.4 Labor Arrangements

The final organizational issue examined at the sites visited involved labor arrangements and relationships at nuclear construction projects. Because the project team visited only union-staffed projects, the team cannot comment on the impact or influence of union vs. non-union vs. open shop arrangements. However, an important insight concerning labor arrangements was gained from the three projects visited: a carefully structured agreement with union organizations is very helpful in controlling project costs and schedules.

An example of this structured agreement is found at Site 2, where a no-strike, no-lock-out agreement was negotiated with unions. Under this agreement, work is disrupted minimally if a dispute occurs, and there is no risk of other union groups slowing or halting work when a dispute arises with one particular union. Accordingly, project costs and schedules have been virtually unaffected by work stoppages or slowdowns. In seven years of construction, only nine days have been lost to labor disputes.

Size and duration of most nuclear construction projects create an advantage in negotiating these structured agreements. Craft unions and locals have a strong incentive to enter into long-term agreements and to abide by project-specific work rules and procedures when they can guarantee large numbers of their members steady work for many years to come.

The complexity of nuclear construction projects may also be an advantage in negotiating labor agreements. At Site 3, for example, the utility has established extensive training programs for craft workers and supervisory

personnel, enabling union members to acquire new skills. While obviously bene-
fiting the quality of the work done on the project, such programs also enhance
the skills and hence future marketability of union members at the utility's
expense.

Finally, these labor agreements give the construction projects stability.
Labor-management negotiations occur before the project begins and therefore
tend to stay outside the construction site itself. All project contractors are
bound by the agreement, assuring equity among classes of workers and similar
work environments and rules for all. These factors can minimize disputes and
prevent the harmful effects of work stoppages and slowdowns.

C.4.5 Organizational Issues--A Summary

From the four organizational issues examined by the project team, several
insights are relevant to the assurance of quality in nuclear power plant construction. First, while the major roles in a nuclear construction project can
be executed through various organizational arrangements, all roles must be
played by strong and effective occupants.

Second, all objectives of a nuclear construction project appear to be
enhanced by the owner's establishing a project team. Normal organizational
departments and channels are not as effective as a project team, with established authority and lines of communication to the construction personnel.

Third, because of differences between nuclear construction and other types
of construction, experience is crucial in building a nuclear power plant. If
the utility/owner lacks experience, others should be hired to act on its
behalf. Experience helps in selecting competent project participants and in
anticipating and dealing effectively with emergent problems. Both these functions make important contributions to project success.

Finally, a nuclear construction projects' size, duration, and complexity
allow adoption of long-term, labor agreements. Such agreements can benefit not
only cost and schedule but the quality of work performed.

C.5 INSTITUTIONAL ISSUES

Nuclear power plant construction projects are defined by the contractual
obligations established as well as by the organizational arrangements that
govern their interactions. At the same time, a larger institutional environment exists in which these projects and their owners are regulated, financed,
and find competition. A full examination of the impact of the institutional
environment surrounding nuclear power plant construction projects was well
beyond the scope of this study. However, at each project visited, two institutional issues were examined: 1) state pubic utility commission (PUC) policies
toward nuclear power plant construction efforts; and 2) various types of ownership arrangements for nuclear power plant construction projects. The reasons
for examining both of these issues and the insights gained from the site interviews, secondary sources, and other project activities are described below.

C.5.1 State PUC Policies Toward Nuclear Construction Efforts

States have almost exclusive responsibility for determining the rates that utilities may charge for the costs of constructing new generating facilities.[a] The state PUCs are responsible for determining when (and whether) costs of new plants are to be passed on to consumers by including such costs in the utility's rate base. If a utility is to regain its investment in a nuclear construction project, then it must be aware of its PUC's policies on construction programs. Similarly, a PUC's policies potentially have significant impact upon the initiation, progress, and completion of a nuclear construction effort.

Rate Base Approaches to Costs of New Plant Construction

Over the years, the PUCs of the various states have developed generally uniform rules concerning when costs, incurred by utilities for constructing new generating facilities, can be passed on to customers as part of the rates charged for power usage. Recently, utilities and PUCs have responded to the economic pressures of nuclear power plant construction by advocating or adopting changes in these rules.

Historically, a plant is not included in a utility's rate base until it is placed in service, i.e., until it becomes "used and usable." Typically, a utility's rate base includes the original or historic cost of bringing the plant into service. Many jurisdictions have allowed some small amount of this cost to be offset by "contributions in the aid of construction." These contributions are non-refundable amounts that utilities have charged customers for installing abnormally costly or extensive facilities, before such plant is placed in service (Howe and Rasmussen 1982). The major offset to the cost of a plant in service in the rate base is accumulated appreciation on the plant.[b]

Construction Work in Progress. Presently, much controversy exists on whether construction work in progress should be included in the rate base. Construction work in progress (CWIP) is the investment in the plant under construction. A recent nationwide survey of privately owned utility companies concluded that, historically, approximately three-fourths of the commissions allowed all or part of CWIP in the rate base, and that currently 27 commissions, approximately half, allow all or some portion of CWIP in the rate base (Edison Electric Institute 1983).[c]

(a) An exception is sales of electricity between utilities (wholesale sales), which are regulated by the Federal Energy Regulatory Commission under the Federal Power Act (FPA) of 1935, 16 U.S.C. §§ 791a et seq. These wholesale sales account for approximately 10% of the "firm" power sales of electricity in the U.S. (U.S. House of Representatives 1983).

(b) There are other offsets to the cost of a new plant, including refundable customer advances for construction, certain deferred income taxes resulting from accelerated depreciation, and pre-1971 income tax credit and customer deposit (Howe and Rasmussen 1982).

(c) The survey covers companies operating in all states except Nebraska, which does not have any investor-owned electric utilities, and the District of Columbia.

The debate regarding CWIP has focused upon several issues: the unpredictability of nuclear plant construction costs; fairness to rate payers; and financial hardship to utilities undertaking large and costly new plant construction (Edison Electric Institute 1983; U.S. House of Representatives 1983). Thus, the Edison Electric Institute (1983), an association of electric companies, concludes that including CWIP in the rate base represents "sound regulatory policy that has been shown to benefit both rate payers and utilities" (p. 1). The House Committee on Energy and Commerce (H.R. 555), on the other hand, has concluded that CWIP is not favorable to consumers and has moved to prohibit the Federal Energy Regulatory Commission (FERC) from allowing CWIP to be included in the rate base of regulated utilities (U.S. House of Representatives 1983). State legislatures, expressing similar sentiments, have also moved to prohibit inclusion of CWIP in the rate base (Edison Electric Institute 1983).

Although nuclear construction projects have figured prominently in the CWIP debate, the focus has been on the total costs of such projects. Little attention has been given to identifying and disaggregating unnecessary or unwarranted costs stemming, for example, from quality assurance breakdowns. Instead, advocates to the debate regard either all costs or no costs as unreasonable.

Construction Cost "Phase In". A similar concept to that of including CWIP in the rate base is that of phasing construction costs into the rate base before a new plant is placed into service. Such "phase-in" plans include paybacks to consumers in the form of lower rate increases after the plant is constructed and in service. Connecticut has revised its law on rate-base treatment of electric plants under construction to allow "phase in" of costs associated with two nuclear generating facilities, Millstone 3 and Seabrook 1, before they are completed.[a] This legislation requires the utilities requesting "phase in" to show that serious financial difficulties are being encountered by the utility or are likely to be encountered unless "phase in" is allowed (State of Connecticut 1983).

Like CWIP, the "phase-in" approach allows the utility to collect funds from rate payers for the new facility before it is placed in service. However, the Connecticut "phase-in" legislation, for example, requires that the funds are to be paid back within the same amount of time after the facility was completed as was allowed for collecting such funds from rate payers before the facility was "used and useful for public service" (State of Connecticut 1983). CWIP differs in that payments made by customers are returned to them over the full useful life of the facility. The 1983 application of Connecticut

---

(a) The New York Public Service Commission is reviewing a similar "rate moderation" plan for the Shoreham nuclear plant (telephone interview with Jack Treilsen, New York Public Service Commission Rate Section, November 8, 1983). In a separate proceeding, the New York Public Service Commission is also reviewing the prudence of utility management decision-making regarding Shoreham.

Light and Power Company to Connecticut's Department of Public Utility Control suggests that consumers be refunded through lower rates within less than three years after Millstone 3 is placed in service (Furland 1983).

As with CWIP, reimbursing a utility for quality-related construction cost overruns has not been the focus of debate in the "phase-in" approach. Instead, the focus has been on the financial condition of individual utilities as it has been affected by nuclear construction projects. Proponents view "phase in" as a reasonable and necessary financial assist to a utility; opponents see such plans as unreasonable. Investigating individual cost items or types of costs has had no place in such a discussion, to date.

Influence of PUC Policies on Sites Visited

Because of the potential impact of a PUC's policies on financing a nuclear project and its effect on the financial integrity of the utility itself, this aspect of the project's environment was examined. At each site visited, utility staff were asked what impact, if any, state PUC policies had on 1) the initial decision to build the plant; 2) subsequent decisions on project organization and progress; and 3) the contracting process generally and, in particular, the requirements placed upon contractors and vendors.

At Sites 1 and 3, a single PUC had rate setting jurisdiction. At Site 2, four state PUCs had rate setting authority. However, none of the interviewees figured PUC policies prominently in their decisionmaking for the nuclear project. Instead, in their initial decisions to build the plants, all sites indicated that the major factor involved assessing projected load requirements and existing capacity.

State PUCs were not reported as significant when major decisions were made during the course of the projects, despite the fact that two of the sites had themselves halted project construction for significant periods. At Site 1, the institutional factor that most influenced the self-imposed work stoppage was the NRC. At Site 3, on the other hand, the project was stopped early to secure adequate financing. Also, none of the sites reported being influenced by PUC policies in their contracting practices and requirements. Rather, the utilities' own contracting styles and preferences as well as various NRC and code requirements appeared to have the greater impact on these project decisions.

Although a PUC's policies have a potential impact, the sites visited appeared to have been little affected by state regulators. This situation may be a function of the particular PUCs and utilities visited, or it may typify the relationship between nuclear construction projects and their PUCs, generally. To examine this, several PUCs were contacted to determine their policies toward the construction phase of nuclear power plants.

PUCs' Historic Position Toward Nuclear Power Plant Construction

Telephone contact was made with 24 PUCs involving states in which currently operating nuclear power plants are located. Each PUC contacted was asked if any of the initial construction costs of operating nuclear plants had been excluded from the rate base and, if so, which costs and why. The study team hoped the survey would reveal the extent that quality-related breakdowns were considered in PUC cost disallowance decisions.

Of the 24 PUCs contacted, only 6 instances were reported where any initial construction costs had been disallowed. This was out of a possible 52 operating nuclear projects reviewed for rate-base treatment. In cases where costs had been disallowed, they were generally small amounts ($1-2 million) when compared with the total cost of most nuclear plants. In addition, construction cost disallowances typically involved special circumstances, e.g., warranty litigation between the utility and major contractor, or the propriety of rate-payer reimbursement for a plant visitors' center. Specific conclusions about breakdowns in project quality assurance, management, or oversight did not appear to be considered in any of the PUCs' decisions to disallow construction costs.

PUC treatment of construction costs appears to differ from their position on CWIP and on operating and maintenance expenses. In these latter areas, several of the same PUCs that had never disallowed construction costs had taken action on CWIP and/or disallowed replacement fuel costs or maintenance expenses. PUC disallowance of operating or maintenance costs appears to be based primarily on its conclusion that utility management had been "imprudent," "improvident," or "unwise." Several of the PUCs indicated that they have not disallowed construction costs because they are not convinced that current methodologies can accurately determine whether costs should be disallowed. This is less true in the operational phase of nuclear plants where PUCs report feeling on firmer ground in reviewing the propriety and prudence of costs incurred.

Recent Developments in PUC Scrutiny of Nuclear Power Plant Construction Projects

The traditional rate-base treatment of nuclear construction costs by PUCs appears to account for their reported significance by the projects visited. However, recent developments in several states suggest that the traditional position of PUCs toward nuclear power plant construction costs is changing and in directions that could significantly impact such projects. Several of these developments are summarized below.

New York's and New Jersey's Incentive Rate of Return Approach

The New York Public Service Commission (hereafter "PSC") has adopted an innovative approach to including one nuclear plant's construction costs in the rate base of participating utilities. This approach is known as the Incentive Rate of Return (IROR). The New Jersey Board of Public Utilities has adopted a similar plan, known as the "Incentive/Penalty Revenue Requirement Adjustment Plan" (New Jersey Board of Public Utilities 1983).

The New York Commission's adoption of IROR resulted from its decision on an inquiry into the cost implications of continued construction of the Nine Mile Point No. 2 nuclear station. New York's inquiry was initiated in response to public and PSC staff concerns regarding repeated increases in construction cost estimates and repeated extensions of the estimated completion time of the station. The New Jersey plan was proposed by the participating utilities to ensure continued funding for the project.

The New York PSC adopted IROR from among several options under consideration. Another option included a shutdown of the project to prevent alleged severe financial and economic implications for both the participating utilities and the affected rate payers. The PSC found that continuing with construction, which had been under way for some time, was the best option, but only if there was IROR to provide some assurance to rate payers that construction costs would not continue to escalate and that the completion schedule would not continue to slip.

The New York plan provides an incentive for the co-tenant utilities to complete construction on or before the scheduled date and at a cost which is at or below the PSC's target cost. Similarly, a disincentive exists for exceeding the "target cost" set.[a] If the completion cost exceeds the target cost, only 80% of such excess costs may be included in the co-tenants' rate base. The remaining 20% will not be passed on to consumers. There is an incentive for completing the station at less than the target cost, since 20% of any cost underrun from such target costs will be allowed into the rate base under the IROR plan (State of New York 1982).

New York's IROR approach allows for the target cost to be modified upward or downward upon request, given "extraordinary events" (State of New York 1982). The PSC also limited any IROR-induced reduction in the return on common equity, applicable to prospective investments in the station, to no more than one-half the normal rate of return (State of New York 1982).

Two of the seven Public Service Commissioners dissented, arguing that the target cost was set too high, that application of the 20% constant sharing factor allocated too great a risk to rate payers because it failed to take into account different tax treatments afforded investors. The dissenters pointed out that several events besides increased construction costs could render the station uneconomic and contended that no meaningful risk sharing could result under the plan unless the "extraordinary events" under which target costs could be modified were clearly delineated (State of New York 1982).

Standard and Poor's, commenting during PSC proceedings, stated that IROR "could have the harmful effect of weakening investor confidence in these utilities and subjecting them to risks, over and above those inherent in the heavy nuclear construction program, particularly since the NRC looms as an immense, exogenous variable" (Standard and Poor's Corporation 1983, p. 20). The

------

(a) The PSC set the "target cost" following public hearings and contested case proceedings before an administrative law judge. The figure was arrived at following review of time and cost estimates submitted by the co-tenants, by PSC staff, and by an independent consulting firm retained at the PSC's request and expense.

dissenters argued that such a "risk premium" would in turn lead to higher rates. On the other hand, both the New York and New Jersey decisions conclude that these plans would not adversely affect the utilities' ability to attract outside financing at reasonable rates.

The New York dissenters also argued that, by including only capital and interest costs in the target cost, the IROR approach would provide incentives for the co-tenants to cut necessary capital expenditures with resultant higher operation and maintenance expenses. Finally, the New York dissenters contended that IROR was legally questionable on two issues. First, the IROR purposed to bind future commissions that would determine the actual rate base of the co-tenants when construction is completed and secondly, the incentives/disincentives depart from what has legally been considered in the past to be a just and reasonable rate of return.

At present, whether New York's IROR has affected the quality of construction at Nine Mile Point No. 2 positively or negatively is not known. The New Jersey decision is also too recent to have produced any discernable effect upon construction quality at Hope Creek 1. Some have argued that IROR plans have the potential to negatively affect construction quality because they place special emphasis on time and costs rather than on quality considerations. Others welcome the scrutiny such plans introduce to nuclear project construction costs. In any case, the adoption of such plans reflects a more proactive PUC position than has been the case historically.

Other PUC Decisions on Nuclear Construction Costs. The Ohio and California PUCs also have recently taken action on the construction costs of nuclear plants under their jurisdiction. In November, 1982, the Ohio PUC decided that only 25% of construction costs associated with the Zimmer plant should be included in the rate base under Ohio's CWIP allowance (State of Ohio 1982). The plant was 75% complete when the order was issued and was expected to be in service by 1975 at a cost of $235 million. The total cost is now expected to be $1.7 billion and a start-up date is still uncertain.

During 1981, the Ohio commission continued to permit inclusion of Zimmer in the rate base despite an NRC report that a widespread breakdown in implementing the Zimmer QA program had occurred. The plant had been included based on assurances that no more breakdowns would occur. In this proceeding, the owners argued that because the plant was 75% complete, the plant should be included on the basis of a state statute that allowed costs to be entered into the rate base when the plant was at least 75% complete. The PUC, however, exercised its discretion to include only 25% of the total cost associated with Zimmer in the CWIP allowance because the plant would not be providing service as soon as was expected in previous proceedings, wherein CWIP allowances were set at higher levels (State of Ohio 1982). The commission denied a request by a consumer group for a management audit of construction of Zimmer but did not bar the possibility of a future audit.

The California PUC recently allowed only a very limited rate increase for the San Onofre 2 nuclear plant. The PUC is planning a lengthy investigation of the reasonableness of construction costs. Unit 2 has been on-line for some time and is running close to full capacity. A spokesman for the PUC stated, "...for the time being rate increases related to plant costs will be limited to match decreases in rates associated with fuel savings produced by the plant. Rate increases for plant costs that are beyond fuel savings will be held in abeyance pending review of the prudency of construction costs." The PUC will also review San Onofre Unit 3, which is due on-line shortly. (Energy Daily 1983).

These developments reflect once again a new PUC position of active involvement in investigating the prudence of management decisions made during the construction of nuclear power plants. The effects that these and other PUC decisions may have on the quality of projects currently under construction, however, are unknown.

The Relationship of the NRC to State Regulation of Nuclear Construction Projects. PUC positions on nuclear construction projects was examined partly to determine the extent to which the NRC was sharing, or could expect to share, responsibility for construction quality assurance with state regulators. Some recent PUC action and subsequent litigation in Florida may indicate the limits/possibilities of shared federal/state regulatory action.

The case involved a forced outage at Florida Power Corporation's Crystal River 3 plant. The issue was whether planning and supervision of a work activity involving the use of a test weight device was deficient. The PUC first ruled that the planning and supervision of the project was inadequate and that Florida Power Corporation must bear the responsibility for the replacement fuel costs. The PUC found that 55 days of the forced outage were attributable to a dropped test weight, which corresponded with replacement fuel costs of $11,056,000, plus interest. Florida Power Corporation then appealed the decision to the Florida Supreme Court.

The Supreme Court reversed the PUC's decision and remanded the case to the PUC for reconsideration. The court stated that the PUC had relied excessively on an NRC notice of violation and a Nuclear General Review Committee (NGRC) report. The court reasoned that use of the documents was analogous to using evidence of subsequent repairs and design modifications to show that the original design was faulty. The court independently reviewed the record and held that the test weight incident was not, per se, safety-related. The court further ruled that the NRC and NGRC reports were issued after the incident, and hindsight should not be the basis for the PUC's decision.

On remand, the PUC re-examined the entire record and decided that an independent basis for disallowing the costs did exist. The PUC ruled that they could rely on the NRC and NGRC reports as secondary sources of information for their conclusion. The PUC's review states that the basis for finding imprudent management was that Florida Power Corporation lacked a formal plan or written firm directives specifying procedure in this type of situation. Additionally, supervision of the work activity was lacking by management. Therefore, whether or not the work should have been considered "safety-related", the PUC ruled that the work was not adequately planned or supervised (State of Florida Public Service Commission 1983).

The PUC stated that the Florida Supreme Court exceeded the normal scope of review, that is, whether substantial evidence supported the PUC's finding. Instead, the court found it necessary to reweigh the evidence and conclude that both the PUC and the NRC were wrong that repair work was safety-related. In its defense, the PUC cited other states that have disallowed costs that were over and above the costs of efficient operations. These states included Arkansas, New York, Iowa, and Virginia.[a] In addition to these states, Ohio has also disallowed operating expenses. According to the Ohio PUC, several million dollars of every rate increase is disallowed on the Davis Besse plant because of its poor operating history.

The Florida litigation suggests that while PUCs may be willing to follow the NRC's lead and undertake special scrutiny of a utility where the NRC has found problems, state courts may not view such a relationship favorably. Thus, while there may appear to be a basis for parallel or complementary actions on the part of PUCs and the NRC with respect to the quality of nuclear plant construction projects, this may only develop to the extent that their respective missions are viewed as complimentary.

Recent State PUC Actions--An Overview

PUCs, such as those in Ohio and Florida, have actively investigated the prudence of management decisions. Ohio investigated a plant's management during the construction phase; and the Florida PUC investigated management decisions during operation of the plant. Although only a few state PUCs have disallowed costs incurred during construction, several other states have, or are considering, disallowing imprudently incurred operating expenses.

Two states, New York and New Jersey, have adopted a wholesale approach to reviewing construction costs. The IROR approach, which does not involve active examination of individual construction management decisions, affords some protection to consumers from further rate increases. Both New York's and New Jersey's PUCs state that their approaches do not involve relinquishing the Commission's authority to review and disallow imprudently incurred construction costs when the plant is completed (New Jersey Board of Public Utilities 1983; State of New York 1982).

The judicial system has become more involved in examining PUC decisions to disallow costs arising from imprudent management decisions. The Florida Supreme Court, for example, is examining more carefully PUC decisions that may result in disallowing costs. The court is looking at whether the experts' data, i.e., the NRC notice of violations and the NGRC report, are sufficient basis for a PUC ruling to deny recoupment of costs incurred by the utility. In

---

(a) Florida Power Corporation has appealed and this case is, again, before the Florida Supreme Court.

one case, the Florida Supreme Court distinguished between the NRC's and the PUCs' primary function. In the Court's view, the NRC's primary function relates to safety. A PUC judgment on the prudence of management decisions must rely on information directly related to such a decision. NRC safety violation reports were not viewed as an appropriate basis for such a PUC decision.

C.5.2 Project Ownership Arrangements

The second institutional issue examined at the construction sites visited was the project ownership arrangement. This issue was examined to determine what impact the ownership arrangement had on the construction effort and, in particular, what benefits certain ownership configurations might have for assuring quality in construction projects.

Of the three nuclear projects visited, Site 1 is joint-venture owned and financed primarily by a small private utility, with participation of a small, rural electric cooperative. Site 2 is a joint venture involving five separate utilities (four investor-owned and one public cooperative) in four states. Site 3 is also a joint venture of four utilities (one investor-owned and three public utilities). This project is dominated by the largest owner, the investor-owned utility, which owns more than 50% of the project. In addition, the investor-owned utility at Site 3 is the subsidiary of a larger holding company, introducing further ownership variety.

Because of the differences among the sites, the benefits of various project ownership arrangements could be examined from the point of view of those interviewed. However, the effects of such arrangements on construction quality could not be assessed objectively. Nevertheless, changes in ownership arrangements, particularly those resulting in enhanced coordination, have been generally regarded as positive developments for the nuclear industry [Jaskow and MacAvoy 1975; International Energy Associates Limited (IEAL) 1979].

Because various ownership arrangements are used in the U.S. nuclear industry, the current arrangements and the statutory and regulatory parameters that shape them were examined. This examination was aided by insights gained at the sites visited.

Current Ownership Arrangements in the Nuclear Industry

The three basic types of electric utility ownership in the United States are investor-owned, government-owned, and cooperative. Investor-owned utilities comprise about 84% of the nation's generating capacity and annual electric power production. Government-owned utilities comprise 13.6% of the U.S. generating capacity, of which municipalities are the most frequently encountered public owners (IEAL 1979). Cooperatives generate comparatively little of the nation's electric power (3%), and only one currently operates a nuclear plant (Osborn et al. 1983).[a]

(a) According to this study, several cooperatives own shares in investor-owned nuclear projects. For example, approximately 40 cooperatives own shares in the financially troubled, publicly owned Washington Public Power Supply System nuclear projects.

Investor-owned utilities vary in size and are organized in several different ways. Some utilities are owned directly by shareholders and some are subsidiaries of holding companies. These holding company arrangements also differ. For example, some parent companies sell power to the public, while others have no such role (with the subsidiary utility handling the sale of power). The largest nuclear generating system in the country, Commonwealth Edison, owns 7 units, comprising 10% of the nation's nuclear generating capacity. Of the 69 U.S. nuclear power plants operating in June 1979, 60 were investor-owned (IEAL 1979).

Within this environment that supports a variety of ownership arrangements, utilities appear to have recognized the importance of coordinating the planning and operation of generating facilities, as well as other facilities, to achieve, for example, more rational investment planning and to minimize dislocations caused by power outages (Breyer and MacAvoy 1973). Observers of the nuclear industry also have noted the potential advantages of increased coordination (IEAL 1979; Gilinsky 1983).

Reviews of relevant statutes and regulations, literature, and information collected during the site visits suggest that coordination is not precluded by existing legal or economic considerations. Neither, however, are there clear incentives (particularly within statutory and regulatory frameworks) for increased coordination. Therefore, despite the possible advantages, increases in coordination are not expected without a compelling impetus, such as might be provided by new legislation. Whether such an impetus is appropriate, however, remains something of an open question.

The Statutory and Regulatory Parameters of Project Ownership Arrangements

Several state and federal laws and regulatory agencies affect the financial and/or ownership arrangements of utilities with nuclear generating projects. Most of the laws and the agencies charged with enforcing these laws are concerned with utilities generally, although certain procedures for enforcing antitrust laws are found in the Atomic Energy Act and are specifically related to the licensing of nuclear power generating facilities.

The number and variety of existing ownership arrangements suggest that these laws and regulations have not prevented formation of varied, viable ownership arrangements for nuclear power plant construction. None of the construction sites visited mentioned the existence of these laws and regulations as a significant obstacle to the project's success. Two sites visited (Sites 2 and 3) indicated that antitrust concerns had been a factor encouraging joint project participation. Nevertheless, while the Federal Power Act provides the Federal Energy Regulatory Commission (FERC) with authority to increase the amount of coordination and efficiency in the industry, this authority has not been broad enough to force changes in project planning and management by individual utilities. Thus, while no insurmountable obstacles to coordination are present, there are also no real incentives in the legal and regulatory system for increased cooperation in project planning and management.

In some foreign countries, nuclear plant ownership and regulatory arrangements differ dramatically from those in the U.S. For example, France has standardized nuclear power plants and just one operating company (Gilinsky 1983). Plants in the Japanese nuclear industry, also highly centralized, experience fewer automatic scrams than do U.S. plants (Dircks 1983). Nevertheless, foreign practices are not detailed here because they involve major changes in industry structure that are generally considered unlikely to be implemented in the U.S. (IEAL 1979; Johnson et al. 1976; Osborn et al. 1983).

The Effect of Federal Antitrust Laws on Project Ownership Arrangements. Through Section 105 of the Atomic Energy Act the NRC is charged with three forms of responsibility for enforcing federal antitrust laws.[a] The NRC must enforce antitrust judgments reached elsewhere, report any apparent antitrust law violations to the Attorney General, and follow the procedure outlined in Section 105 of the Act to solicit the views of the Attorney General on possible antitrust implications of a construction permit application.

The antitrust provisions of Section 105 have been cited as a source of costly delay in the licensing of new nuclear power generating facilities (IEAL 1979). The vast majority of antitrust reviews under Section 105 have resulted in agreements among the utility or utilities, the Department of Justice, and the NRC staff for resolving antitrust concerns, usually in the form of license conditions (Johnson et al. 1976).

Historically, where the NRC's licensing reviews have involved antitrust concerns, the issue has been access to the generating capacity of the plant, rather than the procurement of the design, construction, or supply of components for nuclear plants. Thus, license conditions that have arisen because of antitrust concerns have been grouped into the following four categories (Johnson et al. 1976):

1. Unit Access - involves arrangements for outside utilities to use a nuclear facility.

2. Transmission Services - involves agreements about services to be provided by the applicant to facilitate access.

3. Coordination - involves requirements for such things as emergency and scheduled maintenance support and participation in joint planning and development.

4. Contractual Provisions - involves requirements that the applicant delete discriminatory or restrictive conditions from its contracts, including restrictions on interconnections and coordination agreements, power pool membership, and use or resale of power.

Johnson et al. (1976), authors of this categorization, suggest that if, for example, breeder plants were to be clustered in "nuclear energy centers" resulting in much greater generating capacity than the nuclear plants currently being constructed, special antitrust problems could arise. However, they further suggest that "licensing conditions could probably be worked out to assure equitable access by smaller utilities" (Johnson et al. 1976, p. 51). Nevertheless, Johnson et al. (1976) speculate that such a clustered development might lead to a more complex and time-consuming antitrust review process than that experienced today by utilities with single plant proposals. It seems certain that antitrust laws would have to be addressed by any legislation or initiative providing the impetus for increased coordination in the nuclear industry.

The Effects of Other Federal Legislation. In addition to the NRC, the Federal Energy Regulatory Commission (FERC) and the Securities and Exchange Commission (SEC) play roles in regulating public utility ownership arrangements. The SEC enforces the Public Utility Holding Company Act of 1935. This legislation led to the breakup of several corporate empires that held diverse utility assets in widely separated states and that had been effectively outside the control of state public utility commissions (Breyer and MacAvoy 1973).

It has been suggested that the Public Utility Holding Company Act and SEC review have impeded mergers of public utilities through stock acquisition (Breyer and MacAvoy 1973). However, the thrust of governmental policy appears to be in favor of pooling among individual utilities, not mergers of utility ownership (Breyer and MacAvoy). Furthermore, the Act specifically encourages mergers within the utility industry which would rationalize the production and generation of electricity (Breyer and MacAvoy 1973; 15 U.S.C. 79z-4).

The FERC, formerly the Federal Power Commission, is authorized by Title II of the Federal Power Act (15 U.S.C. 791 et seq.) to "divide the country into regional districts for the voluntary interconnection and coordination of facilities for the generation, transmission and sale of electric energy" [16 U.S.C. 824(a)]. Under its power to regulate interstate commerce, the FERC has asserted jurisdiction over nearly all U.S. generating and transmitting electric facilities because of the existing degree of interconnection among facilities (Breyer and MacAvoy 1973).

The immediate practical effect of such jurisdiction is that companies, including those that are primarily engaged in intrastate commerce, must now obtain FERC approval before entering into mergers and certain security transactions, submit information that the FERC requests, and subject interstate wholesale electricity rates to supervision by the FERC (Breyer and MacAvoy 1973). Although the FERC has authority over almost all utilities, its efforts have been to promote voluntary interconnection within the industry, rather than to compel interconnection or to seek additional legislative authority for compulsory pooling, interconnection, and planning of future generating projects.

(a) The Sherman and Clayton antitrust laws are made specifically applicable to licensees by Section 105, 42 U.S.C. §2135.

Site 3 is a project undertaken by a subsidiary of a large parent firm that is registered as a holding company under the Public Utility Holding Company Act of 1935. This site is an example of a utility and a parent firm that are subject to regulation by the SEC under the Act. Also, certain aspects of the holding company and subsidiary's operation are subject to regulation by the FERC under the Federal Power Act, as discussed above. The other sites visited are similarly subject to FERC regulation. In addition, virtually all significant utilities in this country are subject to state regulation of wholesale or retail rates charged for power.

Insights into Ownership and Management Arrangements: The Effects of Size and Market Power

The construction project management arrangements as well as the utilities' size and nuclear experience varied at the three sites (see Section C.4). Site 1 was being constructed by a relatively small utility and a rural electric cooperative with no prior nuclear experience. Site 2 was being undertaken jointly by several investor-owned and public utilities, also without prior nuclear experience. Site 3 was also a joint project undertaken by several small public utilities and the subsidiary of a large parent firm, registered as a holding company under the Public Utilities Holding Company Act of 1935. This subsidiary (and another subsidiary of the parent firm) had previous nuclear experience.

The site visits suggested that these project ownership arrangements are feasible. However, the sample of sites was small and it is not practical to draw conclusions concerning different ownership and management arrangements. For example, comparisons among investor-owned, government-owned, and cooperative ownership arrangements cannot be made since the site visits were restricted to investor-owned utilities or dominated by such entities.

The site visits did suggest that the presence of the utility or its agent in the marketplace can impact the project. The subsidiary utility acting as the agent for the owners' group at Site 3, closely linked to a large holding company, was in a position to effectively negotiate with contractors, suppliers, and vendors for the goods and services necessary to a successful, high-quality project. The advantage of this association with a major parent company is, in the words of one utility executive, "all about tomorrow." The holding company and the utility are not only contracting for a nuclear project today, they will also be contracting for construction and maintenance projects for years to come. Furthermore, the utility's position is supported by prior experience in nuclear construction projects, providing familiarity with the marketplace and increased knowledge and expertise that can benefit the project. Procurement and contracting are thereby facilitated, as is the expertise necessary to secure satisfactory performance on the procurements and contracts.

The joint owners at Site 2, without prior nuclear experience, took a different approach to meet its goals in the marketplace. They established a major A-E firm as their agent. The A-E, with its well-established systems for evaluating and auditing suppliers' and contractors' bids, products, and performance, has its own considerable market presence. Although the owners lacked nuclear experience, they did not suffer from unfamiliarity in the marketplace. Rather, they used the A-E's experience to the benefit of their project.

Size also is an issue in determining the economic viability of a particular construction project. A small utility beginning a necessarily complex and costly nuclear project can find that the costs and investment in the construction project far exceed the utility's net assets. For this reason, economic decisions within holding company systems may be typically made by the holding company, considering the overall system rather than the operation of the particular subsidiary involved in the project (Breyer and MacAvoy 1973; Osborn et al. 1983).

One of the reasons that most nuclear generating capacity in the United States is owned by investor-owned utilities may be that many investor-owned utilities are larger than government-owned or cooperative companies. Therefore, the investor-owned utilities may have resources that other utility companies lack to invest in nuclear projects (Osborn et al. 1973). Joint ventures and holding companies may also provide necessary support and back-up for nuclear projects, as at Sites 2 and 3 (Osborn et al. 1973). Thus, pooling resources may represent one vehicle for increasing coordination within the industry and for enabling initiation and continuation of a nuclear project that might otherwise be fiscally, if not managerially, impossible. However, for managing the project a joint venture requires an effective arrangement that avoids the difficulties often linked to management by committee (Breyer and MacAvoy 1973).

C.5.3 Institutional Issues--A Summary

Nuclear power plant construction projects are affected to some extent by the larger institutional environments in which these projects and their owners are regulated, financed, and compete. Two aspects of this institutional environment were examined at the sites visited: 1) state PUC policies toward nuclear power construction efforts; and 2) various types of ownership arrangements of nuclear power plant construction projects.

The utilities visited indicated that possible PUC disallowance of construction costs associated with quality problems has not been a significant consideration in utility decisionmaking. This attitude reflects the fact that in the past PUCs have been relatively uncritical of new plant construction costs proposed for inclusion in the rate base. However, recent activity by certain PUCs, such as those of New York, Florida, and California, creates a potential for a significant deterrent to a laissez-faire owner attitude toward contractors of new generating facilities.

While this potential trend may or may not result in better utility management of quality-related construction problems at nuclear projects, such rate scrutiny by PUCs can seriously affect the financial health of utilities, as is true of rate regulation policies generally. The negative side of the trend toward PUC disallowance of quality-related construction cost overruns, then, is that it may increase the risk of undertaking and completing nuclear stations to the point utilities may find otherwise justified power generating projects to be uneconomical. Thus, the impact of a more active PUC posture toward nuclear construction efforts remains unclear. Further examination of state regulatory policy on the quality of nuclear construction projects, and of the NRC's relationship to that policy, is needed.

On the second issue examined, nuclear generating facilities being built or in operation in the United States today reflect a wide variety of plant ownership arrangements also found in the electric utility industry generally. Statutory and regulatory parameters shaping project ownership arrangements in the U.S. include federal and state antitrust laws, the Federal Power Act, and the Public Utility Holding Company Act. These parameters do not appear to have prevented the development of a great variety of project ownership and management arrangements, nor are they likely to prevent further efforts at coordination in the industry. However, antitrust laws could delay formation of more consolidated ownership arrangements in the U.S. Furthermore, positive regulatory incentives for further coordination or consolidation within the industry appear to be lacking.

Increased coordination may be desirable in ownership arrangements. While some individuals have reviewed utility ownership arrangements and project management issues (Breyer and MacAvoy 1973), careful empirical examination of many aspects of utility and project ownership arrangements and their relationship to project outcomes is lacking (Osborn et al. 1983). The limited site work undertaken here, when combined with additional site work at government-owned or cooperative utility companies, or the study of different construction management arrangements (such as one undertaken by a single, large utility company), could begin to identify some of the relative strengths and weaknesses of different types of ownership and management arrangements. Through such additional study, it might also be possible to determine the appropriate vehicle for advocating increased coordination within the industry, assuming that additional investigation offered further evidence of the merits of coordination.

C.6 REFERENCES

Atomic Energy Act. U.S. Code. Title 42, Section 2135.

Bauman, D. S., P. A. Morris and T. R. Rice. 1983. An Analysis of Power Plant Construction Lead Times, Vol. 1: Analysis and Results. Prepared by Applied Decision Analysis, Inc. for Electric Power Research Institute, Palo Alto, California.

Borcherding, J. D. and D. F. Garner. 1981. "Work Force Motivation and Productivity on Large Jobs." Journal of the Construction Division, Proceedings of the American Society of Civil Engineers. 107(C03):443-453.

Breyer, S. and P. W. MacAvoy. 1973. "The Federal Power Commission and the Coordination Problem in the Electrical Power Industry." 46 So. Cal. L. Rev. 661.

Cavanagh, J. B. 1974. "Risks and Remedies--The Nature and Extent of Liability." In Risks and Remedies in Government Contracting. Proceeding of a Briefing Conference held in Los Angeles (American Bar Association).

Cibinic, J., Jr., and R. C. Nash, Jr. 1981. Cost Reimbursement Contracting. George Washington University, Washington, D.C.

Connecticut State. June 1983. Public Act No. 83-239.

Edison Electric Institute. 1983. State Commission Policy on Construction Work in Progress in Rate Base. Edison Electric Institute, Washington, D.C.

Federal Power Act, U.S. Code. Title 16, Sections 791a et seq. (1935).

Federal Procurement Regulations, 41 C.F.R. Chapter 1, Parts 1-2 and 1-3.

Florida Power Corporation v. Public Service Commission, 424 So. 2d 745 (1982).

Furland, E. J. July 1983. Testimony on behalf of applicant. Docket No. 83-07-15.

Gilinsky, V. May 25, 1983. Statement of Commissioner Victor Gilinsky before the Committee on Environment and Public Works, Subcommittee on Nuclear Regulation on Nuclear Licensing Reform.

House Committee on Energy and Commerce. H.R. 555, 98th Congress, 1st Session.

Howe, K. M. and E. F. Rasmussen. 1982. Public Utility Economics and Finance. Prentice-Hall, Inglewood Cliffs, New Jersey.

International Energy Associates Limited (IEAL). 1979. "Institutional Options for Nuclear Power Generation in the United States." Prepared for the U.S. Department of Energy, Washington, D.C.

Johnson, L. J. et al. 1976. Alternative Institutional Arrangements for Developing and Commercializing Breeder Reactor Technology. Rand Corporation, New York, New York.

Joskow, P. L. and P. W. MacAvoy. May 1975. "Regulation and the Financial Condition of the Electric Power Companies in the 1970s." American Economic Review. 65(2).

New Jersey Board of Public Utilities. August 12, 1983. In the Matter of Utility Construction Plans; Hope Creek Inquiry. Docket No. 8012-914-IPRRA, Decision and Order.

New York State, Public Service Commission. April 16, 1982. Proceeding to inquire into the financial and economic cost implications of constructing the Nine Mile Point No. 2 nuclear station. Opinion No. 82-7, Case 28059.

Ohio State, Public Utilities Commission. November 5, 1982. Opinion and Order

Case Nos. 81-1058-EL-AIR and 82-654-EL-ATA.

Osborn, R. N. et al. 1983. _Organizational Analysis and Safety for Utilities with Nuclear Power Plants._ Vol. II, NUREG/CR-3215, U.S. Nuclear Regulatory Commission, Washington, D.C.

Public Utility Holding Company Act of 1935, _U.S. Code._ Title 15, Section 79z-4.

Standard and Poor's Corporation. March 29, 1983. _Comments on Incentive/Penalty and Risk-Sharing Mechanisms._ Submitted in New York Public Service Commission, Case 28059, Opinion of the Dissent.

State of Florida Public Service Commission. July 13, 1983. In _Re: Investigation into Federal Shutdown of Florida Power Corporation's Crystal River No. 3 Unit._ Order on Demand, Docket No. 780832-EU.

State of New York, Public Service Commission. August 31, 1983. "Staff Requirements--Continuation of Evaluation of Implications of Salem Event." Memo from W. J. Dircks, NRC Executive Directors for Operations. Opinion No. 82-7.

The Business Roundtable. 1982. _Contractural Arrangements--A Construction Industry Cost Effectiveness Project Report._ Report A-7.

The Energy Daily. Wednesday, September 14, 1983. "California PUC Trips Up Utilities Over San Onofre 2 and 3 ..."

Theodore Barry and Associates. 1979. _A Survey of Organizational and Contractual Trends in Power Plant Construction._ Theodore Barry and Associates, Los Angeles, California.

U.S. Code. Title 15, Sections 79 through 79z-6.

U.S. Code. Title 16, Sections 791 et seq.

U.S. Code. Title 16, Section 824(a).

U.S. Code. Title 16, Section 824a(v).

U.S. House of Representatives. 1983. _Construction Work in Progress Policy Act of 1983._ Report No. 98-35, Washington, D.C.

Virginia Electric and Power Company v. Division of Consumer Counsel. 220 VA 930 (1980).

C.7 BIBLIOGRAPHY

C.7.1 Sources Cited

Bauman, D. S., P. A. Morris and T. R. Rice, An Analysis of Power Plant Construction Lead Times, Vol. 1: Analysis and Results, prepared by Applied Decision Analysis, Inc., for Electric Power Research Institute, Report of Research Project 1785-3, published by Electric Power Research Institute, February 1983, at p. 4-6.

Breyer, S., and P. W. MacAvoy, "The Federal Power Commission and the Coordination Problem in the Electrical Power Industry," 46 So. Cal. L. Rev. 661 (1973).

Borcherding, John D., and Douglas F. Garner. "Work Force Motivation and Productivity on Large Jobs," Journal of the Construction Division, Proceedings of the American Society of Civil Engineers, Vol. 107, No. CO3, September 1981, pp. 443-453, at pp. 445-446.

The Business Roundtable. Contractural Arrangements--A Construction Industry Cost Effectiveness Project Report, Report A-7, October 1982, at p. 26.

Cavanagh, John B. "Risks and Remedies--The Nature and Extent of Liability," pp. 20-30 in Risks and Remedies in Government Contracting, proceedings of a Briefing Conference presented by the American Bar Association Public Contracts Section, Federal Bar Association, and National Contract Management Association, in Los Angeles, CA; published by American Bar Association, 1974, at p. 21.

Cibinic, John, Jr., and Ralph C. Nash, Jr. Cost Reimbursement Contracting, Washington, D.C.: George Washington University, 1981, at Chapter 5, pp. 170-217 and pp. 254-273.

Edison Electric Institute, Special Report No. 1, State Commission Policy on Construction Work in Progress in Rate Base, published by Edison Electric Institute, August 5, 1983.

Howe, Keith M, and Eugene F. Rasmussen, Public Utility Economics and Finance, Inglewood Cliffs, NJ: Prentice-Hall, 1982, at Chapters 2 and 3.

International Energy Associates Limited, "Institutional Options for Nuclear Power Generation in the United States," prepared for the U.S. Department of Energy, November 30, 1979 (hereafter cited as IEAL Report), at 41.

Johnson, Leland, J., Edward W. Merrow, Walter S. Baer, and Arthur J. Alexander, Alternative Institutional Arrangements for Developing and Commercializing Breeder Reactor Technology, report prepared under Grant No. OEP 76-01895 from the National Science Foundation, Santa Monica, CA: Rand Corporation, November 1976, at Chapter 5, pp. 61-74.

Joskow Paul L. and Paul W. MacAvoy, "Regulation and the Financial Condition of the Electric Power Companies in the 1970s," American Economic Review, Vol. 65, No. 2, May 1975.

Osborn, R. N., J. Olson, P. E. Sommers, S. D. McLaughlin, M. S. Jackson, M. V.

Nadel, W. G. Scott, P. E. Connor, N. Kerwin and J. K. Kennedy, Jr., "Organizational Analysis and Safety for Utilities with Nuclear Power Plants," Vol. II, NUREG/CR-3215, U.S. Nuclear Regulatory Commission (July, 1983).

Theodore Barry and Associates, A survey of organizational and contractural trends in power plant construction, Theodore Barry and Associates, March 1979, at pp. IV-1.

C.7.2 Sources Relied On But Not Cited

Allen, Bruce T., and Arie Melnik. The Market for Electrical Generating Equipment, East Lansing, MI: Michigan State University, 1973.

Asbury, Joseph G., Ronald O. Mueller, and Jarilaos Stavron. "Rate of Return Regulatory Policy Alternatives and Their Effects," Public Utilities Fortnightly, June 9, 1983, pp. 42-48.

"CWIP in Rate Base: The Financial Needs Standard," Public Utilities Fortnightly, February 18, 1982, pp. 65-69.

Ford Foundation. Energy: The Next Twenty Years, report by a Study Group, Cambridge, MA: Ballinger Publishing Co., 1979.

Ford Foundation. Nuclear Power Issues and Choices, report of the Nuclear Energy Policy Study Group, Cambridge, MA: Ballinger Publishing Co., 1977.

Frout, Robert R. "A Rationale for Preferring Construction Work in Progress in the Rate Base," Public Utilities Fortnightly, May 10, 1979, at pp. 22-26.

Heinmann, Fritz F. "Nuclear Litigation from the Supplier's Perspective," in Nuclear Litigation, New York, NY: Practicing Law Institute, 1979.

Hughes, William R. "Scale Frontiers in Electric Power," in William M. Capron (ed.), Technological Change in Regulated Industries, Washington, D.C.: The Brookings Institute, 1971, at pp. 44-85.

Hunt, Raymond G. "Concepts of Federal Procurement: The Award Fee Approach," Defense Management Journal, Second Quarter 1982, at pp. 8-17.

Johnson, Johnny R. "Construction Work in Progress: Planning for the Rate Case," Public Utilities Fortnightly, August 2, 1979, pp. 15-21.

Kalbe, A. Lawrence. "Inflation-Driven Rate Shocks: The Problem and Possible Solutions," Public Utilities Fortnightly, February 17, 1983, pp. 26-34.

Kelley, Doris. Electric Utility Industry--Nuclear Power Annual Power Plant Review. Merril Lynch, Pierce, Fenner & Smith, Inc., May 1983, pp. 1-36.

Keys, W. Noel. Government Contracts in a Nutshell, St. Paul, MN: West Publishing Co., 1979.

McBride and Wachtel. Government Contracts: Cyclopedic Guide to Law, Administration, and Procedure, 1963, Chapters 19 and 20.

Model Procurement Code for State and Local Governments. The American Bar Association, 1979.

Model Procurement Code for State and Local Governments: Recommended Regs. American Bar Association Section of Public Contract Law, 1980.

Muhs, William F., and David A. Schauer. "State Regulatory Practices with Construction Work in Progress: A Summary," Public Utilities Fortnightly, March 27, 1980, pp. 29-31.

Nash, Ralph C. and Cibinic, John, Jr. Incentive Contracting. Federal Procurement Law, George Washington University, 1966.

National Regulatory Research Institute. Commission-Ordered Management Audits of Gas and Electric Utilities, report prepared for U.S. Department of Energy, July 1979.

National Regulatory Research Institute, The Measurement of Electric Utility Performance, Preliminary Analysis, report prepared for the U.S. Department of Energy, December 1981.

Restrick, John K. "The Nuclear Contract," in Nuclear Litigation, New York, NY: Practicing Law Institute, 1982.

Sartonius, Scott. New Nuclear Plants: Will "Phase-Ins" Be Necessary?, New York, NY: Salomon Brothers, Inc., December 14, 1982.

Sichel, Werner and Thomas G. Gies. Application of Economic Principles in Public Utility Industries, Ann Arbor, MI: University of Michigan, 1981.

Simon, Louis P. "What's Really Wrong with the Nuclear Power Industry: An Operator's Viewpoint," Public Utilities Fortnightly, November 19, 1981, pp. 26-30.

Smith, Bruce A. Technological Innovation in Electric Power Generation 1950-1970, East Lansing, MI: Michigan State University, 1977.

Stevenson, J. D. and F. A. Thomas. Selected Review of Foreign Licensing Practices for Nuclear Power Plants, NUREG/CR-2664, U.S. Nuclear Regulatory Commission, April 1982.

U.S. Department of Energy. The Future of Electric Power in America: Economic Supply for Economic Growth, Report of the Electricity Policy Project, DOE/PE-0045. Washington, D.C., U.S. Department of Energy, June 1983.

York, Stanley and J. Robert Malko. "Utility Diversification: A Regulatory Perspective," Public Utilities Fortnightly, January 6, 1983, pp. 3-8.

APPENDIX D

PROGRAMS OF OTHER AGENCIES, INDUSTRIES, AND FOREIGN COUNTRIES FOR THE ASSURANCE OF QUALITY

PRINCIPAL CONTRIBUTORS

E. W. Brach (a)   P. L. Hendrickson (b)
M. G. Patrick (b)   J. F. Nesbitt (b)
W. J. Apley (b)   B. W. Smith (b)

PREPARED BY

M. G. Patrick

March 1984

Prepared for
Division of Quality Assurance, Safeguards,
  and Inspection Programs
Office of Inspection and Enforcement
U.S. Nuclear Regulatory Commission
Washington, D.C. 20555

Pacific Northwest Laboratory
Richland, Washington 99352

(a) Nuclear Regulatory Commission
(b) Pacific Northwest Laboratory

EXECUTIVE SUMMARY

The purpose of the study addressed in this appendix is to assist in the formulation of the long-term direction of NRC's assurance of quality (AOQ) policies and programs. This study is consistent with the direction provided the NRC by Congress in the FY 1982-83 Authorization Act (Public Law 97-415, Section 13) to study alternatives for improving the quality assurance (QA) and quality control (QC) in the design and construction of nuclear power plants.

This study has three objectives:

• conduct a review of the AOQ programs and practices of the U.S. nuclear power industry; of selected private industries and associated regulatory agencies; and a limited review of the foreign nuclear power industry

• identify some AOQ program aspects and practices of these industries applicable to improving the U.S. nuclear power industry

• determine where changes may be appropriate to improve the NRC AOQ program requirements and practices.

In addition to the NRC program, the AOQ programs of five other U.S. government agencies and of six foreign countries were studied. The following domestic programs were examined:

• the Federal Aviation Administration (FAA) program as applied to the manufacture of large commercial transport aircraft

• the Department of Energy (DOE) program as applied to a government-owned nuclear reactor project, the Fast Flux Test Facility, and a nuclear project for enrichment of uranium, the Gas Centrifuge Enrichment Program

• the National Aeronautics and Space Administration (NASA) program as applied to the aerospace industry

• the U.S. Navy (USN) program for shipbuilding under the Department of Defense

• the Maritime Administration (MarAd) program for commercial shipbuilding under the Department of Transportation.

The foreign nuclear programs examined were those in Canada, the Federal Republic of Germany, France, Japan, Sweden, and the United Kingdom.

The study was conducted by reviewing published information on each of the programs selected for study and supplementing this review with information obtained from interviews with representatives of the FAA, the DOE and the NRC, both at Headquarters and at selected regional offices. Limited interviews were also conducted with the NASA Washington, D.C., staff. Published information and interviews with those in the private sector organizations corresponding with these government agencies were also utilized.

One of the principal investigators in this study also participated in the case studies effort relative to the nuclear power plants; therefore, information obtained from the case studies was also utilized in this study.

The reviews of the foreign nuclear programs were based almost entirely on publicly available information. Subcontractors with experience in the countries of interest conducted these reviews. There were also limited contacts with foreign nationals in developing the necessary information.

The studies of the shipbuilding programs in the United States, both Naval and commercial, were conducted entirely through reviews of publicly available documents.

Several findings from this study are considered worthy of more in-depth study leading to their potential adoption in the NRC's assurance of quality program. These are the following:

- The NRC should consider requiring that plant designs be well advanced prior to initiating construction activities. Design requirements should include the completion of safety, reliability, and availability analyses including failure mode and effect analyses, and fault-tree and hazard analyses. The analyses should be integrated with quality assurance and should be completed prior to the initiation of construction. This recommendation is based upon findings from the DOE, NASA, FAA, foreign nuclear, and shipbuilding programs.

- The NRC should consider requiring establishment of a QA system that prioritizes quality efforts. Systems and components should be assigned to the various priority grades on the basis of the safety, reliability and availability analyses. This recommendation is based upon findings from the DOE, NASA, and shipbuilding programs.

- The NRC program should require "readiness reviews" during nuclear power plant construction. These reviews might involve plant designers, construction managers, owner-operators, and (possibly) NRC staff and should be required at key points in the project beginning with "design ready for construction." It may be useful to have additional reviews at selected key milestone points. This recommendation is based upon findings from the DOE, NASA, and shipbuilding programs.

- The NRC should study ways to better integrate NRC inspection functions with system design reviews, test program reviews, and test program evaluations. This recommendation is based upon findings from the USN, FAA, DOE, and NASA programs.

- Consideration should be given to expanding the NRC's vendor inspection program. The licensee should continue to be held fully responsible for vendor-supplied items. Necessary enforcement actions relevant to vendors could be applied to the licensee. The NRC should consider supporting, perhaps through the Institute of Nuclear Power Operations (INPO), continued development of a data bank on performance of and problems with vendor-supplied components. These data should be analyzed and the results published periodically. This recommendation is based upon findings from the NRC, FAA, and USN programs.

- The NRC should expand its inspector training program to increase emphasis on "how to inspect." Such a program should concentrate on such areas as conducting inspections, use of time, and interpersonal skills and should include specific guidance on identifying possible indicators of developing problems. This recommendation is based upon findings from the USN program.

- The NRC should consider requiring inspections of nuclear power plants by independent inspecting agencies. This recommendation is based upon findings from the foreign nuclear programs.

- The NRC should re-examine its posture on quality assurance to emphasize to the licensees that quality and the assurance of quality are responsibilities of overall management rather than responsibilities of the QA/QC organizations. This recommendation is based upon findings from the DOE program.

ACKNOWLEDGMENTS

The authors wish to express their sincere appreciation to the following individuals and organizations for their help in this study:

- The staffs of NUS, Inc., Gaithersburg, Maryland, and Battelle Institute e.V., Frankfurt, Germany, for their studies of the nuclear programs in foreign countries, and to N. C. Kist and Associates for specific information on the programs in Japan and Sweden.

- The staff of Comex, Inc. for drawing upon its extensive U.S. Navy experience in assisting in the efforts to characterize the shipbuilding programs.

- The staffs of the Federal Aviation Administration (Washington, D.C. and Seattle, Washington); the Department of Energy (Germantown, Maryland; Oak Ridge, Tennessee; Portsmouth, Ohio and Richland, Washington); the Nuclear Regulatory Commission (Bethesda, Maryland, Atlanta, Georgia, Arlington, Texas and Walnut Creek, California); the Boeing Co. (Seattle, Washington); Systems Development Co. (Oak Ridge, Tennessee and Portsmouth, Ohio); Stone and Webster, Inc. (Portsmouth, Ohio); Union Carbide Co. (Oak Ridge, Tennessee); and Westinghouse Hanford Co. (Richland, Washington) for their courteous and extreme cooperation in sharing information about their respective programs.

CONTENTS

TABLES

FIGURES

APPENDIX D

PROGRAMS OF OTHER AGENCIES, INDUSTRIES, AND FOREIGN COUNTRIES FOR THE ASSURANCE OF QUALITY

D.1 INTRODUCTION

This appendix reports the results of a study of the assurance of quality (AOQ) programs of five other U.S. government agencies and of NRC counterparts in six foreign countries. Section D.1 presents introductory material on the study's background, purpose and objectives, and technical approach. Conclusions and findings are presented in Section D.2. Section D.3 gives the significant findings from each program, and Section D.4 summarizes the studies of the domestic and foreign programs. References are provided in Section D.5. Appendix D contains two similar terms, "Assurance of Quality" (AOQ) and "Quality Assurance" (QA). The term "Quality Assurance" has been commonly used in recent years to connote a rather specific, single element in an overall management and/or regulatory process to provide both requisite quality and the assurance that it has been attained. Since this appendix addresses both the QA element and other related elements of these processes, the term "Assurance of Quality" has been used to distinguish between the overall process and the narrower, more specific part represented by "Quality Assurance."

D.1.1 Background

The complexity and extent of problems that have been identified in the past few years at some of the commercial nuclear power plants under construction in the U.S. have caused concern regarding the quality of the design and construction of these plants. Analyses of the experience at problem sites have identified three primary problem areas: 1) failure of the project management team to provide adequate management controls to prevent a significant breakdown in quality from occurring; 2) failure of the owners' quality assurance program to detect the breakdown in a timely manner and to obtain the appropriate corrective action; and 3) failure of the NRC's programs to recognize the true extent and nature of the problems (Dircks 1982).

In response to these problems, the NRC developed several initiatives aimed at bringing about effective improvements in the programs to assure quality. As a part of this overall effort, the NRC initiated a long-term review for continuing evaluation of quality and QA problems related to design, construction, testing and operation of nuclear power plants. Also included in this review is the evaluation of potential solutions to these problems and their impact on the adequacy of QA policies and programs.

D.1.2 Purpose and Objectives

The purpose of the study addressed in this appendix is to assist in the forumulation of the long-term direction of NRC AOQ policies and programs. This study is consistent with the direction provided the NRC by Congress in the

FY 1982-83 Authorization Act (Public Law 97-415, Section 13) to study alternatives for improving the quality assurance and quality control in the design and construction of nuclear power plants.

This study has three objectives:

- conduct a review of the AOQ programs and practices of the U.S. nuclear power industry; of selected private industries and associated regulatory agencies; and a limited review of the foreign nuclear power industry

- identify some AOQ program aspects and practices of these industries applicable to improving the U.S. nuclear power industry.

- determine where changes may be appropriate to improve the NRC AOQ program requirements and practices.

The scope of this study is limited to design-, construction- or fabrication-related activities of the industries and programs selected for review. Follow-on operational activities were not studied.

D.1.3 Technical Approach

At the initiation of this effort, an assessment plan was prepared to provide guidance in carrying out this study. This plan established a methodology for selecting industries to be studied, the content of the various reviews, and the format of the final report.

An important element of this study is the selection of the industries and programs to be examined. One organizational category of interest is nuclear endeavors that are not under NRC jurisdiction. This category includes the Department of Energy (DOE), the U.S. Navy (USN), and nuclear programs in foreign countries. A second organizational category of interest is nonnuclear endeavors that involve highly complex technology that requires high-quality standards in design and manufacture and that strives for low probability of failure because the consequences of failure may be substantial. This category includes aircraft manufacturing regulated by the Federal Aviation Administration (FAA), shipbuilding under both the USN and the Maritime Administration (MarAd), and spacecraft under the National Aeronautics and Space Administration (NASA).

Included in these categories are two subcategories. One is represented by situations where a government agency is the owner and/or operator of products or facilities generally produced by the private sector under contract to the government. These include the DOE, NASA, and the USN part of the shipbuilding industry. The second subcategory is characterized by those instances of private sector endeavors being regulated by a government agency. Aircraft manufacturing and commercial shipbuilding are examples of this subcategory. Foreign nuclear programs reviewed include both government and private ownership and operation of nuclear power plants.

The following domestic programs were studied:

- the Federal Aviation Administration (FAA) program as applied to the manufacture of large commercial transport aircraft

- the Department of Energy (DOE) program as applied to a government-owned nuclear reactor project, the Fast Flux Test Facility, and a nuclear project for enrichment of uranium, the Gas Centrifuge Enrichment Program

- the National Aeronautics and Space Administration (NASA) program as applied to the aerospace industry

- the U.S. Navy (USN) program for shipbuilding under the Department of Defense

- the Maritime Administration (MarAd) program for commercial shipbuilding under the Department of Transportation.

The programs have been studied to the extent that each can be characterized with respect to its AOQ features and activities and to identify specific elements that may have potential application to the NRC program. Each program was studied in sufficient depth to gain a good understanding of the total program to adequately analyze those particular features deemed pertinent to the NRC program. No attempt was made to evaluate the effectiveness of these outside programs.

Each program was studied by reviewing the publicly available information describing the program, including legislation, regulations, guides and miscellaneous instructions. The literature review was supplemented by interviews with representatives from the government agencies and by interviews with pertinent private sector representatives involved in the DOE, FAA and NRC endeavors. There were also limited contacts with NASA representatives. The characterizations of the USN, MarAd, and foreign nuclear programs are based entirely upon literature reviews, except for limited discussions between subcontractors and a few people in the foreign countries. In the case of the USN and MarAd programs, an experienced, expert consultant assisted in developing the program characterizations.

The studies of the foreign nuclear regulatory programs were conducted primarily by subcontractors selected on the basis of their already existing knowledge of the programs, their geographical locations, and their ability to overcome language differences. The programs in West Germany, France and Sweden were studied by Battelle Institute e.V. located in Frankfurt, Germany. The programs in the United Kingdom, Japan and Canada were investigated by the NUS Corporation, including their Japanese subsidiary, JANUS. Assistance in studying the programs in Sweden and Japan was provided by N. C. Kist and Associates.

The reviews of the foreign nuclear programs were based almost entirely upon publicly available information. These reviews were supplemented with subcontractor knowledge of these programs. Because the information obtained in this way is limited, it may be desirable that some of the foreign nuclear programs be selected for a more in-depth study at a later date.

It was also important to characterize the NRC program for assuring quality in design and construction of nuclear power plants, in order to properly consider adopting features from other programs. The abundant literature available on the NRC program was reviewed and supplemented with interviews of officials in the NRC's Inspection and Enforcement Office and with interviews of staff in regional offices for Regions 2, 4 and 5.

D.2 CONCLUSIONS AND FINDINGS

There are several significant differences among the programs investigated in this study:

- the nature and extent of the interfaces between the government sector and the private sector differ

- the incentive systems for achieving quality vary

- in some cases, the major thrust for quality needs arises from safety considerations; in others, from a need for reliable performance; however, safety and reliability are frequently closely intermixed.

Each of the programs reviewed in this appendix operates within its own "cultural ambience" and such differences profoundly affect the resulting program for assuring quality. This is particularly evident in the foreign nuclear programs.

In spite of such differences, there are also identifiable areas of commonality. One example is that all of the programs studied are quite dynamic. Although each of the programs has experienced its own evolutionary process and some are much older than others, changes aimed at improving the effectiveness of the QA programs are ongoing.

One of the observations from this study is that the FAA, NASA, USN, and MarAd shipbuilding regulatory programs are directed towards industries that have evolved as specific entities. These are, respectively, the aircraft manufacturing industry, the aerospace industry, and the shipbuilding industry. Each of these industrial sectors obtains equipment, materials and services from other industrial sectors. Design and fabrication are normally performed by industrial sectors that have evolved generally in parallel with the corresponding regulatory programs. In contrast, the NRC program is directed towards regulating the "nuclear industry," which has never evolved as a specific industrial entity in the traditional sense. The design and construction of nuclear power plants is accomplished as an offshoot activity from several traditionally established industries, each with its own historical methods of doing business. These are the electrical utilities, the architect-engineers, the major power plant equipment suppliers, and the construction industry. Implementing the NRC program in these industries has required major changes in traditional practices. Furthermore, the NRC program is directly applied to each utility that chooses to build a nuclear power plant with the stipulation that the requirements be passed on to others.

One consequence of the complex institutional arrangement for building nuclear power plants has been that major changes in long-established ways of doing business have been imposed across a large number of business-management interfaces. It is beyond the scope of this study to pursue such a complex issue to the point of developing recommendations; however, it is reported here as an issue that emerged from the study of other programs and is deserving of further study.

Although significant differences exist between the NRC's AOQ program and the other programs reviewed, some elements of the other programs may be applicable to the NRC program.

The major findings discussed in this appendix were derived from studies of the various individual programs. It must be emphasized that these studies were limited in scope to general concepts. Therefore, these findings should be viewed as features worthy of consideration by the NRC for its assurance of quality program rather than as features that should be immediately adopted.

In formulating these findings, consideration was given to the institutional differences that exist between the NRC and between the outside programs reviewed. For example, the relationship between the government and the private sector is of a regulatory nature in some cases (FAA, NRC, MarAd) and a contractual nature in others (DOE, NASA, USN). Other intrinsic aspects of the programs studied include cultural differences, as observed in the foreign nuclear programs, and a national commitment to developing the product, as observed in the USN shipbuilding, NASA, and foreign nuclear programs.

Findings are categorized below by Design, Assurance of Quality Programs, Program Reviews, Vendors, Inspection Programs and Craftsmanship.

D.2.1 Design

The NRC should consider requiring that plant designs be well advanced prior to initiating construction activities. Design requirements should include the completion of safety, reliability, and availability analyses including failure mode and effect analyses, and fault-tree and hazard analyses. The analyses should be integrated with QA and should be completed prior to the initiation of construction. This recommendation is based upon findings from the DOE, NASA, FAA, foreign nuclear, and shipbuilding programs.

D.2.2 Assurance of Quality Programs

The NRC should consider establishing a QA system that prioritizes levels of quality efforts. Systems and components should be assigned to the various priority grades on the basis of the safety, reliability and availability analyses discussed under "Design" above. This recommendation is based upon findings from the DOE, NASA, and shipbuilding programs.

D.2.3 Program Reviews

The NRC should consider adopting the following three recommendations, which relate to program reviews:

1. The NRC program should require "readiness reviews" during nuclear power plant construction. These reviews might involve plant designers, construction managers, owner-operators, and (possibly) NRC staff and should be required at key points in the project beginning with "design ready for construction." It may be useful to have additional reviews at selected key milestone points. This recommendation is based upon findings from the DOE, NASA, and shipbuilding programs.

2. The NRC should study ways to better integrate NRC inspection functions with system design reviews, test program reviews and test program evaluations. This recommendation is based upon findings from the USN, FAA, DOE, and NASA programs.

D.2.4 Vendors

Consideration should be given to expanding the NRC's vendor inspection program. The licensee should continue to be held fully responsible for vendor-supplied items. Necessary enforcement actions relevant to vendors could be applied to the licensee. The NRC should consider supporting, perhaps through the Institute of Nuclear Power Operations (INPO), continued development of a data bank on performance of and problems with vendor-supplied components. These data should be analyzed and the results published periodically. This recommendation is based upon findings from the FAA, USN, and foreign nuclear programs.

D.2.5 Inspection Programs

The NRC should consider adopting the following inspection-related points:

1. The NRC should expand its inspector training program to increase the emphasis on "how to inspect." Such a program should concentrate on such areas as conducting inspections, use of time, and interpersonal skills and should include specific guidance on identifying possible indicators of developing problems. This recommendation is based upon findings from the USN program.

2. The NRC should consider requiring inspections of nuclear power plants by independent inspecting agencies. This recommendation is based upon findings from the foreign nuclear programs.

D.2.6 Other

The NRC should re-examine its posture on assurance of quality to emphasize to the licensees that quality and the assurance of quality are responsibilities of overall management rather than responsibilities of the QA/QC organizations. This recommendation is based upon findings from the DOE program.

D.3 SIGNIFICANT FINDINGS FROM EACH PROGRAM

The intent of this study was not to evaluate the other programs studied but, rather, to focus on identifying features with potential for improving the NRC program. In general, these were features that were viewed as positive factors in their respective programs by the administrators of those programs. This section discusses the significant findings from each of the programs to provide a basis for the major findings presented in Section D.2.

D.3.1 FAA Program

The portion of the FAA program that was reviewed is that relating directly to the design and manufacture of large, commercial transport aircraft. The following five items are considered to be significant findings from this program relative to the NRC program.

1. The FAA closely reviews and monitors all of the design, fabrication and flight testing of prototype airplanes. This involvement includes flight tests by FAA pilots. It is only after these flight tests of prototypes that the first FAA certificate, Type Certificate, is issued for a new aircraft model. Both designated engineering representatives and designated manufacturing and inspection representatives are utilized extensively throughout this process to supplement the FAA's resources. These representatives are industry employees, individually certified by the FAA to conduct certain review and inspection activities on behalf of the government. This practice reflects the very substantial FAA effort in this phase of producing a new airplane.

2. The FAA reviews and approves all of the manufacturer's QA/QC, work performance, and testing procedures prior to issuing a Production Certificate. This permits the manufacturer to produce replicated aircraft following the FAA issuance of a Type Certificate for that model.

3. The aircraft manufacturer is held responsible for safety and quality but the FAA accepts some responsibility for the certification program being properly conducted.

4. The FAA issues certificates to vendors supplying parts for airplanes but holds the prime manufacturer responsible during the manufacturing process, including enforcement actions being applied to the prime

manufacturer rather than to the vendor. After airplanes are in service, the vendor certification program is of greater significance and the FAA interfaces more directly with the vendors. When supplied parts cannot be fully inspected after delivery, more attention is devoted to the vendor's plant.

5. Although it is not required by the FAA, aircraft industry practice requires that mechanics sign off completed work prior to QC inspection. Some items are also signed off as acceptable by FAA inspectors or the designated representatives.

D.3.2 DOE Program

Two DOE projects were studied. One is the Fast Flux Test Facility (FFTF) at Hanford, Washington. Operated under the control of the Richland Operations Office of DOE, the FFTF was constructed in the 1970s. The basic element of this facility is a fast reactor that achieved operational testing in 1980. The other project is the Gas Centrifuge Enrichment Program (GCEP), which is currently under construction near Portsmouth, Ohio, under control of the Oak Ridge Operations Office. The significant findings from these projects are as follows.

1. Both projects used a prioritized quality assurance program. At the FFTF three and later four levels of quality assurance were established. At GCEP two basic classifications used are "routine" and "special." A relatively standard QA program is applied to routine items and specific quality assurance action plans are prepared for special items which may incorporate additional variations, depending upon the established degree of importance.

2. Thorough design reviews were conducted on both projects. In both cases steps were taken to assure that all potentially impacted interests were represented in the design review process.

3. At GCEP the QA efforts are combined with a specific systems engineering effort early in the design process. This includes the use of failure and effects modes analyses and reliability, availability, and maintainability analyses. The developed listings of critical items from these analyses provide a basis for determining the extent of the graded QA/QC to be applied.

4. Both projects have used a form of "readiness reviews" prior to initiating the next or new project phases or activities. In both projects, care has been taken to include the plant owner, engineering design, construction management, and operations interests in these reviews.

5. Both projects have emphasized that quality is a line management responsibility rather than the responsibility of the quality assurance organization. In other words, there has been a major effort to integrate quality assurance into the overall management process of both projects.

D.3.3 NASA Program

To date the study of the NASA program has been primarily limited to a review of the available literature. Based upon this limited review, however, the following findings have been identified as significant.

1. NASA applies an extensive "systems approach" to safety and reliability considerations. This incorporates, for example, risk-of-failure analysis, failure modes and effects analysis, single-failure-point analysis, criticality analysis and hazards analysis, using systems engineering techniques to identify the critical items for application of more stringent QA/QC controls. This systems approach is initiated early in the design phase and is ongoing throughout a project.

2. NASA requires that detailed designs be essentially completed prior to starting fabrication.

3. NASA's contractors are required to establish appropriate QA/QC programs and these programs are closely monitored by NASA.

4. NASA uses detailed, in-depth readiness reviews at predetermined stages of the project. Among other things, these reviews verify that any and all changes and discrepancies have been properly addressed and dealt with.

D.3.4 Shipbuilding Program

The study of the shipbuilding industries was based entirely upon review of publicly available information. The USN programs studied involved the design and construction of both nuclear and nonnuclear ships. In this instance, the USN is the owner, the operator, and the regulator.

In the case of the MarAd programs, the ships are designed, built and operated by private organizations. However, there is extensive financial participation by the federal government in constructing these ships. Consequently, a single government agency, the Department of Transportation, simultaneously promotes and regulates the design and construction of the vessels. The significant findings from both the shipbuilding programs are as follows.

1. A close and cooperative relationship has developed, apparently successfully, between the builders, buyers, regulators and standards-setting organizations.

2. In both the USN and the MarAd programs, the fabricating contractor is held responsible for the assurance of quality, with a significant inspection overview effort by USN inspectors for Naval vessels, and by U.S. Coast Guard inspectors for Maritime ships.

3. Designs for ships are reviewed and approved by the responsible federal agency before construction begins. For the Maritime ships,

design reviews by the American Bureau of Shipping may be accepted or supplemented by U.S. Coast Guard reviews.

4. Design and fabrication is performed by a relatively small number of shipyards with a work force considerably more stable than in the general construction industries.

5. The standardization of ship design is a major policy of the Maritime industry and the USN.

6. The USN uses a graded QA system to identify critical equipment, systems and/or material.

7. The USN provides specific guidance for its personnel and its contractors to prevent or detect deliberate malpractice and fraud.

8. The USN has developed a data bank for analyzing the performance of vendor-supplied components.

.3.5 NRC Program

The NRC program for assuring quality in the design and construction of nuclear power plants was not studied in great depth. The objective was to investigate the NRC program sufficiently to have a good understanding of the program as a basis for considering specific features identified in the other programs. This understanding is important in determining those features that deserve further investigation for potential adoption by the NRC. This less-than-in-depth study did, however, identify the following findings considered to be significant when considering the applicability of findings from the other programs to the NRC program.

1. The NRC holds the licensee (the utility) totally responsible for quality and safety in the design and construction of nuclear power plants.

2. The "nuclear industry" does not exist in the United States as a specific entity in the traditional sense. Therefore, the regulation of this industry has been more difficult because it has required bringing about significant major changes in traditional methods and practices of several industries that continued to perform other types of work. These include the utilities, the architect-engineers, the equipment suppliers, and the constructors. The regulatory process is therefore applied to offshoots of several established industries by focusing on one of them (the utilities) and requiring that the regulations be passed on to the others (i.e., the vendors, contractors and suppliers).

D.3.6 Nuclear Regulatory Programs in Other Countries

This section identifies the findings considered significant from the studies of the AOQ programs for building nuclear power plants in six other countries. These findings are identified below by the country of interest.

Canada

The significant findings of a study of the AOQ regulatory program in Canada are as follows.

1. A graded approach is used with five levels. The level is determined based upon an evaluation of six factors (design complexity, design maturity, manufacturing complexity, item or service characteristics, safety, and economics).

2. The regulatory process is a joint effort between national and provincial governments which relies on technical expertise of the utility, except for critical pressure components.

3. The emphasis in design and construction is in establishing quality engineering rather than documentation of existing practices. The term "quality engineering" refers to the management decision process which ensures that all parties involved communicate with each other and clearly understand requirements and objectives throughout the design and construction process.

4. Suppliers are qualified by the utility before a contract award, and the Canadian Standard Association has initiated a qualification program.

Federal Republic of Germany

The significant findings from a study of the assurance of quality regulatory program in West Germany are as follows:

1. The regulatory process, including the setting of rules, is conducted in more of a collaborative mode than an adversarial mode between government and industry.

2. The utilities in West Germany contract with a single organization for the total design and construction of a nuclear power plant on a turnkey basis. The contractor therefore bears full vendor's liability.

3. The onsite inspection functions to assure compliance with regulatory requirements are performed by independent, not-for-profit organizations, Technische Uberwachungs-Vereine (TUVs). These are organizations which have a long history of providing inspection services in a number of business and industrial areas, and they are accepted as highly competent and trustworthy.

4. The control measures and inspections are predominantly hardware- or product-oriented. "Supplier Certificates" and "N" stamps are not used, but the suppliers of equipment and plants must show to the inspection authority's satisfaction that standards are met.

5. In addition to a safety report, the applicant for a license (the utility) must provide "factual statements enabling the examination of the reliability and expert knowledge of the persons responsible for the erection of the installation and the management and control of its operation as well as factual statements enabling the examination of the requisite knowledge of all persons working on the installation." (From the License Procedure Ordinance, "AVerfVO.")

France

The signifcant findings from the study of the assurance of quality regulatory program in France are as follows:

1. The light water reactor power plants in France are designed and constructed under a turnkey arrangement with Framatome, a government-owned corporation that designs the plants, manages the construction and provides the nuclear steam supply system for the utility, which is also government-owned.

2. A series of three standardized nuclear power plant designs have been licensed. Additional licensing considerations for each plant are restricted to consideration of siting issues.

3. The onsite inspection activities on behalf of the government are by private individuals or small associations. These inspectors have not only been qualified by the government and certified, but individually take an oath of office and therefore function as government deputies.

4. The single utility, which operates all of the light water reactor plants, has developed a sophisticated information system to gather data on operating experience. These data are used as a basis for improvements in plant designs and components.

Japan

The significant findings from the study of the assurance of quality regulatory program in Japan are as follows:

1. The government agency, Ministry of International Trade and Industry (MITI), and licensees have a mutual trust and cooperation based upon a stated common goal of safe operations. MITI has also licensed an independent nonprofit organization, the Japan Power Plant Inspection Institute (JPPII), which is

funded by users to perform inspections of welds and hardware. When JPPII performs an inspection, no additional inspection is performed by MITI. MITI inspections are primarily programmatic.

2. The QA practices emphasize the inspections and records rather than the system. Certain inspections are required by law.

3. The current system does not include regulatory criteria for QA, but the Japan Electric Association has published QA guidelines. MITI established a QA Investigation Committee in 1980, which recommended a QA program similar to those in the U.S. (10 CFR 50 Appendix B) and Europe.

4. ASME Stamp Accreditation has been used in Japan since 1973. MITI established a Committee for Nuclear Accreditation under the JPPII which is an agency authorized to inspect nuclear power plant components on behalf of MITI. The Committee has discussed the introduction of an accreditation system similar to ASME "N" stamps and establishment of a third-party agency to conduct surveys and audits.

Sweden

The significant findings from the study of the assurance of quality regulatory program in Sweden are as follows:

1. The program for constructing nuclear power plants in Sweden has taken advantage of replicated basic designs.

2. The government regulatory agencies have relatively small staffs and rely heavily upon reviews and inspections performed by a nonprofit, government-owned, third-party organization. This organization reviews designs, inspection plans and work procedures, and inspects hardware.

3. A "hold point" system is utilized by the independent inspection agency at specific points in the construction program. The third party must approve designs, inspection plans and work plans and procedures before construction is allowed to proceed with specific activities.

United Kingdom

The significant findings of a study of the assurance of quality regulatory program in the United Kingdom are as follows:

1. A "hierarchical system" is used in which the extent of responsibility and authority, and the lines of communication, are clearly defined starting from the licensee through the main contractor and finally to the smallest supplier. Although any higher-order organization may

audit QA/QC practices of any lower organization, an organization is only accountable to the organization immediately above it in the hierarchy.

2. The site license is granted only after design intent and safety prin-ciples and the construction design description are judged suffi-ciently complete that construction can proceed with small risk of significant changes being subsequently required for safety reasons.

3. QA/QC procedures approval by the Nuclear Installations Inspectorate (NII) is a license condition.

4. Inspection and testing of major items may be carried out by the licensee's own inspection organization or by recognized independent inspecting agencies, but the arrangement requires NII approval.

5. NII inspectors visit each site to witness tests and examine test records, and NII consents are required at various major steps bebore construction proceeds further.

6. There are four grades of QA requirements normally employed, namely: "Q" - highest grade for safety class plant items; "N/S" - important to safety and "N/O" - important to operational reliability items; "N/E" - lower class items which still require significant design engineering; and N/-" - the lowest class of off-shelf, mass-produced items.

D.4 SUMMARY OF ASSURANCE OF QUALITY (AOQ) PROGRAMS

This section provides a brief summary of each of the AOQ programs studies that resulted in the findings identified in Section D.3. Time restraints pre-cluded the NRC staff from forwarding the summary descriptions to all the govern-ment agencies for their review, comment, and correction. As a result, inaccuracies may exist in these summaries. If warranted, corrections to these summaries will be made in future revisions or supplements to this report.

D.4.1 FAA Assurance of Quality (AOQ) PROGRAM

This part of the study focused on the Federal Aviation Administration's (FAA) program for assuring quality in the design and manufacture of large com-mercial transport aircraft. To obtain program information, publicly available documents on this program were reviewed. FAA staff in Washington, D.C., and staff in the Transport Airplane Certification Directorate Office, Seattle, Washington, also were interviewed. Staff of the Boeing Co. also were inter-viewed, including a Designated Engineering Representative (DER) and a Desig-nated Manufacturing Inspection Representative (DMIR). Finally, limited obser-vations of aircraft manufacturing work in progress were conducted at Boeing.

The Assurance of quality (AOQ) program being applied to the licensing and certification of large commercial aircraft is all-inclusive in that it addresses all aspects of design, material fabrication, assembly and tests.

Like nuclear reactors, aircraft involve highly complex technology and require high-quality standards in their design, construction and operation.

Aircraft are complex structures that are fabricated of many lightweight systems located in limited space. The aircraft must perform in a wide variety of environments for many years. Aircraft safety demands not only a design that is tolerant of failure, but also careful production that is of the highest quality and excellent maintenance following manufacturing.

The Federal Aviation Act of 1958 authorizes the FAA to issue certificates for aircraft in the interest of safety. Section 603.A of the Act addresses the requirements for a "Type Certificate" (design). The requirements for a "Production Certificate" (production) are covered in Section 603.B, while Section 603.C states the general requirements for an "Airworthiness Certificate" (license for operation). Essentially, these sections of the Act address the safety of or the assurance of quality for the aircraft.

D.4.1.1 Organization and Responsibilities

The responsibilities for an airworhtiness program in the FAA involve both headquarters and field operations. Headquarters is responsible for establishing rules, issuing directives, and distributing guidance publications. Field operations are responsible for the receipt of applications, examination, certification, surveillance, and enforcement.

Title VI of the Federal Aviation Act sets forth the responsibilities of the basic certification processes. The interested party files an application, and the FAA makes a finding and issues certificates as well as any regulatory corrective action necessary. The FAA is also responsible for certificate amendment, suspension, and revocation. The administrator is given the responsibility to issue minimum standards, rules, and regulations as well as the use of various kinds of airworthiness inspectors. The responsibilities and duties of the industry require "air carriers to perform their services with highest possible degree of safety in the public interest."

Policy and guidance responsibilities are retained at FAA headquarters, while the field offices develop and implement programs. Airworthiness programs are carried out by four regional directorates located in Seattle, Washington (commercial transport aircraft); Ft. Worth, Texas (rotary aircraft); Kansas City, Kansas (general aircraft); and Boston, Massachusetts (engines and propellers). The Seattle office has responsibility for review and oversight of commercial transport aircraft design, production, determination of airworthiness, and maintenance throughout the world.

The directorates were established to perform technical policy and airworthiness project management for the aircraft certification programs. The directorates of the regional offices report to the Administrator. The directorates, while assigned specific policy and programmatic responsibilities, are also responsible for implementation of the airworthiness programs within their respective geographical boundaries.

D.4.1.2 Certification Program

In issuing certificates for aircraft, the FAA is responsible for exercising its powers and performing its duties to reduce or eliminate the possibility

of accidents in air transportation. This section discusses the three certification programs employed by the FAA to assure the quality of large commercial
transport aircraft: the Type Certificate, the Production Certificate and the
Airworthiness Certificate.

Type Certificate. The first step in the FAA's certification of an aircraft is design approval or Type Certification. The Type Certificate is an FAA
approval of an aircraft design based on engineering review of reports,
drawings, and data, and on flight tests and tests of materials and parts. The
FAA review during the Type Certificate process is very detailed and includes a
design review of basically all parts and pieces of the aircraft. The Designated Engineering Representative (DER) activities are a very integral part of
this FAA review process. Designated Manufacturing Inspection Representatives
(DMIR) also provide support in the Type Certification program during the production of prototype aircraft or parts for testing. The DMIR provides conformity inspection assistance to the FAA during the long proto-typing process which
precedes the Type Certificate.

Type Certificates are issued for complete aircraft, but they may also be
issued for components such as engines and propellers. Basically, the FAA
defines the minimum safety standards to be met, and the applicant develops,
defines, analyzes, tests and shows compliance with the requirements to obtain
design approval. Before a Type Certificate is issued, the FAA evaluates the
applicant's compliance by design review, inspection of prototype fabrication,
and performance of flight tests.

All of the activities leading to a Type Certificate are monitored closely
by the FAA. The FAA has prepared and issued a handbook (Order 8110.4 Type Certification) to guide and assist all personnel in performing their responsibilities and in efficiently accomplishing the assigned tasks.

Production Certificate. After the conditions of the Type Certification
program have been met, the Production Certification phase begins. To obtain a
Production Certificate, the the manufacturing facility and process for the
replication of a Type Certificated aircraft, including the manufacturing quality control system, must be approved by the FAA. In issuing the Production
Certificate, the Administrator can inspect and require any tests of the aircraft, aircraft engine, propeller, or appliance as is needed to assure that
each unit has been manufactured adequately according to program specifications. If the Administrator approves production duplicates of the aircraft,
aircraft engine, propeller, or appliance for which a Type Certificate has been
issued, then a Production Certificate is issued, authorizing the production of
such duplicates. In the Production Certificate program, the Administrator may
set the duration of the certificate and any other terms, conditions, and limitations required in the interest of safety.

Assuring the adequacy of the production system involves various levels of
FAA quality control surveillance. FAA inspectors review and approve the company's manufacturing, QA/QC and testing procedures and processes.

Before the Production Certificate is issued, the FAA requires applicants
to demonstrate that a QA system will be established and maintained so that each

plane produced will meet the design provisions of the applicable Type Certificate. Each applicant for a Production Certificate must also submit to FAA for approval the following information:

- a statement of QA organization and responsibilities

- a description of inspection procedures for materials, parts, and supplies

- a description of the methods used for production inspection

- an outline of the materials' review system

- an outline of the system for informing QA inspectors of manufacturing changes

- a chart showing the location of inspection stations

- information on delegation of inspection authority to subsidiary manufacturers.

A holder of a Production Certificate must allow the FAA to make any inspections it desires (including suppliers) to assure compliance with the above requirements.

The FAA has considerable involvement in and control over a manufacturer's processes through the requirements of Production Certificates. Also, the FAA exercises a similar control over various suppliers and vendors who are considered an extension of the manufacturer or prime contractor. However, most enforcement actions resulting from problems associated with vendors or suppliers are applied to or through the prime manufacturer of the aircraft. FAA involvement occurs primarily through the manufacturer, the holders of a Type or Production Certificate, through Parts Manufacturer's Approval (PMA), or through a Technical Standard Order Authorization (TSOA). When parts supplied to a manufacturer cannot be adequately inspected after delivery, the FAA may inspect them at the supplier's location.

The PMA and TSOA are used primarily when parts are supplied for repair or modifying aircraft in service. In both cases, the FAA issues an approval (license) for the manufacture of certain parts to an approved design after FAA approval of the process and the QA/QC program and procedures. These approvals give the FAA the prerogative to inspect and audit the facilities, products, and processes of a manufacturer and his suppliers.

Order 8120.2A, Production Approval and Surveillance Procedures, was prepared to guide personnel in accomplishing FAA's responsibilities for the evaluation, approval, and surveillance of the production activities of manufacturers and their suppliers producing products, parts and appliances in accordance with Code of Federal Regulations 14 CFR, Part 21.

Airworthiness Certificate. The third and major part of FAA's certification of an aircraft is the original Airworthiness Certification program. An

Airworthiness Certificate from the FAA is required for a U.S. registered aircraft to operate. Basically, the registered owner of any aircraft may file an application for an Airworthiness Certificate with the Administrator. The Airworthiness Certificate is issued after the Administrator finds that the aircraft conforms to the Type Certificate, and if that aircraft is found to be in condition for safe operation after inspection. The Administrator can set the duration of the certificate, the type of service for which the aircraft may be used, and any other terms, conditions, and limitations that are required in the interest of safety. Each certificate is registered by the Administrator and can include any information that the Administrator feels is necessary.

The FAA has issued Order 8130.2B, entitled Airworthiness Certification of Aircraft and Related Approvals, which contains procedures and instructions for personnel involved in issuing Airworthiness Certificates and related approvals.

D.4.1.3   Program Implementation: Designated Representatives (DR)

To ensure that the design and fabrication of a new airplane meets all regulatory requirements, the FAA is assisted by specified independent persons who also may be employees of the aircraft manufacturers. In accordance with the Civil Aeronautics Act of 1938, examinations and reports could be accepted from properly qualified private persons in place of those made by government employees. In 1950, Congress passed bills authorizing the delegation of certain functions to properly qualified private persons (designated representatives). These functions include the examination, inspection, and testing necessary for issuing certificates in accordance with properly established standards.

These Designated Representatives (DR) review the design and fabrication processes to ensure compliance with all aspects of the regulations. In 14 CFR, Part 183, Representatives of the Administrator, the requirements are described for designating private persons to act as representatives of the Administrator in examining, inspecting, and testing persons and aircraft prior to the issuing of airman and aircraft certificates. In addition, it states the privileges of those representatives and prescribes rules for exercising those privileges. The review of the Designated Representatives program focused on the following two types:

- Designated Engineering Representative (DER). Individuals designated to approve engineering information. Order 8110.37, DER Guidance Handbook, identifies the policies, procedures, technical guidelines, and limitations of authority for DERs. This information is amplified in Chapter 5 of Order 8110.4, Type Certification.

- Designated Manufacturing and Inspection Representative (DMIR). Individuals designated to issue original airworthiness, export, ferry, and experimental certificates. The qualifications, appointment, responsibility, authority, etc. of DMIRs are identified in Chapter 8 of Order 8130.2B, Airworthiness Certification of Aircraft and Related Approvals.

The DER principally supports the FAA in issuing Type Certificates. The DMIR principally supports the FAA in issuing Airworthiness Certificates, and, when necessary, the DMIR provides support for a Type Certificate during the production of prototype aircraft or parts for testing.

The DER and DMIR are authorized to perform certain examinations, inspections, and tests on behalf of the FAA. Depending on the specific limitations in their designation, they also provide FAA approval (sign-off) or recommend approval by the FAA. DER activities focus on insuring compliance with the requirements of the FAA regulations, whereas the DMIR principally ensures that aircraft and components are manufactured according to FAA-approved designs, specifications, and QC programs.

Designees are usually nominated by the applicant (aircraft manufacturer) and are appointed by the FAA regional director after the director reviews their personal and professional qualifications and experience. Once appointed, they are delegated by the FAA Administrator, through the regional office, to represent the FAA in helping to determine that the aircraft complies with the relevant requirements of the regulations. In this capacity, designees are bound by the "...same requirements, instructions, procedures, and interpretations as FAA employees..." (FAA 1967). While designees perform considerable work for the FAA, the agency reserves for itself the approval of the following necessary elements in the certification process:

- the regulatory process
- analytical criteria to be used
- major design philosophy affecting safety
- all fault-type safety analyses
- all test proposals
- witnessing of all major tests
- all major flight testing
- all in-service safety problems
- aircraft flight manual
- QC manual
- surveillance of production facilities
- production certification of facilities and QC functions.

D.4.1.4 Industry's QA Program

As part of this study of FAA quality assurance programs, staff at one plant were interviewed and work was observed. Each of the QA functions that constitutes the foundation of the QA/QC system in the manufacture of large commercial transports is briefly described in Table D.1 (FAA 1976).

In reviewing the industry program it was noted that fabrication is tracked by a very detailed operations and inspection record. After a particular item of work has been completed, the record is initialed or stamped by the person performing the activity. Then, inspections are performed by company employees who are required to verify by formal record that the product meets the established standards. This record signifies who performed each task and that the inspector stands behind the proper performance of the work. Also, articles are tagged or stamped with marks that identify the individual inspector and ensure that only inspected and accepted items are used in the finished product. For

TABLE D.1. Descriptions of Quality Assurance Functions (FAA 1976)

1. TECHNICAL DATA CONTROL--Assures that only the latest approved drawings, drawing change notices, engineering data, etc. are available to production and inspection personnel and that obsolete drawings and data are promptly removed from the production and inspection areas.

2. MANUFACTURING PROCESSES--Provides for selecting and controlling procedures to ensure that all characteristics affecting safety will be inspected and that products or processes conform to approved design data where specific operations such as machining, riveting, welding, etc. are performed.

3. SPECIAL PROCESSES--Controls all processes and services such as welding, heat treatment, bonding, plating, casting, forging, etc. where the material being processed undergoes any physical, chemical, or metallurgical transformation and the conformance to specifications cannot be verified by external visual inspection.

4. INSPECTION/IDENTIFICATION--Ensures that only articles and processes that have been accepted and that conform to approved design data are used in the product. Items are identified with stamps or marks traceable to qualified individuals.

5. NONDESTRUCTIVE INSPECTION--Establishes requirements for inspection methods used to determine conformity to the design data through or by a means which will not have a detrimental effect on a part. Example: Magnetic particle, ultrasonic radiographic, etc.

6. TOOL AND GAUGE CONTROL--Establishes control of precision weight and measuring devices (tools, scales, gauges, fixtures, etc.) used in fabricating and inspecting parts, assemblies, and complete products to assure conformity to type design data.

7. SUPPLIER CONTROL--Encompasses the purchasing, testing, and acceptance of all materials, parts, and services furnished the manufacturer from an outside source, including proprietary items.

8. TESTING--Assures that all functional components and/or assemblies are subjected to tests that will ensure that the product will perform its intended function safely.

9. MATERIALS REVIEW--Identifies system of control for withholding, evaluating and disposing of all materials, parts, etc. that do not conform to engineering design data.

10. STORAGE AND ISSUANCE--Assures proper protection and prevention of damage and deterioration of materials, parts, assemblies, etc. that have passed inspection while awaiting use. Also assures that only articles current with applicable design changes are released for incorporation in the product.

TABLE D.1. (contd)

11. AIRWORTHINESS CERTIFICATION--Identifies system for evaluation of the com-
    pleted article or product and related documents to assure that all
    required inspections and tests have been satisfactorily performed and that
    it is in a condition for safe operation.

12. SERVICE DIFFICULTIES--Establishes a system for recording, investigating,
    determining cause, and assuring corrective action on all known or reported
    failures, malfunctions, or defects.

example, suitable "acceptable," "rework," or "rejection" stamps are placed on
articles subjected to heat treatment, welding, riveting, soldering, hardness
tests, laboratory analysis, and other tests. It should be noted that the sign-
off of completed work by the mechanic who did the work is not required by the
FAA, but is reportedly an aircraft industry practice.

D.4.2 Department of Energy (DOE) Assurance of Quality (AOQ) Program

The DOE, its predecessor organizations Energy Research and Development
Administration (ERDA) and the Atomic Energy Commission (AEC), and contractor
organizations and laboratories have been developing, constructing, and operating nuclear reactors and other nuclear facilities for some four decades. They
have developed and applied many methods and practices for safely carrying out
these activities. Many of the accepted and proven practices of nuclear technology, such as the nuclear application of QA and engineering standards, were
pioneered in these endeavors. For this reason, two DOE nuclear projects were
selected for review in this study to determine whether there were attributes of
the DOE Program for assurance of quality which may be transferable to the NRC.
The first DOE project reviewed, the Fast Flux Test Facility (FFTF), is a
reactor facility that achieved initial start-up in 1980. The other one, the
Gas Centrifuge Enrichment Plant (GCEP) for uranium enrichment, is currently
under construction. The assurance of quality program for each project is
discussed separately in Sections D.4.2.2 and D.4.2.3. The following section
gives a brief overview of the DOE organization and responsibilities for nuclear
programs.

D.4.2.1 Background

The Atomic Energy Commission was disbanded by the Energy Reorganization
Act of 1974 and replaced by NRC and ERDA. Section 107(a) of the Energy Reor-
ganization Act states that the nuclear functions of ERDA will be subject to the
Atomic Energy Act of 1954. All functions of ERDA were transferred in 1977 to
DOE by the DOE Organization Act. DOE is basically subject to the same direc-
tives regarding safety and AOQ in the Atomic Energy Act as the NRC.

The Atomic Energy Act has no specific language addressing AOQ. Indirectly, however, the act empowers DOE to regulate AOQ to protect health and
safety and to minimize danger to life and property. A basic purpose of the
Atomic Energy Act is to encourage widespread use of atomic energy, but only to

the extent that its use is consistent with the health and safety of the public. Section 161(b) of the Atomic Energy Act requires DOE as follows:

Establish by rule, regulation, or order, such standards and instructions to govern the possession and use of special nuclear material, source material, and by-product material as may be deemed necessary or desirable to promote the common defense and security or to protect health or to minimize danger to life or property.

DOE was formed in 1977 to centralize responsibility for national energy policy and to continue and expand the energy research and development that was transferred from the Energy and Research Development Administration (ERDA). The DOE Organization Act of 1977 placed the operation of government-owned nuclear plants and the independent safety overview function in a larger organization. Normally, these programs were administered through an agency headquarters group, and facilities were operated by a contractor at the site. A DOE field office, located on or near the site, provides close oversight of the programs.

The organizational placement of nuclear energy activities in DOE can be characterized as decentralized. Although essentially all duties of the Assistant Secretary for Nuclear Energy are nuclear related, other major nuclear activities have been assigned to the Assistant Secretaries for Defense Programs; Environment Protection, Safety, and Emergency Preparedness; International Affairs; and to the Director of Energy Research.

The management of the DOE nuclear programs (including the FFTF and GCEP is generally administered through three organizational tiers depicted in Figure D.1.

The DOE field organization and project officers have overall responsibility and authority for defining and assuring effective implementation of required quality assurance (QA) activities to be established and implemented on DOE programs by contractors under their direction. Any order or standard that DOE adopts can readily be made applicable to the activities of its contractors simply by inserting an appropriate applicability clause in the contract.

D.4.2.2 Fast Flux Test Facility

In addition to reviewing pertinent project documents for the DOE's FFTF, interviews were conducted with DOE headquarters staff in Germantown, Maryland, and in the Richland Operations Office (RL), Richland, Washington. Staff at the Westinghouse Hanford Corporation (WHC), which operates the facility, were also interviewed.

Background. The FFTF is a 400 MW (thermal) sodium-cooled fast neutron flux reactor designed for the irradiation testing of fuels, materials and components for fast breeder reactors.

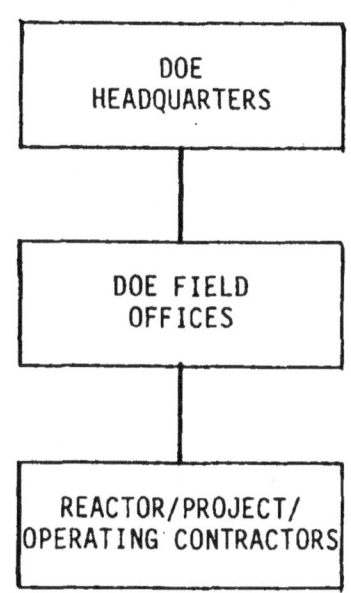

FIGURE D.1. Organizational Tiers in Nuclear Field Programs

In the FFTF, fuels and materials are exposed to conditions typical of those expected in future reactors and, in some cases, to conditions beyond anticipated plant conditions to explore safety margins, to extend fuel technology and to build confidence in the design of future power plants. The FFTF was initially started up in 1980 and recently completed the third of its planned cycles of operation. Performance to date has exceeded DOE's expectations and program milestones.

The FFTF was designed and constructed to meet NRC license requirements, although it is not licensed. A Preliminary Safety Analysis Report (PSAR) with the required section on QA was submitted to AEC's Division of Reactor Licensing in September 1970. Since the FFTF was owned by the AEC, the Commission had the prerogative to proceed with the FFTF program even though the Regulatory Review was not complete. The initial or limited work authorization was obtained in July 1971, and the final construction authorization came in May 1973. A report to the Advisory Committee on Reactor Safeguards (ACRS), dated December 29, 1971, and prepared by AEC Division of Reactor Licensing, stated: "Our review of the Quality Assurance Program indicates that it meets the intent of Appendix B, 10 CFR Part 50, and that adequate quality control is available at the site to assure quality in the safety-related structures."

Organization and Responsibilities. Through nearly all of the design and construction phase of the FFTF project, the responsibility for the AOQ program was delegated directly from AEC/ERDA/DOE headquarters to the prime contractor, Westinghouse Hanford Co. (WHC). Currently, the delegation of such responsibilities is from DOE Headquarters through the DOE Richland Operations Office (RL) to WHC as the prime operating contractor. The prime contractor developed the details of the AOQ program and how it was to be applied to the

FFTF project. To assure the implementation and oversight of quality responsibilities for the DOE Richland Operations Office (including FFTF), DOE RL Order 5700.1, Quality Assurance (Fremling 1980), contains the following responsibilities and authorities for QA in the DOE RL:

- **Director, QA.** The director develops and coordinates the RL QA program, assures that each contractor establishes an appropriate QA program according to the order's basic requirements, and assures, together with the affected RL program or project manager, that each contractor establishes an adequate QA plan for each program or project. The director also audits the RL and contractor QA activities to evaluate their effectiveness and selectively reviews contractor component and material contracts and purchase orders to assure optimum use of available offsite inspection services. The director attends periodic forum meetings with contractor QA management to review mutual QA practices and problems and to coordinate and standardize practices and procedures where appropriate. Finally, the director conducts appraisals of contractor QA activities to assure compliance with applicable requirements.

- **Division Directors and Project Managers.** These directors and managers determine when special considerations require a QA plan that may vary from the program required by this Order. They also verify that the contractor has identified appropriate QA requirements for individual systems, components, materials, processes, and services and that QA requirements have been considered in conceptual stages of construction projects. Finally, they verify that appropriate quality requirements are established in project design criteria and that contractor QA plans are effectively implemented.

- **Director, Construction Division.** This director determines when it is appropriate to assign the responsibility for the items to a prime contractor and formally delegates such responsibility and monitors the performance of the operating contractor. The director reviews and evaluates preliminary activities and plans for construction projects to verify that QA plans are appropriate and ensures that appropriate QA requirements, in accordance with this Order, are included in construction project contracts. The director also reviews, evaluates, and assures that QA activities are effectively implemented and reviews and approves key design and testing documents and plans for construction projects. Finally, the director reviews and evaluates the satisfactory completion of all required construction and testing activities before RL accepts a new facility or major modification, and he/she accepts the facilities for the government when all requirements are satisfied.

- **Director, Procurement Division.** This director takes the contractual actions required to support the Directors, Quality Assurance, Construction and Program Divisions and Project Offices in implementing the responsibilities and authorities delegated above.

● Responsibilities and Authorities of Contractors. The contractors develop a generic QA program and implementing procedures for DOE and other government agency-sponsored programs and projects performed in accordance with this Order, and they prepare and implement QA plans for assigned projects and programs. They also verify effective implementation of the QA program and plans for assigned programs and projects and monitor the performance of an A-E and/or construction contractor, as delegated by the Director of the Construction Division.

QA Program. The basic requirements of the QA criteria used on the FFTF project, RDT F 2-2, Quality Assurance Program Requirements, and RDT F 2-4, Quality Verification Program Requirements, are essentially the same as the other recognized criteria given in Appendix B, 10 CFR 50, Quality Assurance Criteria for Nuclear Power Plants and Fuel Reprocessing Plants. The AOQ program that has been applied on the FFTF project throughout its life is all-inclusive (pertaining to the entire facility), yet flexible. This program and the management philosophy established and followed during the design, construction, fuel fabrication, testing, and startup of the FFTF (i.e., that the line organization has the responsibility and that QA is integral to the work) has resulted in a plant that is exceeding its established operational goals.

During the review of the FFTF AOQ program, four QA functions were identified that appeared to be key to the project and that may have applicability to the NRC. These functions are QA classifications, design review levels, data-type QA classifications, and readiness reviews, and are discussed separately below.

● QA Classifications. QA classifications were established for varying levels of effort necessary to provide a controlled system that assures a safe and properly functioning facility or component. This enabled AOQ efforts to be concentrated on the items and systems crucial to the reactor and its supporting facility.

Items, components and systems provided for or used in the FFTF were evaluated by a set of factors and assigned a QA classification or type. These types/classifications were used as guidelines for applying appropriate QA efforts for the various components and elements of the system. During much of the construction phase, three levels of QA were applied. The most extensive application was in accordance with RDT F 2-2, the second was in accordance with RDT F 2-4, and the third represented standard commercial practices. This was later expanded to four levels. The four levels or types of QA classifications and the factors to be considered in assigning QA classifications are shown in Table D.2, extracted from the Westinghouse Hanford QA Manual.

TABLE D.2:   Definition of FFTF QA Classifications
(Hanford Engineering Development Laboratory 1982)

The following definitions are established for the various QA classifications.   They may be used in determining the level of quality assurance effort necessary to provide a controlled system that will assure that the facilities and/or components function safely and properly.

Some factors to be considered in assigning QA classifications include the following:

1. the consequence of the item's malfunction or failure
2. the item's design and fabrication complexity or uniqueness
3. the need for special controls and surveillance over processes and equipment
4. the degree to which functional compliance can be demonstrated by inspection or test
5. the item's history and degree of standardization
6. the difficulty of the item's repair or replacement and the associated cost, including procurement lead time.

DEFINITION OF QA CLASSIFICATIONS

Type I - Applies to items that are highly or moderately complex whose failure can have a direct effect upon operability, performance or safety. Items, which if failed, could cause or fail to prevent an incident affecting health and safety, are also included in this classification.   Typical examples of Type I include reactor core components, fuel handling equipment, and high-level radioactive waste systems.

Type II - Applies to items that are moderately complicated and whose failure can have a significant impact on the validity of development test results, operation, performance or safety.   Items, which if failed, could cause an incident affecting the health and safety of personnel on the Hanford Site are also included in this classification.   Typical examples of Type II include radiation monitors, pressure-retaining components, and HVAC equipment for contaminated zones.

Type III - Applies to items of standard or customized design of a unique but simple nature whose consequences of failure are unlikely to be severe, and/or which can be readily controlled through simple inspections or tests. Typical examples of Type III include standard electronic equipment and air sampling monitors.

Type IV - Applies to activities or items with minor consequences of failure, whose quality is adequately assured through undocumented examination by the requester, such as temporary buildings, roads and fences, and commercial tools.

- **Design Review.** All FFTF design contractors and suppliers, as part of their QA programs, were required to submit a plan for performing independent design reviews, for submitting design review reports, and for assuring resolution of problems revealed by the design reviews. Formal design reviews were identified in the design contractor's or supplier's project planning documents, and the meetings were scheduled sufficiently in advance to avoid unnecessary delays in major milestones. Design reviews were to be regarded as contract "HOLD" points; for example, as a prerequisite to requesting approval of a design package to be released for fabrication.

During the early stages of a design development, the FFTF cognizant engineer responsible for developing a design was required to review the preparation of design drawings, preferably at the design contractor's office, and to verify the understanding of the basic approach and the content of the design packages. The design contractor was required to notify WHC when a design had reached the agreed-upon state for a meeting (Preliminary Design Review).

All principal design documents, such as specifications, drawings, and analyses, were reviewed and evaluated by the cognizant design contractor, before release, to verify the completeness and adequacy of design criteria and contract requirements. When a document was submitted, FFTF project management was required to verify consistency with the functional requirements of the system design descriptions and with the specific requirements of the design application and submit comments (when appropriate) to the originator.

Each design contractor was required to define a system for selecting design review participants, for defining the design and data to be reviewed, for stating proposed objectives and the agenda, and for identifying the review chairman, and time and place of the meetings. After the design contractor determined that a design had reached a point requiring project approval, the responsible design organization was required to call the scheduled formal design review meeting.

The cognizant design contractor or supplier was responsible for taking appropriate action to ensure that all action items resulting from the design review are promptly and adequately resolved.

The design review system was integrated with the design release control system. Design review plans provided for successive reviews and corresponding release depending upon the Data Type Classification. The lowest level was suppliers, next was the cognizant Design Contractor, then the FFTF Project, and finally DOE. The most sensitive design terms were reviewed by each level and required DOE approval. The least sensitive could be reviewed and approved by a supplier. This is discussed further below.

- **Data Type Classifications and Releases.** The basic FFTF procedures and requirements for data review and initial project release are discussed in this section. To determine the level of approval required for initial project release, all FFTF Principal Design Documents were divided into the following data types:

- Data Type 1 included key controlling documents and drawings, such as Safety Analysis Reports, System Design Descriptions, plot plans, piping and instrument diagrams, major assemblies, and general plan arrangements.

- Data Type 2 included documents and drawings such as engineering studies, design or stress reports, quality control procedures, piping and mechanical layouts, radiation zoning, control logic diagrams, and instrument locations.

- Data Type 3 included documents and drawings such as detailed design drawings, and other supporting design documents not classified as Data Type 1 or 2.

- Data Type 4 included supplier drawings or other documents not otherwise classified as Data Types 1, 2, or 3.

All changes to principal design documents were subject to a categorization program to determine the impact of the change and the appropriate level of approval required for project release. The originator of the document change made the first determination of the impact level. Three impact levels were used in the classification. For impact level 1, the originator had to obtain the FFTF cognizant engineer's and DOE's approval. Impact level 2 required approval of the FFTF cognizant engineer while impact level 3 changes could be approved and released by the design contractor.

- Readiness Reviews. Another AOQ or program management function used at the FFTF is that of a project review board or readiness review. In a new or modified facility or system, the coordination of many elements and attention to every detail is required to assure that it is ready to proceed to the next step safely and effectively. Project or readiness review boards have been used at the FFTF since 1976. They were applied on system startup tests and are now being used during reactor shutdowns and startups. Guidance for the current readiness review process at FFTF is given in detailed procedures. Those procedures basically direct that a readiness review be conducted to document line management's certification of the readiness of 1) the FFTF plant, 2) the operating staff, and 3) the support groups to conduct startup and operation following a schedule outage. In addition to FFTF line management, personnel from outside the FFTF plant organization are on the Review Board. Each review is specific and addresses the status of the Operations and Test Plan, the Reload Design Report, refueling documentation, significant plant repair and maintenance activities, major plant changes, engineering system readiness assessments, reactor and industrial safety issues, technical specification/procedural changes, and the plant's transition into the operational mode.

In addition, special emphasis topics may be included in a review at the discretion of the Review Board. Other assessments of a broader nature, such as the status and quality of engineering instructions, are performed as part of the system of routine audits conducted by FFTF plant, safety, and quality assurance organizations.

D.4.2.3 Gas Centrifuge Enrichment Program

This part of the study addresses the QA programs followed by DOE and associated organizations on the design, construction, testing, and startup of the Gas Centrifuge Enrichment Plant (GCEP) and related development facilities. In addition to reviewing the publicly available documents about the GCEP program, interviews were conducted with DOE staff and some of their contractors at the Oak Ridge Operations Office (ORO), Oak Ridge, Tennessee, and at the GCEP construction site near Portsmouth, Ohio. The Centrifuge Program Development Facility (CPDF) was also visited.

Background. Construction of the GCEP production facilities near Portsmouth, Ohio, began in the spring of 1979. A "cascade" of production machines is scheduled to go into operation in the spring of 1984. Certain aspects of the fabrication and construction activities have been reported to be ahead of schedule. In 1982, the CPDF was placed in operation. Although identified as a development facility, the CPDF is a large structure and its startup was the culmination of an involved engineering and construction effort. The project was completed ahead of schedule and under budget, and the QA program that was applied during the CPDF project was felt to be a positive factor in putting a workable facility in operation.

The QA program or system being applied on the GCEP is integrated into the management realm where the QA elements are combined with other management requirements. A series of documents has been developed to provide general requirements on and specific instructions for establishing and executing the various management aspects during the design, fabrication, construction, installation, startup, operations and maintenance of structures, components and systems of the GCEP.

Organization and Responsibilities. The DOE field offices have the overall responsibility and authority for assuring that the required QA activities of contractors under their direction are implemented. Thus, the field offices have a direct relationship with organizations such as Union Carbide or Stone and Webster, which have prime contracts with DOE. For other suppliers, manufacturers, or contractors under contract to a prime contractor, DOE QA personnel have contact only through the prime contractor. The DOE can and does authorize prime contractors to administer DOE contracts, including the QA functions, with other organizations.

Some DOE-ORO staff members who are involved in AOQ functions administratively report to the Quality Reliability Division under the Office of Assistant Manager for Safety and Environment. Others functionally report to the various divisions within an operational office such as Office of Assistant Manager for Enriching Operations and Development. The majority of DOE-ORO staff involved with the various GCEP QA functions are permanent personnel identified as professional QA or nuclear engineers. A member of the QA division is assigned to

GCEP QA on a full-time basis; however, those in the operations and development office handle QA functions along with other engineering and program management assignments.

The responsibilities and authorities for QA policy coordination and overview and for developing, implementing, or evaluating QA activities in support of design and construction of DOE programs at Oak Ridge (including GCEP) are contained in OR 5700.6 Quality Assurance - ORO Site Implementation Plan:

- Contracting Officers and Contracting Officers Representatives (COs/CORs) for AE and Construction. Provide contactors with QA requirements, assessments, and plans for implementation. Obtain contractors' comments and contributions to assessments and plans for follow-on participants. Provide copies of assessments and plans to the Director, Q&R Division, for comment and concurrence before approval. Report significant quality problems and unusual occurrences. Obtain participation of cognizant operating contractor personnel during design and construction to identify potential problems with satisfactory performance in service.

- Director, Procurement and Contracts Division. Assure that contracts contain provisions for AOQ of materials and services. For procurement contracts exceeding $1,000,000, obtain concurrence of the Director, Q&R Division, on requirement for AOQ.

- Director, Quality and Reliability Division. Manage the ORO QA program. Establish QA policy for implementation by ORO program and project divisions. Develop and provide specific guidance for application of QA to all ORO programs and projects, except weapons components and assemblies. Review and approve selected contractor policies and plants for QA. Maintain surveillance of contractor activities and assure compliance. Perform management appraisals to verify adequacy and effectiveness of contractor QA programs; coordinate appraisals with and utilize resources of other cognizant DOE organizations, as appropriate. Investigate significant quality problems, identify quality-related issues, and cause corrective actions to be taken by responsible contractor organizations through COs/CORs and project managers.

GCEP QA Program. The QA program that has been developed at Oak Ridge Operations (ORO) and that is being applied on GCEP programs is an integral part of the project's planning and management activities. QA is included in a "systems" approach from the start of a particular activity. The systems approach addresses the quality, safety, reliability, operability, and maintainability of all components, equipment, and processes involved. Each architect-engineer (AE) or contractor must have a formal program for deliberately and systematically assuring the performance of equipment and facilities. Each of these programs must 1) show management support and concern of QA, 2) emphasize prevention of major problems, 3) provide the means for all employees to understand their roles, and 4) provide a basis for measuring the effectiveness of QA.

A main aspect of the GCEP QA program is the required evaluation of failure consequences as well as the probability of failure of a component, equipment, or process. This procedure provides the means to establish the criticality of an item within a system and the relationship of that system to the project and permits the concentration of QA activities where needed most.

If the risk of failure is high or unacceptable, special attention to prevent failure is required and specific QA actions are prepared to reduce the risk to an acceptable level. Formal planning is required to prevent potential quality problems when the risk of failure is not acceptable. These plans assure adequate considerations of actions to prove quality of development, design, procurement, fabrication, construction, operation, or maintenance and to find quality problems in time to minimize their impact.

Another aspect of the GCEP systems QA program that is considered to be beneficial is the requirement that all participating groups, including AEs, take part in the early planning and participate in all the various phases of the project.

The schematic in Figure D.2 depicts some of the QA elements which are fundamental to the ORO QA program. This figure shows that QA or the assurance of quality is used in a much broader sense than the NRC traditional use of QA requirements: the ORO approach incorporates the QA elements into the overall management of the project.

QA Program Implementation GCEP QA methodology and responsibilities are specified in ORO-EP-105, GCEP Quality Assurance Requirements. Overall responsibilities and authority of project participants are defined in ORO-EP-103, GCEP Project Management Plan, and ORO-EP-116, System Engineering Management Plan. The Deputy Manager for Enrichment Expansion Projects is responsible for establishing and executing the QA Program and assigning parts of it to other organizations, although he retains responsibility for overall program effectiveness.

The basic elements of ORO-EP-105 are as follows:

- Each project participant must have a formal program for assuring quality of equipment and facilities.

- Concern for quality must be visible and should receive management attention comparable to that given to costs and schedules.

- To maximize effectiveness, the QA program must be selectively applied to emphasize prevention of major problems.

- The program must include provisions that assure that each employee clearly understands this role in providing assurance of quality.

- To provide a basis for judging the effectiveness of the QA program, the costs of significant quality problems must be documented and presented to appropriate levels of management.

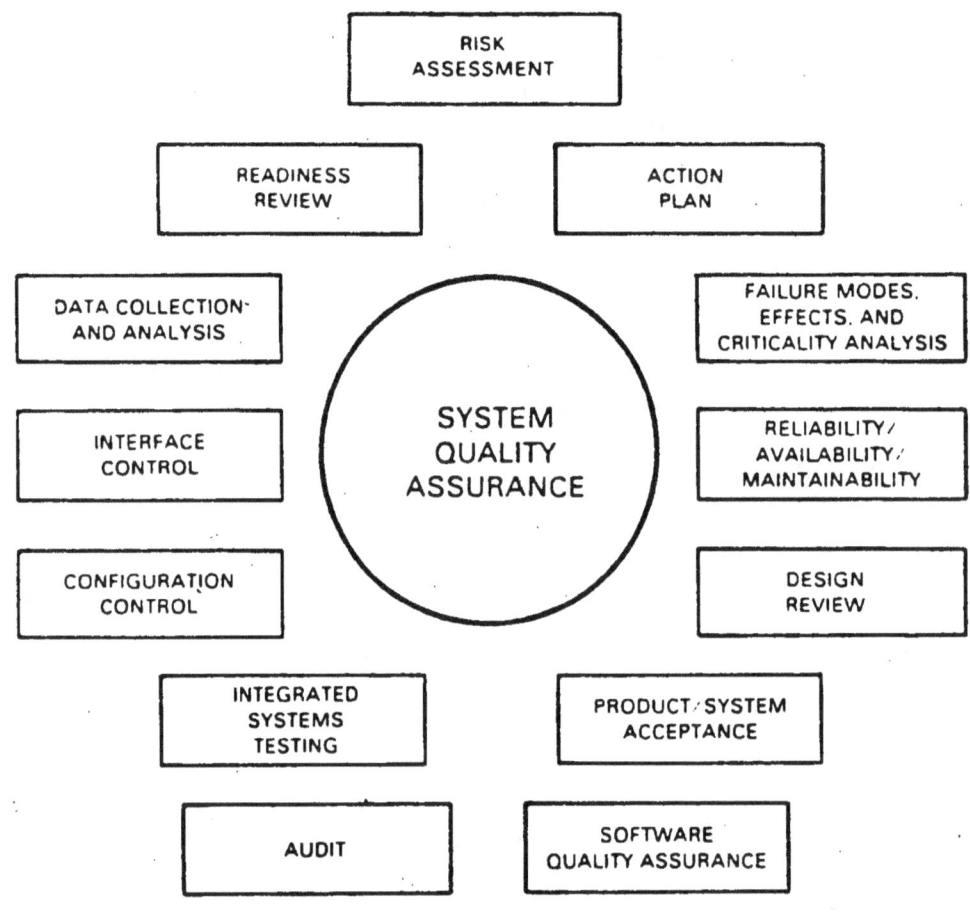

FIGURE D.2. Quality Assurance Elements

The system engineering function for the GCEP covers all of the system requirements for definition, analyses, verification, technical reviews, and other system efforts, including QA, necessary to assure the optimum balance of performance, safety, costs, and scheduling. The system engineering function will support the design, installation, startup, and operational phases of the GCEP. The principal objectives of the system engineering function are as follows:

- to assure that the system requirements of the GCEP process are adequately specified and documented and that due consideration and emphasis is given to all aspects of the project

- to provide system analyses of the designs as they progress to assure that system requirements are met and that GCEP interfaces are compatible

- to assist in defining programs for the necessary and sufficient verification of GCEP systems

- to integrate reliability, maintainability, logistics, safety, producibility, and other related specialties into a total system effort.

QA Classification. At ORO, quality is defined as "fitness for intended use." Accordingly, the GCEP project's basic approach to quality assurance is to assure that the plant's equipment and facilities will be of a quality consistent with their importance to plant operation, reliability, and safety. Therefore, a formal evaluation is required for each system or subsystem to determine the consequence of failure of equipment and facilities. Those performing the assessments are required to consider the effects of failure on safety, environment, cost, schedule, and plant reliability. If it is determined that a failure will have significant consequences and the risk of failure is unknown or unacceptable, or that the consequences of failure are so severe as to be unacceptable under any circumstances, regardless of the probability of occurrence, the item assessed is classified as "special" and a specific Quality Assurance Action Plan (QAAP) is prepared. The requirements of the QAAP are tailored to address the specific area of concern to assure that the equipment or component will function as intended and required. Items determined not to be special are classified as "routine" and come under the basic QA Plan.

QA Program Requirements. The requirements for application of QA by the GCEP contractors are identified in ORO-EP-105, GCEP Quality Assurance Requirements. This document identifies the quality responsibilities and programs for each contractor to implement. ORO-EP-105, GCEP Quality Assurance Requirements, is also structured in two distinct categories: 1) basic QA program requirements, and 2) supplementary control systems in support of quality assurance. The basic requirements apply to all major participants who are DOE prime contractors and are the primary concern of QA personnel. The supplementary controls are administered by appropriate line management.

The basic requirements as identified in ORO-EP-105 are incorporated into a QA program at the earliest practicable time consistent with GCEP schedules. Personnel from all participating groups, including quality assurance and operations, contribute to the program plans. Operations personnel are involved in the review of design and construction activities, and management of the participating organizations review the status and adequacy of the program parts.

Quality Assurance Action Plans (QAAP) are prepared by the responsible design organization for those items identified as special by quality assessments. The implementing organization is responsible for developing the plan and procedures required to accomplish the actions specified in the QAAP; for items established as routine in the QA assessment, participants are required to take appropriate steps to identify and prevent quality problems.

A Construction Critical Items List is made for each design package released or certified for construction. This list is included in the construction package.

Participants document their basic QA program in a manual or program plan and keep it current. Each manual or plan is reviewed and approved by management or upper-tier contractors.

Supplementary Control Systems in Support of QA. In ORO-EP-105, supplementary controls are grouped in the categories of design, procurement, manufacturing/fabrication, testing, and construction/installation. Under design, control requirements are noted for QA-related systems that are selectively applied based on the nature and scope of the work as well as the importance of the items and services being provided. Included are such elements as assurance that design requirements are correctly translated into design documents, interface control, design verification, design change control, and document control. ORO-EP-105 also provides appropriate quantitative and/or qualitative criteria in the form of instructions, procedures, or drawings; corrective action; quality records; and QA audits.

The procurement controls include procurement document control that provides for a supplier QA program, basic technical requirements, source inspection and audits, documentation requirements and lower-tier procurements. It also includes control of purchased material, equipment and services; corrective action; quality records; and audits of procurement activities.

For manufacturing/fabrication, the elements of instructions, procedures, and drawings; document control; identification and control of materials, parts, and components; control of special processes; inspection; and control of measuring and test equipment are noted. The elements of handling, storage and shipping; inspection, test and operating status; nonconforming items; correcti action; quality records; and QA audits are covered.

Testing includes the elements of test control; instructions, procedures, and drawings; document control; failure analysis; records; and QA audits.

Under construction/installation, the elements included are instructions, procedures and drawings; document control; control of purchases, material, equipment and services; identification and control of material, parts, and components; control of special processes, construction and installation inspection; and acceptance testing. The elements of control of measuring and test equipment; handling, storage and shipping; inspection and test status; nonconforming items; corrective action; quality records; and QA audits are also included.

The responsibility and authority to produce reliable GCEP machines and systems are delegated to contractors. From a QA/QC viewpoint, the construction management contractor, Stone and Webster, or the A-E overviews and monitors activities of their subcontractors, as well as other construction contractors under contract to the DOE. DOE Portsmouth contracts for and overviews the activities of contractors such as Stone and Webster, while DOE-ORO audits the QA activities of DOE Portsmouth and Stone and Webster.

Integration by DOIT Teams. A management team concept has been established and placed in effect at GCEP for work execution and control including quality assurance. This system is outlined in PPO-EP-104, GCEP Construction Work Package Execution and Control System (DOIT). It also establishes appropriate interactions among the many GCEP participants.

Working execution and control teams, or DOIT teams, are used to manage defined pieces of work (work packages) so they can be completed cost effectively within prescribed parameters. Currently, at GCEP four DOIT centers are in operation: DOIT-P (process facilities), DOIT-R (Recycle, Assembly, Centrifuge Training and Test Facilities), DOIT-S (site and support facilities), and DOIT-F (feed and withdrawal). Each DOIT center or team includes representatives from each of the principal and responsible participating organizations. These participants and their areas of responsibility are as follows:

> Portsmouth Project Office (PPO) - General Manager
> Architect-Engineer (A-E) - Engineering and Design
> Stone and Webster Engineering Corp. (SWEC) - Construction Management
> Operating Contractor (OC/OCPO) - Startup and Operations

In the embryonic stage, a work package (WP) is equivalent to a design package (DP). During its evolution it may encompass many DPs or combinations thereof. Regardless of evolution, type of contract, work breakdown structure, number, etc., the WP identifies a defined piece of work that is to be accomplished. Each WP includes technical requirements, quality standards, performance period, estimated cost, and how it is to interface with other WPs. The WP identifies a certain portion of the facility design by the A-E and its acceptable state before the Portsmouth Project Office (suboffice of DOE-ORO) can accept it from the constructor and to turn it over to the operating contractor for custodianship.

D.4.3 NASA Assurance of Quality (AOQ) Program

In the review of the National Aeronautics and Space Administration's (NASA) programs for safety, reliability, and quality assurance (SR&QA), extensive use has been made of previous studies on the NASA programs, which examined the possible applicability or transfer of the NASA program concepts to the NRC and nuclear power industry. These studies include Space and Missile Reliability and Safety Programs, prepared for the Nuclear Safety Analysis Center (NSAC), 1981, and the Application of Space and Aviation Technology to Improve the Safety and Reliability of Nuclear Power Plant Operations, prepared for the Department of Energy (DOE), 1980. Additional information on the NASA programs was obtained from published NASA documents and through meetings with NASA Headquarter's SR&QA staff.

The NASA approach for assuring the quality of their space and missile program is generally perceived as very successful. In only 25 years, the U.S. has probed the reaches of outer space, has had men walking on the surface of the moon, and has established a highway to space with space shuttle craft. These space projects have generally required advances in the technical state-of-the-art in many systems and subsystems.

D.4.3.1 Background

NASA was formed in 1958 (by legislation commonly referred to as the Space Act of 1958) and was given a Congressional mandate 1) to restore U.S. technological leadership, and 2) to lead the world into the space age for its peaceful benefits. NASA inherited several field installations from the U.S. Army, Navy, Air Force and the National Advisory Committee for Aeronautics. These sites include the Marshall and Goddard Space Flight Centers, Jet Propulsion Laboratory (California Institute of Technology), Cape Canaveral, Wallops Island, and Langley Research Center (NSAC 1981).

Since its origin, NASA has generally operated as a decentralized agency with the NASA headquarters responsible for the development of policy for field use. The field offices develop programs to implement the NASA policy and tailor the programs to fit and meet the peculiarities of the projects assigned to the NASA field offices. In mid-1961, a separate quality assurance organization was formed. With approximately 90% of the NASA-sponsored work performed by contractors, the QA organization initiated efforts to develop quality program requirement documents that would be used by the NASA Centers or field installations to place QA requirements on the NASA contractors. These early QA documents established basic QA policies that remain in effect today. These policies are as follows:

- Quality is the overall responsibility of the NASA Centers and cannot be delegated.

- Central direction is provided by a QA organization having responsibility and authority in each NASA Center.

- NASA QA requirements are included in NASA contracts.

- Assuring satisfactory performance in developing and maintaining system quality is the responsibility of the NASA procuring installation (NSAC 1981).

In 1967, the Apollo fire more sharply focused NASA attention on the need to assure system safety. The emphasis on systems engineering was increased. The importance and relationship of safety, reliability, and quality was recognized, and the three disciplines were coordinated and integrated. In the late 1960's, revised policy documents were issued to address revised requirements for system safety, reliability, and quality. These documents closely coordinated the safety analysis (such as hazard identification), reliability analysis (such as failure mode and effect analysis), and the quality program requirements necessary to achieve the safety and reliability performance goals.

D.4.3.2 Organization and Responsibilities

The Chief Engineer at Headquarters and the Directors of the various NASA Centers and laboratories all report directly to the NASA Administrator. The NASA headquarters' safety, reliability and quality assurance staff report to the Deputy Chief Engineer, whereas at the NASA Centers, the Director of the Safety, Reliability and Quality Assurance staff generally reports to the Center Director.

The basic responsibilities for planning, developing, conducting, and evaluating NASA programs and other activities to ensure the achievement of necessary levels of safety, reliability, and quality are identified in NHB 1700.1(V1.A), Basic Safety Manual, and NMI 5300.7B, Basic Policy and Responsibilities for Reliability and Quality Assurance. In addition, NASA documents NHB 5300.4(1A), Reliability Program Provisions for Aeronautical and Space System Contractors, and NHB 5300.4(1B), Quality Program Provisions for Aeronautical and Space System Contractors, contain specific reliability and quality program requirements for NASA contractors, while the safety requirements of NHB 1700.1(V1.A) are also applicable to contractors.

The basic NASA policy for SR&QA, as stated in the NASA policy documents, includes two objectives:

- Safety (NHB 1700.1(V1.A))

    1. Avoid loss of life, injury to personnel, damage to or loss of equipment or property, mission or test failure, and undue risk.
    2. Promote safety by instilling safety awareness in all NASA employees and contractors.

    3. Use an organized and systematic approach to identify and control hazards ensuring that safety factors are fully considered from conception to completion of all agency activities.

- Reliability and Quality Assurance (NMI 5300.7B)

    1. Plan and execute NASA activities to achieve levels of reliability and quality that are commensurate with mission objectives and overall life-cycle costs.

    2. Tailor the provisions of the reliability and quality assurance manual to the extent needed and consistent with NASA program planning. Use the NASA Procurement Regulation, NHB 5100.2, Part 14, in conjunction with the Reliability and Quality Assurance Manual for contracted effort.

    3. Define and assign reliability and quality assurance tasks to minimize duplication of resources, make effective use of ground and flight experience, and properly consider interfacing disciplines.

    4. Periodically review and evaluate plans, systems and activities for achieving reliability and quality to ensure that objectives will be met within available technology, funding and schedule constraints.

D.4.3.3 Program Requirements

Each NASA Center has the responsibility to develop and tailor programs to implement the NASA policy requirements for safety, reliability and quality for the programs and project activities assigned to that Center. To characterize the NASA program requirements for SR&QA, the NASA guidance documents for the space shuttle program were selected as examples to illustrate both the coordination and integration of NASA SR&QA. The following statement is an excerpt from the preface of NHB 5300.4(1D-1), Safety, Reliability, Maintainability and Quality Provisions for the Space Shuttle Program:

> This publication establishes common safety, reliability, maintainability and quality provisions for the Space Shuttle Program.
>
> NASA Centers shall use this publication both as the basis for negotiating safety, reliability, maintainability and quality requirements with Shuttle Program contractors and as the guideline for conduct of program safety, reliability, maintainability and quality activities at the Centers. Centers shall assure that applicable provisions of this publication are imposed in lower tier contracts. Centers shall give due regard to other Space Shuttle Program planning in order to provide an integrated total Space Shuttle Program activity.
>
> In the implementation of safety, reliability, maintainability and quality activities, consideration shall be given to hardware complexity, supplier experience, state of hardware development, unit cost, and hardware use. The approach and methods for contractor implementation shall be described in the contractor's safety, reliability, maintainability and quality plans.
>
> This publication incorporates provisions of NASA documents: NHB 1700.1, NASA Safety Manual, Vol. I; NHB 5300.4(1A), Reliability Program Provisions for Aeronautical and Space System Contractors; and NHB 5300.4(1B), Quality Program Provisions for Aeronautical and Space System Contractors. It has been tailored from the above documents based on experience in other programs. It is intended that this publication be reviewed and revised, as appropriate, to reflect new experience and to assure continuing viability.

NHB 5300.4(1D-1) stipulates the NASA approach for SR&QA, and requires the following:

- thorough planning and effective management

- definition of the major safety, reliability, maintainability, and quality assurance tasks and their place as an integral part of the design and development process

- evaluation of hardware safety, reliability, maintainability and quality through analysis, test, review, and assessment

- timely status indication by formal documentation and other reporting to facilitate implementation of the safety, reliability, maintainability, and quality assurance efforts

- compatible requirements among manufacturing, test and operational sites.

The following three sections present an overview of the NASA program requirements for safety, reliability and quality assurance for the space shuttle program.

Safety. According to NASA program requirements, a safety plan must be developed and must include a description of the approach for identifying, eliminating and/or controlling potential safety hazards that could lead to injury, loss of personnel, and damage or loss of flight or ground hardware throughout the program's complete cycle. The safety plan will integrate and describe the relationship of all safety activities. The safety requirements and tasks are to be reflected as appropriate in other program plans.

The basic elements of a NASA safety plan are summarized below (DOE 1980):

- System Safety Analysis. Establish and identify procedures and instructions that will be used to execute all safety analyses. Perform system and safety analyses to assure the following:

  - Safety is to be designed into the product. Known hazardous conditions that cannot be eliminated through equipment design or operation procedures are to be controlled or reduced to an acceptable level. Residual hazards shall be tracked and identified to NASA.

  - Hazard level classifications are to be used to provide a continuous tracking and status of severity to reduce catastrophic and critical hazards to controlled levels within the constraints of risk management.

- System Safety Guidelines and Constraints. Develop and establish system safety guidelines, constraints, and requirements to guide the vehicle system's design, ground support equipment design, and operations planning. These criteria shall satisfy programmatic guidelines and constraints, system safety goals, and other top-level safety objectives.

- Safety and Trade Studies. Specific, inherently hazardous characteristics of the alternatives being considered shall be identified. Rationale shall be documented to support the selected concept and to demonstrate that it includes the optimum safety provisions consistent with program objectives, risk management, performance, cost, and schedule.

- Hazard Analysis. Perform a qualitative hazard analysis to identify hazards and to assure their resolution. Hazards shall be defined and classified by hazard levels. Conduct periodic performance and refinement of hazard analysis and periodic assessment of achieved versus specified requirements. All hazards, including those resulting from failures, irrespective of subsystem or component redundancy, shall be analyzed. In addition to hazards resulting from failures, those emanating from normal or emergency equipment operations, environment, personnel error, design characteristics, and credible accidents shall also be analyzed. Identify and eliminate or control any failures or malfunctions that could independently or collectively present a hazard to interfacing hardware, and assure that normal operation of a hardware, item cannot degrade the safety of interfacing hardware or the total system. Early hazard analysis emphasizing design shall be the baseline of an expanded analysis. The hazard analysis shall be updated as the program progresses, providing continuity and covering the interrelated areas of design, operations, and vehicle subsystem integration.

- Human Engineering. Procedures shall be developed to assure the application of safety-related human engineering principles during design, development, manufacture, test, maintenance, and operation of the system or subsystem to minimize human error.

- Interface with Other Program Functions. Safety shall be coordinated and integrated with other program functions to avoid overlaps and conflicts among the technical disciplines, and to establish an integrated effort. This coordination shall include the delineation of responsibilities, management structures, joint analyses, reporting procedures, feedback of test data and corrective actions, use of failure mode and effects analyses, single failure point summaries, or other analytical techniques to identify hazards.

- Waivers and Deviations. For proposed waivers and deviations, the contractor shall establish a way to analyze the safety impact.

- Hazard Data Collection, Analysis, and Corrective Action. Using existing data systems wherever practical, a system for reporting hazards, data storage, and feedback of corrective action shall be formulated.

- Specifications and Procedures Review. Specifications and procedures for manufacturing, testing, and operations shall be reviewed to assure that these activities do not negate the inherent safety of the design.

- Review of Changes. When changes are proposed for equipment design or procedures, identify and resolve hazards that may be introduced into the system. Residual hazards shall be identified as part of the engineering change evaluation.

- Postflight Evaluation. System safety organizations shall participate in postflight reviews and a safety evaluation shall be made in cases where anomalous conditions are revealed. This safety evaluation will provide guidance in planning future missions and establishing necessary corrective action to reduce hazards.

Reliability. According to NASA program requirements, a reliability plan is to be developed in conjunction with other program plans. Reliability is an integrated part of the design and development process and is to include the evaluation of hardware reliability through analysis, review, assessment, and timely status reporting. The three major elements of the reliability program are reliability management, reliability engineering, and testing [NHB 5300.4(1D-1)].

NHB 5300.4(1A), Reliability Program Provisions for Aeronautical and Space System Contractors, prescribes general reliability program requirements for NASA contracts involving the design, development, fabrication, testing and/or use of aeronautical and space systems and elements thereof. Basically, it stipulates that the contractor will maintain a reliability activity planned and developed in conjunction with other contractor elements. Reliability functions will be an integral part of the design and development process and will include the evaluation of hardware reliability through analysis, review, and assessment. The contractor will provide, maintain and implement a Reliability Program Plan that describes how compliance will be ensured with the specified reliability requirements of engineering, design, failure mode analyses, testing and reliability assessments.

A summary of some of the major elements of the NASA reliability program is given below (DOE 1980).

The reliability management task involves the identification of a reliability organization that has unimpeded access to top management including main line and program managers ... [NASA requires] each contractor to conduct audits of their internal reliability and those of his suppliers ... [to] evaluate progress and effectiveness and ... determine the need for adjustments or changes in activities.

Each major contractor must ensure that the reliability of system elements from subcontractors and suppliers meets the requirements of the overall system. The level of reliability is tailored to the supplier.

The reliability engineering tasks involve development of reliability design criteria for each subsystem, a system for receiving and concurring on design specifications and changes, and assuring that no subsystem or component specifications violate reliability design criteria ....

The most in-depth analysis and example of reliability engineering disciplines comes in the area of establishing a system for conducting Failure Mode and Effects Analysis (FMEA) and the control of the results of this analysis, which are in the form of Critical Item Lists (CILs) of single failure points ....

NASA Reliability Engineering establishes the fundamental requirement for contractors supplying major space components to prepare design FMEAs at the lowest levels of system definition required to support potential uses, e.g., testing, failure reporting and corrective action, and preparation of mandatory inspection points. FMEAs must be performed to the 'black box' level and within the 'black box' to pursue all critical functions ... The FMEA includes an integration of all flight hardware, including government furnished equipment and essential launch ground equipment.

Contractors must support the internal and supplier's design reviews at the system, subsystem, and component levels as well as NASA design and readiness reviews. These reviews include the preliminary Design Review (PDR) which covers the system concept; the Critical Design Review (CDR) which is accomplished when the design is about 90% complete and components are ready for fabrication; the Design Certification Review (DCR) which is accomplished by NASA Headquarters; and, finally, the Flight Readiness Review (FRR) which determines that the equipment is ready for flight.

In summary, the NASA reliability technique includes:

● A well organized and managed reliability program.

● Defining and implementing tasks that prevent problems early in the program.

● Establishment of programmatic controls with required formatted documentation.

● Establishment of key points in the program to check and review progress and problems.

● Strict attention to detail by all organizations.

Quality Assurance. As for safety and reliability, a quality plan is to be developed in conjunction with other program plans. The elements of the quality plan are somewhat similar to the elements of 10 CFR 50, Appendix B, which is required for nuclear power plants. The NASA quality program as outlined in NHB 5300.4(1D-1) and NHB 5300.4(1B) is to do the following:

● demonstrate recognition of the quality aspects of the contract and an organized approach to achieve them.

● ensure that quality requirements are determined and satisfied throughout all phases of contract performance, including preliminary and engineering design, development, fabrication, processing, assembly, inspection, test, checkout, packaging, shipping, storage, maintenance, field use, flight preparations, flight operations, and post-flight analysis, as applicable.

- ensure that quality aspects are fully included in all designs and are continuously maintained in the fabricated articles and during operations.

- provide for the detection, documentation, and analysis of actual or potential deficiencies, system incompatibility, marginal quality, and trends or conditions which could result in unsatisfactory quality.

- provide timely and effective remedial and preventive action.

Also, the contractor will prepare, maintain, and implement a Quality Program Plan that describes how the contractor will ensure compliance with cited quality requirements. The Quality Program Plan will be submitted as required by the Request for Proposal or Contract. The plan format shall be readily identified with each cited requirement. The plan shall cover all quality program activities for the time period or phase authorized, be updated periodically and resubmitted, as specified in the contract, and serve as the master planning and control document.

NHB 5300.4(1B), Quality Program Provisions for Aeronautical and Space System Contractors, identifies the quality program requirements for NASA aeronautical and space programs, systems, subsystems and related services. Basically, the contractor will maintain an effective and timely quality program planned and developed in conjunction with all other contractor's functions necessary to satisfy the contract requirements.

D.4.3.4 Program Implementation

NASA Centers are to invoke the requirements of the reliability and quality assurance manual to the extent required and consistent with program planning in procurements of aeronautical or space systems, launch vehicles, spacecraft, associated ground support equipment or elements thereof to ensure the required high quality of materials, parts, components and services; to design reliability into aeronautical and space systems; and to prevent degradation of the design's reliability through the succeeding steps from fabrication to end use. Because their programs require delivery of only small numbers of each system, operate under tight schedules, and require high reliability in the first, as well as subsequent systems, NASA has developed and implemented a program wherein contractors and suppliers use a thoroughly disciplined, systematic approach to safety, reliability, and quality.

NASA requires that engineering designs be essentially completed (90 to 100%) and reviewed prior to starting fabrication work. Further, in-depth, detailed "readiness reviews" are conducted at key points in a program before proceeding with the next phases or steps. These reviews assure that all changes and discrepancies have been properly addressed and resolved.

D.4.3.5 Coordination of Programs

The emphasis of NASA on safety, reliability, and quality assurance programs appears to stem from the definite commitment to coordinate and integrate these programs to achieve the common overall program objective--a safe, reliable product with the necessary level of quality to meet program performance objectives. In DOE (1980) there is a discussion on the NASA system safety approaches and reasoning or rationale behind these approaches. Listed below are a few of these features that appear to be applicable to the "entire systems" approach used by NASA to coordinate and integrate the safety, reliability and quality assurance programs and plans.

- The complexity of systems, subsystems, and components under extreme and varying environment and application conditions places heavy demand on safety systems. The inherent complexity of the NASA flight hardware systems demands technical and analytical techniques of considerable sophistication to identify and solve problems.
- The need to focus considerable attention on the safety considerations arising out of total systems effects cannot be discovered by considering portions of the system independently.

- Assure that the safety aspects of the mission under normal conditions and under mission failure conditions are adequate.

- Know the hazardous characteristics of the system, including operation under all environmental conditions during design, manufacture, test, transportation, storage, and operation. "System" includes the hardware, flight and ground support equipment/electrical support equipment, the facilities, and the procedures that are used to operate and test the system.

- Eliminate, insofar as possible, these hazards. If the hazards cannot be eliminated, take all practical steps to control them. These steps include both hardware and software considerations.

- Recognize that the management responsibility for achieving system safety flows along program organizational lines.

- Keep in mind that the desired results from system safety activities are to minimize risks to the maximum practical extent and apply the knowledge of these risks to management decisions. Also, assure an understanding at all management levels as to the risks being incurred by testing, transporting, or operating the system or portions of the system.

D.4.4 U.S. Shipbuilding Assurance of Quality (AOQ) Programs

This section discusses the AOQ programs for both U.S. Naval shipbuilding and commercial shipbuilding. For each, the program requirements and implementation are described.

This study was based totally on publicly available information obtained from a comprehensive review (including computerized literature data base searches) of pertinent references. Sources for these references included the Naval Sea Systems Command Library, the Naval Sea Systems Command Directives, the Defense Logistics Agency, the National Technical Information Service, the U.S. Department of Transportation Library (including both Maritime Administration and U.S. Coast Guard Material), and the American Bureau of Shipping Library.

However, to validate more completely the material presented here, as well as to expand the material collected so that it describes in greater detail how the programmatic aspects actually work, an outside group with technical experience on U.S. commercial and naval shipyard operations (COMEX) reviewed and expanded the material.

D.4.4.1 U.S. Naval Shipbuilding AOQ Program

The U.S. Naval shipbuilding program involves both nuclear and non-nuclear ship construction. Such construction uses many of the same kinds of materials, construction techniques, and skills used in the civilian commercial nuclear industry. The potential for hazard to the general public and a strong governmental involvement closely relate the two programs.

Background. Before 1960, no formally established quality assurance program existed in Naval or private shipyards. In November of 1960, the Bureau of Ships (now Naval Sea Systems Command - "NAVSEA") published an instruction that formally established a quality assurance program in the shipyards. A Quality Assurance Division was formed in the Production Department (which is primarily responsible for all phases of ship construction in the yard) partly by bringing together existing functions, including inspection and test sections, laboratory functions, and the welding engineers. In the Nuclear Power Division that was set up in some shipyards, there was also a responsibility for quality control functions for all operations involving nuclear power.

In 1966, the publication of a revised edition of the Standard Regulations established a mandatory Quality and Reliability Assurance Department. From 1966 to 1975, various instructions, notices and publications addressing the assurance of quality and reliability were promulgated by the Defense Department, NAVSEA, Naval shipyards, private shipyards and commercial vendors. By 1975 every Naval shipyard and all private shipyards performing work for the Navy had quality control and assurance instructions and manuals. Areas such as the nuclear propulsion program or areas of specific interest or having special, more rigorous requirements or problems had their own instructions, which amplified these basic manuals and directives.

Quality Program Organization and Requirements. The AOQ program for U.S. Naval ship construction is based on Title 32 of the Code of Federal Regulations (National Defense) DAR Section XIV, Procurement Quality Assurance. This defines the government function by which it determines whether a contractor has fulfilled its contract quality and quantity obligations. The contractor is responsible for controlling product quality and for offering to the government for acceptance only those supplies and services that conform to contract requirements. When required, the contractor also must maintain and furnish substantiating evidence of this conformance.

The organization responsible for technical requirements (e.g., specifications, drawings and standards) prescribes inspection, testing, or other contract quality requirements that are essential to assure the integrity of products and services (32 CFR). Systematic control of manufacturing processes by the producer is also an essential prerequisite for assuring the quality of such items (32 CFR). However, criteria for applying contract quality requirements can be dependent on each item's character, importance, and application.

The general framework for the regulations currently governing the assurance of quality program for the U.S. Naval ship construction program, both nuclear and non-nuclear, is shown in Figure D.3.

Three military standards/specifications form the implementing basis (32 CFR) for Department of Defense assurance of quality programs: MIL-STD-109B, Quality Assurance Terms and Definitions; MIL-Q-9858A, Quality Program Requirements; and MIL-I-45208A, Inspection System Requirements.

- MIL-STD-109B Quality Assurance Terms and Definitions. The intent of this standard is to ensure that the Department of Defense quality assurance organizations are able to implement policies based on a commonality in language.

- MIL-Q-9858A Quality Program Requirements. This specification is applicable to the Department of the Army, the Navy, the Air Force, and the Defense Supply Agency. It requires the establishment of a quality program by all contractors furnishing equipment, systems, subsystems, and/or services to the Department of Defense. Commonly referred to as "MIL-Q," this document allows the Supervisor of Shipbuilding, Conversion, and Repair (SUPSHIP) organizations to direct the contractor to establish a quality control and assurance program for a specific procurement in excess of standard contractual obligations. When invoked, MIL-Q requires the contractor to establish the programs and requires the Government Representative, in this case SUPSHIP, to approve and monitor the program.

The complexity of such a program varies, depending upon the work being performed by the contractor. For example, private shipyards engaged in construction of nuclear submarines typically have quality assurance organizations and programs at least as sophisticated as those of Naval shipyards. On the other hand, a private yard engaged only in constructing or repairing small auxiliary vessels such as tugs and barges would not need nearly as complex an organization to satisfy MIL-Q. In April 1965, the Assistant Secretary of Defense (Installations and Logistics) published the Quality and Reliability Assurance Handbook (H 50), which provides general guidance to personnel responsible for evaluating a contractor's quality program when Military Specification MIL-Q-9858A is invoked in the contract.

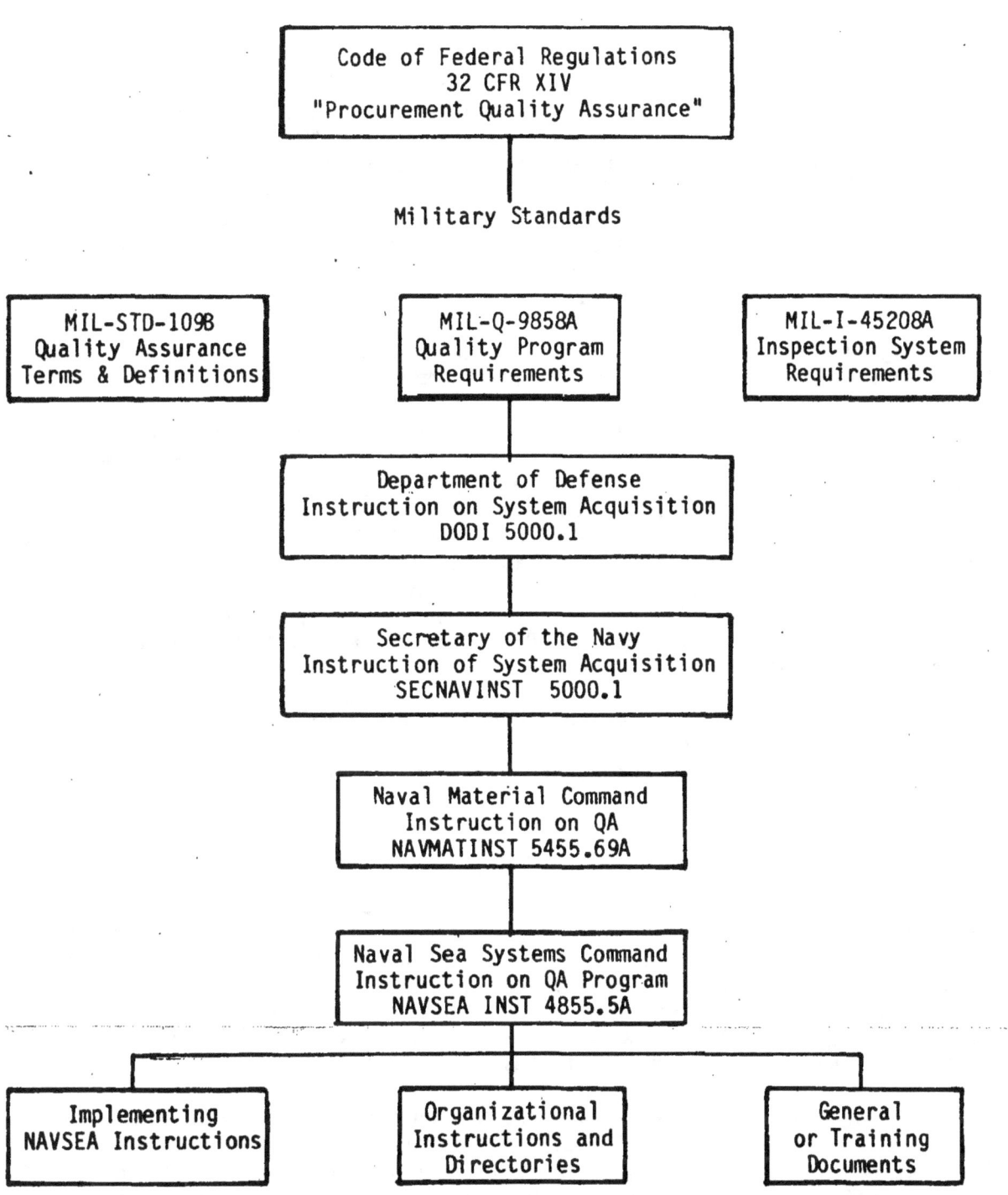

FIGURE D.3. Regulations for the Assurance of Quality Program for the U.S. Naval Ship Construction Program

- MIL-I-45208A - Inspection System Requirements. This specification establishes requirements for the inspection and tests that the contractors must perform to substantiate product conformance to drawings, specifications and contract requirements and to all inspections and tests required by the contract. These requirements are in addition to those inspections and tests set forth in applicable specifications and other contractual documents. Commonly referred to as "MIL-I," this document is similar to MIL-Q in intent, use and assignment of responsibility. Like MIL-Q, the contractor's efforts to satisfy MIL-I requirements varies with the specific procurement. Again, the contractor must satisfy SUPSHIP that compliance has been achieved and is being maintained adequately. MIL-Q and MIL-I interrelate in that in satisfying the requirements of MIL-Q, a contractor may have also satisfied MIL-I requirements.

Program Implementation. The AOQ program is directly administered by the Naval Material Command, Naval Sea Systems Command, and the Naval shipyards.

Within the Navy, the Naval Material Command implements the overall procurement QA program in accordance with NAVMATINST 4355.69A. This document is identical to that used for the Defense Supply Agency (DSAM 8200.1). Army (AR 702-4), Air Force (74-15) and Marine Corps (MCOP 4855.4A). The Deputy Chief of Naval Material for Reliability, Maintainability and Quality Assurance (MAT 06) is responsible for AOQ programs regarding the acquisition of naval material, and has four components reporting to him: (1) Program Assessment Division (MAT 061), (2) Reliability and Maintainability Division (MAT 062), (3) Quality Assurance Division (MAT 063), and (4) Manufacturing Technology Division (MAT 064).

The Naval Sea Systems Command reports to the Naval Material Command and serves as the coordinator of shipbuilding, conversion, and repair for the Department of Defense; and coordinator of ship repair and conversion of the Department of Defense/Department of Commerce.

In July 1975, the Naval Shipyard Quality Program Manual, NAVSEA 0900-LP-083-0010, was promulgated (NAVSEA 1982). This manual established the minimum quality program requirements for constructing, converting, modifying, overhauling, and refurbishing Naval ships and craft. This generic document addresses general responsibilities, technical data, work instructions and authorizations, procurement quality control, material control, process controls, metrology and calibration, inspection and verification, corrective action, preventive action, audits, and training. By mid-1977 the provisions of this manual had been implemented in Naval and private shipyards.

Although the requirements for quality programs are reasonably well consolidated in NAVSEA 0900-LP-083-0010, countless amplifying documents and instructions exist that are more specific, detailed or tailored to the specific needs of ship operators, ship or system types. Separate programs that are distinct from the overall Navy AOQ exist, although they are generally consistent with the overall objective of ensuring safe, reliable output from research and development and operational activities. The SUBSAFE program, for example, is an entire program amplifying the guidelines of the basic instruction to more specifically address submarine safety. Another program, the Naval reactor propulsion program, contains numerous specific quality assurance directives addressing aspects of quality control and assurance ranging from identifying and controlling materials suitable for reactor plant application to controlling safety in reactor system and subsystem operations and testing.

The operational branches of NAVSEA include the Naval shipyards, Naval ship repair facilities, and the offices providing the liaison between the Navy Department and commercial shipyards and repair activities. The liaison offices include the offices of the Supervisor of Shipbuilding, Conversion and Repair, commonly referred to as SUPSHIP. The Naval Shipyard Commander and the Supervisor of a SUPSHIP organization share the responsibility for completing the construction, repair or overhaul mission with the vessel's commanding officer. Frequently, the Naval Shipyard Commander is also the Supervisor of the SUPSHIP office within his Naval District. The administration of the quality program is assigned to the Quality Assurance Officer of the Naval Shipyard and SUPSHIP organization.

The objectives and functions of a Defense Department contract administration activity such as a SUPSHIP office have several major distinctions from those of a Naval shipyard quality assurance organization. The chief difference is in the administration of quality control and assurance procedures. Commercial contractors performing work for the government are required by contracts to assure compliance with the quality requirements of the specific contract being performed. Certain basic minimum quality program requirements must be met for a private firm to be qualified to perform certain types of government work. For the types of work ordinarily performed under SUPSHIP cognizance, these basic guidelines are required by the Master Ship Construction contract (MSC) or the Master Ship Repair contract (MSR). To be eligible to bid for Navy ship construction or repair work, a signator private firm must continuously comply with the provisions of either the MSC or MSR contract regardless of whether the company is presently performing government contract work. Among these provisions are those addressing quality assurance and control.

The SUPSHIP organization acts as the liaison between the government customer and the commercial supplier in all matters including administering the contract and control for quality. The Quality Assurance Office is guided by two basic documents, which have extensive quality assurance supplements:

a. Ship Acquisition Contract Administration Manual (SACAM) - the governing document for use by SUPSHIP in contracting for the construction of vessels.

b. Ship Repair Contracting Manual (Repair Manual) - the governing document for use by SUPSHIP in contracting for ship repair.

Standard Naval shipyard organization for assurance of quality is specified in NAVSEA instruction 5450.14, the Standard Naval Shipyard Organization Manual. Any deviation must be approved by the Naval Sea Systems Command. Besides outlining the organizational structure of Naval shipyards, this organization also describes the duties and responsibilities of staff within the organization.

Using excerpts from NAVSEA Instruction 5450.14, the following subsections briefly discuss the duties and relationships of typical shipyard organizations. First, however, several key observations must be explained to adequately understand how the AOQ function actually works within the Naval shipyard.

- There are parallel, complementary, organizations within the shipyard for non-nuclear and nuclear matters.

- A DOE representative called "Naval Reactors Division of Naval Reactors Representative" is assigned to every Naval shipyard that performs nuclear reactor plant work. This representative reports to the Director, Division of Naval Reactors, U.S. DOE, and provides the Shipyard Commander with an independent review and surveillance of all shipyard work relating to Naval nuclear propulsion matters. The Representative has free access to all elements of the shipyard dealing directly or indirectly with Naval nuclear propulsion. The review and surveillance is not intended to detract from, change or diminish the existing responsibility of the Nuclear Engineering Manager or any other shipyard official. The Naval Reactors Representative is provided suitable and sufficient office space in the shipyard and other administrative support to carry out the assigned function.

- The USN puts the burden of proof for assurance of quality totally on the contractor. USN inspections, while extensive and involving both shipyard and ship's force review, do not substitute for a contractor inspection, and the use by the contractor of independent auditors is encouraged. The intent is to allow the USN inspectors to selectively review phases of the overall program rather than become immeshed in minute details of specific technical areas.

- The USN shipbuilding program requires readiness reviews at the shipyard project level involving ship's force, shipyard departments, contractors, and quality assurance staff at both periodic (e.g., weekly) intervals and workphase points (e.g., pre-hydrostatic testing) (NAVSEA Instruction 5450.14).

- USN shipyard organizational structures have been mandated to ensure that the QA manager and the onsite Naval reactors representative have direct access and responsibility for reporting to senior shipyard management and their respective directorates at Naval Sea Systems Command headquarters.

- A prioritization effort has been made in the USN program to base quality requirements on and to direct audits to the equipment, systems, and/or material which are most critical. A formal mechanism is established for classifying or prioritizing quality efforts to ensure that attention regarding assurance of quality is not placed only on what just happened ("squeaky wheel" syndrome). An important distinction is made, however, that while the range and depth of requirements may change depending on the importance of the component or system, no adjustment is made in the degree of compliance (i.e., requirements must be met).

- The USN has issued strong guidance for detecting and preventing deliberate malpractice and fraud related to construction assurance of quality programs (NAVSEA 1976).

- Careful attention has been taken to ensure that onsite inspectors are not overloaded with administrative responsibilities (i.e., duties are prioritized) such that they do no have the freedom or time to examine problem areas as they arise.

- The USN has established a program to evaluate the quality of contractor products and maintains a computerized data bank of vendor record and component performance information, accessible to both USN and commercial staff. The vendor and component information collected in this data bank is analyzed to identify and track potential items of concern. These analyses have been characterized to look for "warning signs" or other indicators to key inspection staff on aspects or features of assurance of quality programs that need to be monitored in closer detail.

In the Naval shipyard, there are seven individuals whose functions directly encompass the assurance of quality. Their responsibilities for implementation of the quality programs are discussed below.

- Nuclear Engineering Manager. The Nuclear Engineering Manager is responsible to the Shipyard Commander for resolving all nuclear reactor plant technical matters. These responsibilities include the following:

    - testing nuclear reactor plants and integrated propulsion plants on nuclear powered ships

    - advising responsible shipyard officials on quality control and radiological controls of such work, including special fabrication procedures, instructions, proper manning levels, erection and overhaul schedules and sequences, estimates, facilities, and industrial safety and security

    - quality control engineering of nuclear reactor plant work.

- Head, Nuclear Quality Engineering Division. The Head of the Nuclear Quality Engineering Division is responsible during the construction, overhaul, testing, maintenance and refueling of Naval reactor plants for the following:

    - establishing or causing to be established quality control procedures to be used for nuclear reactor plant work

    - analyzing and assessing the quality of reactor plant work; recommending remedial actions to correct and prevent recurrence of errors in workmanship and procedures

- providing information feedback to NAVSEA for improving specifications

- conducting irregular periodic audits of shipyard operations related to nuclear reactor plant quality control matters

- performing the responsibilities of the "governmental inspector" for reactor plant work, as defined in applicable NAVSEA standards.

● **Production Officer.** The production department, headed by the Production Officer, plans, schedules, and supervises all new ship construction work.

● **Nuclear Production Manager.** The Production Officer's nuclear area supervisor is the Nuclear Production Manager, who also has direct access to the Shipyard Commander. Duties include the following:

- exercising line authority as a deputy to the Production Officer for the nuclear aspects of new construction, overhaul, testing, refueling and core loading of concern to the Production Department

- assuring that all nuclear production work is accomplished on time, at reasonable cost, and in accordance with specified technical requirements and good workmanship standards

assuring that lists of production personnel qualified for nuclear work are maintained and concurring in such lists.

● **Quality Assurance Officer.** The Quality Assurance Officer reports directly to the Shipyard Commander and is responsible for the following:

- planning, executing and monitoring a quality assurance program for the shipyard in accordance with applicable quality-assurance criteria and with due consideration to the safety of ships, equipment and personnel

- planning and managing a quality-cost measurement program for the shipyard (prevention, appraisal and failure costs)

- providing guidance, integration, and evaluation of the efforts of the shipyard toward the prevention of product quality degradation

- investigating and evaluating quality problems to determine the fundamental cause, cost, scope, and significance of the problems

- directing a shipyard program to ensure calibration of measuring and testing equipment; maintaining measurement standards and performing calibration

- developing a quality assurance training program for the shipyard

- performing quality assurance functions such as inspection, physical and chemical testing, qualification testing, non-destructive testing; witnessing formal operational tests, as assigned; performing audits of the procedures, conduct, and records of inspections; and performing tests of weight-handling equipment

- making failure mode analyses and process capability studies

- establishing technical requirements for metal fabrication and thermal joining processes

- managing to the shipyard quality assurance audit program, and performing internal audits to determine shipyard compliance with quality requirements

- executing such research, development, test and evaluation programs as are assigned.

- **Chief Quality Assurance Engineer.** The Chief Quality Assurance Engineer reports to the Quality Assurance Officers. Responsibilities include performing and coordinating all activities of the Quality Assurance Office, with the exception of those functions performed by the Nuclear Quality Assurance Manager.

- **Nuclear Quality Assurance Manager.** The Nuclear Quality Assurance Manager exercises line authority as a deputy to the Quality Assurance Officer for the nuclear quality assurance of new construction, overhaul, testing, refueling and core loading of Naval reactor plants. The Nuclear Quality Assurance Manager has direct access to the Shipyard Commander, and is responsible for the following:

  - confirming that nuclear work is performed to specifications and procedures and recording required data to document that the work is performed correctly, including maintenance of documentation files

  - informing the appropriate department heads and the Shipyard Commander of work not being performed to specified requirements or not in accordance with safety practices

  - assuring that adequate planning and scheduling are provided for the nuclear work performed under the responsibility of the Quality Assurance Officer, including assuring that adequate manpower resources and equipment are provided within the Quality Assurance Office to prepare for and perform reactor plant work

- keeping informed of the nuclear work performed under the cognizance of the Chief Quality Assurance Engineer and assuring that the Quality Assurance Officer and the Shipyard Commander are advised of work not performed to required standards

- assuring that lists of Quality Assurance Office personnel qualified for nuclear work are maintained and concurring in such lists

- consistent with the above, seeing that all functions of the Quality Assurance Office concerned with nuclear work are accomplished on time and at reasonable cost.

D.4.4.2 U.S. Commercial Shipbuilding AOQ Program

In areas of complexity, potential for hazard to the general public, and strong governmental involvement, the U.S. merchant marine shipbuilding program can be related to the U.S. commercial nuclear power plant construction program. Vessels include supertankers, combined ore/bulk/oil (OBO) carriers, and liquefied natural gas (LNG) carriers, in addition to containerships, barge carriers and roll-on/roll-off van carriers (Maritime Administration 1982). Because these ships are so large, so complex, and in many cases carry hazardous cargos, assurance of quality in construction is essential.

Background. Private shipyards in the United States employ approximately 175,000 people, about two-thirds of whom are concentrated at 26 major shipyards involved in constructing naval ships and/or major ocean-going or Great Lakes merchant ships (Maritime Administration 1982). The deep-draft merchant vessels being constructed represent the largest and most complex mobile structures manufactured. Their assembly involves nearly every kind of material, draws on the products of almost every industry, employs almost every skill, and is intended to achieve a thirty-year life, which is comparable to that of a commercial nuclear facility. Many of the ships being constructed represent advanced designs that are equivalent to three to five of the older ships that they replace.

Governmental regulatory bodies are involved in the assurance of quality for U.S. merchant vessels for two major reasons:

- concern for the substantial hazard to life and property from commercial vessels in the case of an accident

- involvement of the U.S. Government in Construction Differential Subsidy (CDS).[a] (Although this was applicable to the program studied, the subsidy has reportedly been discontinued except for contracts existing in 1982.)

(a) In 1981 the Maritime Administration (MarAd) awarded CDS contracts to build 83 new merchant ships valued at $4.4 billion; the government's share, including national defense features, was $1.74 billion (Maritime Administration 1982).

AOQ Program Organization and Requirements. The United States Coast Guard (USCG) is responsible for enforcing rules and regulations set down in Title 46 (Shipping) of the Code of Federal Regulations necessary for the safe construction and operation of U.S. flag vessels. The USCG inspects and certifies various tanker, passenger, cargo, and miscellaneous ships prior to their use. The principal legislative authority for these inspection and certification activities are found in 46 CFR 369 and 391. The USCG inspection and certification regulations apply to nearly all large vessels. Smaller vessels may or may not be covered, depending on their size, capacity, and type of power. The USCG Merchant Vessel Inspection Division in the Office of Merchant Marine Safety administers the inspection and certification.

The USCG regulatory structure for each class of vessel is basically similar. Initially, the USCG must approve the plans for a proposed tanker [46 CFR 31.10-5(a)], passenger (46 CFR 71.20-10), or cargo (46 CFR 91.20-5) vessel. When a vessel passes the initial inspection upon completion of construction, the USCG issues a certificate of inspection. Several points relating to assurance of quality in the USCG program are worth noting:

- On a trial trip of each new or converted ship, an inspector is present to observe safe operation (46 CFR 31.10-40).

- It has been conservatively estimated that 9 percent of the total estimated construction costs of a vessel are due to U.S. government regulation. The U.S. government requirements themselves, however, are essentially the same in most cases as internationally recognized standards (Ernst and Whinney 1979, pp. 7-12).

- A survey by Ernst and Whinney in 1979 found that most shipping and shipbuilding companies (>80%) felt that regardless of current regulations, they would continue to perform the same inspections and tests at the same frequency because of their concern for the safety of the crew and ship. Because safety and the assurance of quality were felt to be everyone's concern, including the vessel owner's, mutual understanding and working relationships would be preferred and should be in general more effective than the adversary position that was sometimes felt to exist between the Coast Guard and the industry (Ernst and Whinney 1979, pp. 5-31).

- In the inspection of hulls, boilers, and machinery, the current standards established by the American Bureau of Shipping (ABS) are designated Rules for Building and Classing Steel Vessels. These apply to materials and construction of hulls, boilers, and machinery, except as provided for by other regulations in Title 46, and are accepted as standard by the USCG. The approved plans and certificate of the ABS, or other recognized classification societies for classed vessels, may be accepted by the USCG as evidence of the structural integrity of the hull and the reliability of vessels, except as otherwise specified in 46 CFR 31.10-1.

- Since May 1965, the ABS has been designated as an organization duly authorized to issue the "Cargo Ship Safety Construction Certificate" to certain cargo ships on behalf of the United States of America as provided in the regulations. At the option of the owner or agent of a vessel and on direct application to the ABS, the ABS may issue to a vessel a Cargo Ship Safety Construction Certificate having a period of validity of not more than five years. If the ABS determines that a vessel that was issued a Cargo Ship Safety Construction Certificate no longer complies with the ABS's applicable classification requirements, it will immediately furnish the USCG with all the relevant information to determine whether the USCG will withdraw, revoke or suspend the certificate (46 CFR 91.60-45).

- Before any construction or conversion is started on a vessel, application for the approval of contract plants and specifications and for a certificate of inspection is made in writing to the USCG, and construction or conversion cannot proceed until approval is granted (46 CFR 31.01-20).

- Triplicate copies of contract plans and specifications are forwarded to the Officer in Charge, Marine Inspection, in whose district the construction will take place, for submission to headquarters for approval. However, if the vessel is to be classed, such plans and specifications shall first be approved by a recognized classification society. If the plans and specifications are adequate, they are approved. During construction and upon completion of construction, each vessel is inspected by the Officer in Charge, Marine Inspection, to determine whether it has been built in accordance with the approved plans and specifications (46 CFR 31.10-5).

AOQ Program Implementation. This section discusses the roles of the MarAd, the ABS, the Ship Structure Committee and their interactions with the USCG. However, before discussing the roles of each of these organizations, the following are noted as significant factors in the assurance of quality for the commercial shipbuilding program.

1. A cooperative relationship has been fostered between the builders, buyers, regulators, and standards-setting groups. An emphasis has been placed by the interested parties on maintaining cohesiveness rather than retaining individual freedoms. Involvement of the federal government with industry through the active participation of staff members on standards and codes committees and Memorandums of Understanding has been successful.

2. Both ABS and USCG have a corps of Inspectors/Surveyors adept at shipbuilding practices and interpretation and enforcement of their respective rules. The autonomy of these Inspectors/Surveyors is generally accepted by U.S. shipbuilders without the adversary relationship so common in other programs. The reason for this acceptance, as outlined by Lisanby and Hass (1981), lies in the commercial impact, since denial of certification is accepted by the courts as proof of failure on the part of the shipbuilder so that the commercial loss of the customer can be shifted to the shipbuilder.

3. Standardization of ship design is a major industry policy, which has greatly simplified assurance of quality.

• **Maritime Administration.** The MarAd, an agency of the U.S. Department of Transportation since August 6, 1981, administers federal programs designed to promote and maintain a merchant marine capable of meeting America's requirements for both commercial trade and national defense. From 1950 to 1981, MarAd was part of the Commerce Department.

The MarAd is indirectly involved in assurance of quality, and is mainly an economic and contractual, not a technical, organization whose purpose is to ensure that subsidies (where applied) are not misspent. To accomplish this, it has established policies and procedures for the conduct of subsidy condition surveys (46 CFR 272.2-5). Besides requiring and specifying the extent of surveys necessary to validate subsidies, the MarAd also is involved in developing guidance to assist the maritime industry and in preparing detailed ship specifications. The MarAd's Standard Specifications for Merchant Ship Construction (PB-290,400; January 1979) requires that the contractor submit working plans within 60 days after the award of the contract, including schedules for readiness reviews. These plans identify which reviews are required, who will participate, and what will be reviewed, including change orders. Finally, the MarAd is involved in promoting the U.S. maritime industry through its research and development programs.

• **American Bureau of Shipping.** The American Bureau of Shipping (ABS) is a nonprofit, nongovernmental ship classification society that establishes and administers standards (which it refers to as Rules) for the design, construction and periodic surveillance of merchant ships and other marine structures. Members of the society include naval architects, marine underwriters, ship-owners, shipbuilders, and governmental representatives (including U.S. Department of the Navy, U.S. Coast Guard, and U.S. Maritime Administration). The ABS acts as a self-regulatory agency to the maritime industry, not just in the United States, but in over 90 countries.

The ABS's charter is to continually work to develop and update its rules through a pyramidal committee structure that comprises 19 technical committees and panels, whose members serve without compensation to ensure impartiality. Rule changes are initiated at the special committee or panel level, or by someone in the maritime field. If a special committee (e.g., Materials, Electrical Engineering, Nuclear Applications, Welding, or Operations) recommends that a rule be adopted or changed, such a proposal (depending on application) is forwarded to one of two full committees (Engineering or Naval Architecture).

This second committee will arbitrate whether such action should be taken, then submit their recommendation to the technical committee, which has the final say on each proposal's acceptability. These rules are published in an array of documents that apply to ship construction. The basic implementing document for most major ship construction is Rules for Building and Classing Steel Vessels, which is annually updated. This document is called out in 46 CFR 31.10-1 regarding required USCG inspections accepted as standard by the Coast Guard, except as appropriately noted in the regulations.

The following excerpt from the ABS description of classification describes
how the rules are administered:

The formal classification procedure begins when an official request
for the classification of a ship or marine structure is voluntarily
submitted to ABS. This usually results from an owner specifying a
desire for ABS classification to the shipyard whereupon the shipyard
contracts for classification serves with ABS.

The vessel design is then submitted to ABS for verification that the
plans conform to accept standards of good practice for vessel design
embodied in the 'ABS Rules for Building and Classing Steel Vessels,'
or other various ABS Rules listed earlier. So, in reviewing a given
set of design plans, ABS is comparing them with a compendium of
experience factors and proven scientific principles. In this way,
ABS is able to determine whether the design is adequate in its
structural and mechanical concept and, therefore, acceptable to be
translated into an actual vessel.

To conduct the plan review function, the classification society
employs technical staff surveyors trained in the skills of naval
architecture, marine engineering, and other associated disciplines. These specialists scrutinize the vessel's design to confirm
that the details comply with the standards set forth in the published Rules. Their review may also include sophisticated analytical procedures employing one of the many ABS computer programs.
If the design is found to be not in compliance with the Rules, ABS
amends the plans or notifies the owner or designer of the departures
from the Rule requirements. During the entire review process, ABS
is available for consultations with the owner and designer.

After a design has been reviewed by ABS technical surveyors and
found to be in conformance with the Rules, ABS field surveyors, who
are experienced in the construction of hulls and fabrication of
machinery and components, 'live with the vessel' at the shipyard
from keel laying to delivery. In so doing, they survey construction
to verify that the plans are followed, workmanship is of the best
quality, and the Rules are adhered to in all respects. Field surveyors also witness testing of material, machinery, and components
at manufacturers' plants and fabricators' shops to determine that
they also comply with the Rules. During the entire time of construction, ABS maintains an ongoing dialogue with the owner and
shipyard to make sure the Rules are understood and adhered to and
also to assist in resolving any differences that may arise.

When completed, a vessel undergoes sea trials and an ABS field sur-
veyor attends the trials to verify that the vessel performs according to the requirements as specified in the Rules. In order for a
vessel to be formally classed, a report must be presented to the ABS
Classification Committee. This Committee, composed of prominent

individuals from the maritime industry who serve without compensation, meets twice a month to perform a final review of the vessel's 'credentials.' A vessel found to be acceptable in all respects according to the Rules is then granted ABS classification by that Committee and issued an official ABS classification certificate. In granting class the Committee is saying, in essence, that the vessel is in conformance with the ABS Rules and to that extent is mechanically and structurally fit for its intended service.

An additional item of some importance concerning the ABS involves a Memorandum of Understanding (MOU) signed by the ABS and the USCG in early 1982. As stated in the 1982 ABS annual report, page 3:

One memorandum provides for Coast Guard acceptance of ABS admeasurement and tonnage certification of all U.S. flag vessels. The other, known as MOU II, is an expansion of an earlier Memorandum, known as MOU I, signed in June of 1981, and provides for Coast Guard acceptance of ABS plan review and inspection of various hull and machinery items for new construction of U.S. flag vessels built to the classification requirements of the Coast Guard.

In this regard, it was written into the memorandum that an orderly and deliberate transition will be assured through a phasing-in-process, thereby allowing ABS to augment its resources as necessary and appropriate.

- Ship Structure Committee. To integrate research on marine transportation, a committee involving most of the major participants was formed. As stated by the Booz-Allen Study (1981), the mandate of the interagency Ship Structure Committee (SSC) is to conduct an aggressive research program. This program's objective is, in the light of changing technology in marine transportation, to improve the design, materials and construction of the hull structure of ships and other marine structures by extending knowledge in these fields. Its ultimate purpose is to increase the safe and economic operation of all marine structures. The SSC is composed of one senior official each from the USCG, Naval Sea Systems Command, Military Sealift Command, MarAd, and the ABS. In 1977, the U.S. Geological Survey, which is responsible for the personnel, safety and environmental aspects associated with the offshore oil and mining industry, agreed to participate.

The SSC formulates policy, approves program plans, and directs funds from its member agencies into the research program. Four representatives from different divisions within each agency meet periodically as a Ship Structure Subcommittee to ensure achievement of the program goals and to evaluate the results in terms of ship structural design, construction and operation.

D.4.5 NRC Assurance of Quality (AOQ) Program

This description of the NRC's program for assuring quality in the design and construction of nuclear power plants has been developed by reviewing the available literature and by conducting interviews with NRC staff both at the headquarters of the Inspection and Enforcement Office and at regional offices in Atlanta, Georgia; Arlington, Texas; and Walnut Creek, California. The purpose of this review is to provide a basis for evaluating the transferability of AOQ program features and practices from other industries and agencies to the NRC and the industries involved in building nuclear power plants.

D.4.5.1 Background

The nuclear industry originated with the U.S. Army Engineers' Manhattan District Project in World War II. Shortly after the end of the war, a new government agency, the Atomic Energy Commission (AEC), was formed and the nuclear industry, which at that time involved only the federal government and its contractors, was transferred from the military to the AEC. The expansion into commercial applications by the private sector became possible with the passage of the Atomic Energy Act of 1954. A separate arm of the AEC was established to regulate the private sector in these commercial applications.

The Energy Reorganization Act of 1974 further separated the regulatory function from nuclear energy promotion by forming the Nuclear Regulatory Commission (NRC). This legislation also created the Energy Research and Development Administration (ERDA) to encourage and promote the commercial applications of nuclear energy, with the NRC responsible for the regulatory functions.

Although a few new corporate organizations dedicated to activities in the nuclear field came into being, the major thrust of the commercial industry was carried by existing corporate organizations. These organizations were primarily the electrical utilities, power plant designers (the architect-engineer firms), and their traditional suppliers of central station power plant equipment. The major corporations involved have tended to establish separate divisions or components directed to this new and evolving market place. In general, however, major corporations dedicated primarily to commercial applications of nuclear energy have not evolved in the United States.

The regulatory challenge to the NRC and its predecessor, the AEC, has been formidable. Nuclear technology has evolved very rapidly. In its short history, less than 30 years, the regulatory program and organization have experienced their own evolutionary processes while simultaneously regulating the "nuclear industry." This industry, however, has never existed as an entity in the traditional sense such as the iron and steel industry, the automobile industry or the aircraft industry. By contrast, the "nuclear industry" exists as an offshoot, almost a sideline, of several older, well-established industries, i.e., the utilities, the architect-engineers and the power plant equipment manufacturers. These industries had long been regulated to some extent by codes and standards, public utility commissions, etc. However, the depth and breadth of the NRC regulatory program certainly presented a major change from traditional business and working environments. In essence, fully mature business enterprises with long, well-established methods of operating had to make major (in some cases nearly revolutionary) changes in order to participate in what appeared to be a growing market area. Some of these organizations have made the necessary adjustments much more readily than others. Implementing and maintaining an effective and consistent regulatory program throughout the U.S. under these institutional circumstances has been difficult. This regulatory situation appears much more difficult than, for example, regulation of the aircraft industry, in which the private sector and the regulatory process evolved in parallel. In the latter situation, corporate business traditions and practices evolved much more in concert with the government's regulatory program.

D.4.5.2 Organization and Responsibilities

When the NRC was formed in 1975, the major organizational components were Reactor Licensing, Fuels and Materials Licensing, Inspection and Enforcement, Regulatory Research, and Standards Development. The inspection and enforcement arm included a staff at NRC headquarters and five regional offices.

Criteria for licensee QA programs were developed by the Office of Standards Development. The review of licensees' proposed QA programs was in the licensing components of the Office of Reactor Licensing, and the Office of Inspection and Enforcement was responsible for ensuring that licensees carried out their commitments as approved by NRC licensing and presented in the Safety Analysis Report. The NRC's QA efforts for nuclear power plant construction, therefore, were distributed among the three major organizational divisions.

In 1981, the regional offices were separated from the headquarters Office of Inspection and Enforcement and began reporting directly to the NRC's Executive Director for Operations. In 1982 and 1983, the NRC's headquarters staff was reorganized (in a series of actions), and all QA efforts were assigned to the Office of Inspection and Enforcement. Most of the NRC staff members interviewed felt that this was a very positive step; centralizing QA activities provided a mechanism to expedite the resolution of any differences or disagreements among the various functions within the NRC. The organizational chart for the NRC staff as of January 1, 1983, is shown in Figure D.4.

The headquarters Office of Inspection and Enforcement is now responsible for developing criteria and standards for licensee QA programs, for reviewing licensee QA programs, for licensing (QA issues) and establishing policies, and for defining the program for inspecting licensees by the regional offices to assure that the licensees' programs are carried out. It is responsible for managing major enforcement actions through orders and civil penalties. Further, it recently took on the added responsibility of inspecting and evaluating vendors, designers and suppliers wherever they may be located.

The five regional offices (see Figure D.4) are responsible for executing the established NRC policies and assigned programs relating to inspection and enforcement within their regional boundaries. The regional administrators have the authority to stop any or all safety-related work during the construction and/or operation of nuclear power plants.

In general, the regional offices conduct an inspection program that has been basically defined by the headquarters Office of Inspection and Enforcement. However, the regional offices administratively report directly to the Executive Director for Operations as does the headquarters Office of Inspection and Enforcement.

Since the licenses are issued to the utilities to construct nuclear power plants, the utility is held totally responsible by the NRC. If an enforcement action concerning a construction contractor is deemed appropriate, the action is taken with the utility, not directly with the contractor.

NRC's Relations with Others. The NRC has placed a resident inspector at each of the nuclear power plants under construction. The inspector's efforts are supplemented by periodic visits to the site by regional-office-based inspectors who generally look at specialty areas. The total level of effort is estimated to average about 1-1/2 persons for each reactor unit under construction.

Resident inspectors are provided office space at the site and have ready access to all documents, records and files pertaining to the assurance of quality and the licensee's commitments on quality. The NRC inspectors also observe the work in progress. The basis for their authority, in general, is to assure that the licensee fulfills the commitments made during the licensing process.

The NRC operates with a very high degree of public visibility. For example, individual inspection reports become public information, and extensive public participation occurs in the licensing process, including the various hearings that are conducted. Direct public access to NRC inspectors is provided and encourged.

Resources. As noted earlier, the actual NRC inspection effort for each reactor unit under construction averages the equivalent of about 1-1/2 full-time persons. Nearly all of the inspection staff is made up of engineers. Special multi-week training programs on technical aspects of the inspection job are provided by the NRC, with a one-week course on the fundamentals of inspection.

This normal level of inspection effort is supplemented in some cases by construction assessment teams (CATs) from NRC Office of Inspection and Enforcement staff supplemented by contractor or consultant experts. These teams perform three to four detailed inspection efforts per year. Each inspection covers a four- to six-week period. A typical effort by a CAT amounts to about 14,000 man-hours per inspection.

The staffs of some regional offices were concerned with maintaining high levels of proficiency and adequate numbers of persons in inspection, a concern attributed to competing with industrial organizations for experienced people.

The NRC regional offices each have one or two mobile vans with nondestructive testing capability. The vans can be moved from site to site, which provides some capability to perform independent nondestructive examinations in special cases, generally at sites with major problems. This effort is supplemented by the use of contractors to assist in conducting independent examinations--both nondestructive and destructive.

ORGANIZATION CHART

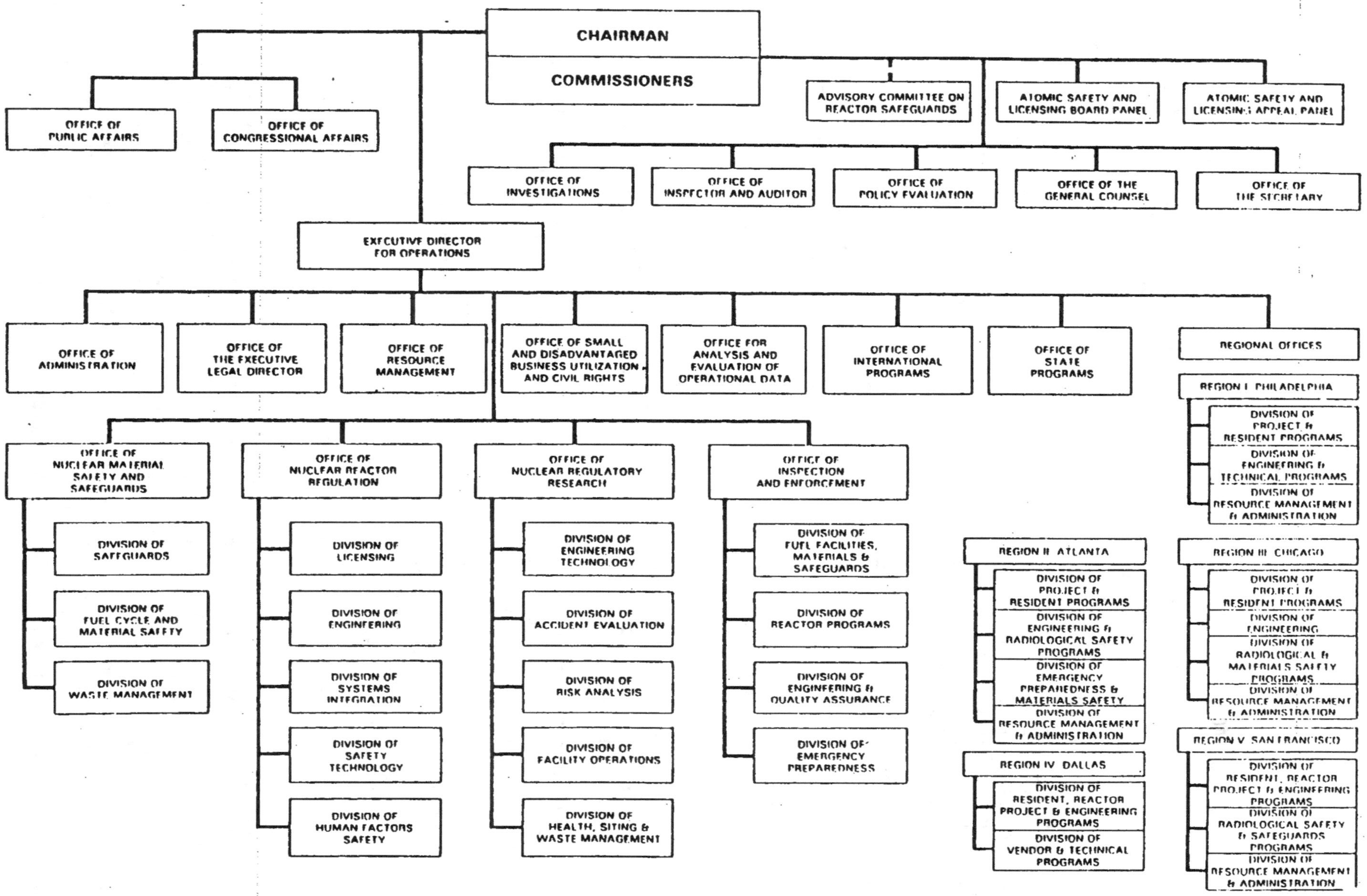

FIGURE D.4. U.S. Nuclear Regulatory Commission

D.63

D.4.5.3 AOQ Program

The major thrust of the NRC's program to assure quality construction of nuclear power plants is directed to the owners/operators of the plants. The utilities that operate the plants include a government-owned corporation (TVA), local government agencies (i.e., public utility districts, cooperatives) and privately owned corporations. In any case, the utility must obtain a permit from the NRC to construct a nuclear power plant. The application for such a permit includes all of the information necessary to analyze safety, siting and environmental issues, and the licensee's program for quality assurance. The QA program for safety-related systems and equipment must meet the requirements of the NRC's QA criteria contained in 10 CFR 50 Appendix B, which is the basis for the NRC's program for quality assurance.

The NRC's Standard Review Plan describes the NRC review of a license application for construction. Chapter 17.1 of the Standard Review Plan outlines in considerable detail the requirements that a licensee must meet in applying for a permit to construct a nuclear power plant. In essence, this requires a description of the QA program that the licensee will implement throughout the design and construction of the plant. This program description becomes the basic commitment by the licensee and is therefore the basis for all following QA inspection efforts.

In addition to the program description, the regional office inspection staff will review the licensee's QA manual and the detailed procedures that are to be applied to the project. The results of this inspection are fed into the application review process.

The inspection program carried out by the five regional offices is in accordance with the Inspection and Enforcement Manual issued by NRC's Office of Inspection and Enforcement. The manual includes many comprehensive and detailed Inspection Modules, ranging from "predocketing" inspection of the licensee's QA manual and procedures to the details of inspecting specific equipment, system and component areas. The modules indicate a minimum frequency for inspections, describe what to look for, and provide checklists of what to look at. They also describe acceptable practices for work in progress. The major thrust of these inspection efforts is to review the documentation and the work being done on a sampling basis to determine if the licensee's program is being carried out effectively and in accordance with license commitments.

The NRC's inspection efforts may result in "deviations," or "violations." A "violation" means that a non-compliance with requirements has been identified. A "deviation" identifies a departure from acceptable, standard practices. The licensee must formally respond to non-compliances by identifying what is being or has been done to correct the item noted and what actions are planned or have been implemented to preclude any further similar occurrences. These required responses are to some extent viewed by the licensees as a form of enforcement penalty because of the resources required to prepare the necessary responses. The corrective actions required may also represent new and unplanned efforts and activities for the licensee and/or its contractors.

In more extreme cases, including those where corrective actions have been ineffective or not adequately implemented so that the problems have continued, or where so many difficulties arose that a major breakdown in the licensee's program has occurred, the Regional Administrator has the authority to stop work. Work cannot be resumed until the regional office has been satisfied that appropriate changes have taken place and there is reasonable assurance that requirements will be met.

Another task force type of effort provided by NRC headquarters is identified as the Integrated Design Inspection (IDI). This is generally done in cases identified as near-term operating license situations. The IDI consists of a detailed review of a sampling of the plant's design. The results of these inspections are incorporated into the review process in preparing for the issuance of an operating license.

NRC regional offices also perform, annually, a Systematic Assessment of Licensee Performance (SALP) for each construction plant site. This is an overall assessment of performance.

In some cases, a regional office forms a task force to conduct a detailed construction assessment effort of selected systems or features at a site to supplement the normal inspection activities.

Incorporated in the inspection effort is a review of the qualifications and certifications of the quality assurance/quality control personnel of the licensee and its contractors to assure that these staffs are properly qualified. The NRC provides technical training of its own inspectors with required minimum grades on written examinations. Annual performance appraisals of NRC inspectors are developed and provided.

In essence, the major focus of NRC's inspection efforts is to assure that the licensee is conducting effective QA and QC programs in accordance with the requirements of Appendix B to 10 CFR 50. This effort consists both of reviewing documentation and procedures and of observing work in progress for a review of the actual implementation of the committed program. The NRC inspection effort itself cannot assure that all design requirements are met in the resulting hardware. The inspecting of hardware and observing of construction work in progress are parts of the NRC's effort to assure that the licensee's QA process is functioning properly.

The NRC AOQ requirements permit the licensee to take a wide variety of approaches in its QA program. One of the major variables is the degree of delegation permitted by the licensee to its contractors. However, the licensee is required to maintain a minimal level of QA activities with ready access to the appropriate high levels of management in the licensee organization. Within this framework, some licensees have chosen to delegate quality control inspections with supplementary QA activities to their construction contractors or construction management contractor. Others have chosen to exercise all of these functions under their own direct management control with, perhaps, supplementary staff provided by a contractor.

All AOQ programs must conform to the criteria in 10 CFR 50, Appendix B. This requires extensive documentation of the program, its procedures, resulting records, and the management and control of the documentation for all activities in plant construction. Independent periodic reviews and audits have become a matter of standard practice and are required as is certification of certain QA and QC staff personnel.

D.4.6 Foreign Nuclear Assurance of Quality (AOQ) Programs

Summary descriptions of the AOQ regulatory programs for nuclear power plant construction in six other countries are presented in this section. The six programs studies are Canada, the Federal Republic of Germany, France, Japan, Sweden and the United Kingdom. The summary descriptions were developed almost exclusively from available literature. Time restraints precluded the NRC staff from forwarding the summary description to the six foreign countries for their review, comment, and correction. As a result, inaccuracies may exist in these summaries. If warranted, corrections to these summaries will be made in future revisions or supplements to this report.

The major efforts on these studies were provided by the NUS Corporation, Gaithersburg, Maryland, and Battelle Institute e.V., Frankfurt, Germany. In their studies of Canada, United Kingdom and Japan, NUS provided the advantages of a staff member in residence in England, a staff member previously employed in the Canadian nuclear program, and staff of their Japanese subsidiary (JANUS). The Battelle Frankfurt Laboratory, in studying Germany, France and Sweden, provided the benefits of their extensive research work in nuclear matters pertaining to the European community. Since the studies were conducted primarily by reviewing the available literature, these organizations were particularly helpful in overcoming the language barriers.

Since both NUS and Battelle have well-established relationships with the nuclear industry sectors in various countries, it was possible for them to supplement the literature review with a few discussions with non-government individuals. The information available on Sweden and Japan was also supplemented with data obtained by a representative of N.C. Kist and Associates whose visit to those countries coincided with the studies.

There are significant differences in the programs of the countries studies, however, there are also common elements. Some of the commonalities are:

° Each has utilized the U.S. NRC's QA criteria, 10 CFR 50, Appendix B, in developing its program.

° Each has utilized the International Atomic Energy Agency (IAEA) established Codes of Practice and Safety Guides for nuclear power plants.

° Each has incorporated the government regulatory functions for nuclear power plants into agencies or departments with cognizance over non-nuclear industries and activities not related to radioactive materials or devices.

The program in each country is discussed in the following subsections.

D.4.6.1 Canada

The Canadian nuclear power program has been producing electricity since 1962. Canada has 14 operating reactors with 7,278 MWe capacity and 10 reactors under construction, for a capacity of 14,469 MWe projected for operation in 1990. Currently, nuclear power plants produce 9.7% of Canada's electricity. The annual load factors are among the highest in the world and have been improving. The high annual load factor (77.1% in 1982) is partially the result of the CANDU pressurized heavy water reactors' being refueled while operating. Canada is currently building reactors with capacities of 516 MWe, 756 MWe and 881 MWe. Generally, four reactors of a given size and type are built at a site.

Organization. The Atomic Energy Control Board (AECB) was created in 1946 to implement and administer the Atomic Energy Control Act of 1946 (amended in 1954). This act, in conjunction with the 1974 Atomic Energy Control Regulations (amended in 1978 and 1979) and the Nuclear Liability Act, governs all nuclear activities in Canada. The AECB reports to the Minister of Energy, Mines and Resources and is composed of five members with a staff of 250 people. The AECB has not issued formal QA regulations for generic nuclear power plants. QA requirements have been imposed as part of the licensing activity for each plant. The AECB staff was reorganized in 1978 with the formation of a Quality Assurance and Standards Division. Formal QA regulations and guidance are being prepared. The AECB power reactor safety criteria and principles are defined in "Licensing and Safety of Nuclear Power Plants in Canada" (AECB 1982).

The licensing process in Canada is the responsibility of the AECB, but because of provincial concerns, the AECB has evolved a "joint regulatory process" that enables all concerned federal and provincial agencies and ministries to participate. The AECB acts as the lead agency. However, the provincial government can veto the proposed construction of a nuclear facility within its borders. A veto only applies to a reactor site, not to an evaluation of plant operation and safety. Additionally, provincial government agencies perform reviews and inspection of pressure-retaining components to verify conformance with ASME and Canadian Pressure Vessel Codes.

Subsequent to site acceptance, application for a construction license is made. Primary documentation supporting the application consists of a Preliminary Safety Report (which includes site characteristics, design description, and preliminary safety analyses), a Quality Assurance Program, and preliminary plans for generation (including staffing and training plants). The AECB staff reviews the supporting documents and, if satisfied, recommends to the Board the issuance of a construction license. This review normally includes consultations with the provincial authorities, the applicant, and the applicants' agents to obtain additional information that may be required.

As construction progresses, the AECB staff meets with the applicant and resolves safety-related problems as they arise. During construction, authorization for acquiring and loading heavy water and fuel is issued by the AECB.

Quality Program Requirements. While formal regulations are still being developed, the AECB has supported the development of national QA standards. The Canadian Nuclear Association (CNA) has issued, under authority of the Canadian Standards Association (CSA), a series of standards for nuclear power plants, with N286 being specific to quality assurance. The standard for quality assurance in manufacturing was developed by the CNA as a general standard, CSA-2299, since the utilities wanted to use it for conventional as well as nuclear equipment. In terms of the principles involved, the standards CSA-Z299 and CSA-N286 are similar to the IAEA Code of Practices and its supporting Safety Guides and to 10 CFR 50 Appendix B. The most significant difference is that CSA-Z299 uses a five-step, graded quality standard for component manufacture and installation. Selection of the appropriate quality program standard is based on the sum of a four-level evaluation of six factors (design complexity, design maturity, manufacturing complexity, item or service characteristics, safety and economics). Canada has also developed standards similar to the ASME Codes, which provide criteria specific to Canadian design and construction characteristics.

In its licensing of nuclear power reactors, the AECB sets basic criteria and requires licensees to design, construct, and operate power reactors to meet those criteria. Besides considering single failures, the licensing process includes analysis of such dual-failure accidents as failure of a process system coincident with failure of a safety system; e.g. occurrence of a large LOCA simultaneously with unavailability of the emergency coolant injection system or impairment of the containment system.

Onsite AECB inspectors monitor compliance with license conditions throughout the construction and into the operating stage. A licensee must submit to the AECB an annual report on operation and maintenance of its nuclear power plant. The report includes a numerical assessment of the reliability of safety-related systems during the reporting period.

To ensure that provincial requirements are met by licensees, the AECB and the provinces have developed a joint regulatory process that is operative from the application stage through construction and facility operation. The AECB licensees are inspected periodically to ensure compliance with license conditions. Inspections may be carried out by AECB staff appointed as inspectors, or by provincial officers also appointed as AECB inspectors on agreement with their provincial ministries or departments. Provincial inspectors so nominated are supplied with an AECB inspector card that provides access to nuclear facilities and users' properties. They inspect according to the AECB regulations and report to the AECB as well as to their home office.

With respect to the design of pressure-retaining components, each province where nuclear power plants are to be located has a Pressure Retaining Component (PRC) Safety Department. The provincial PRC Safety Departments exercise general control over pressure vessel or boiler installation in each of the various provinces of Canada. To the extent that these pressure vessels are related to nuclear facility safety and under the AEC Act, the AECB makes use of provincial expertise to perform design examinations, and fabrication, installation and operational inspections. Provinces rely on the AECB for much work that they cannot cover, and there is joint consideration of all major matters.

Quality Program Implementation. Organizationally, Canada may be considered as a single utility for the purposes of comparison to other foreign countries. Twenty-one of Canada's twenty-four plants are owned by Ontario Hydro, two are owned by Hydro-Quebec, and one is owned by the New Brunswick Electric Power Commission. The utilities are similar, with Ontario Hydro dominating nuclear power plant construction. All three utilities are provincial corporations. The AECB holds the utility solely responsible for construction of a nuclear power plant. Given its limited resources, the AECB relies on the technical expertise of the utility and its vendors in implementing construction criteria developed in the license. Most notable is the reliance on utility inspections of suppliers. A system has been developed whereby suppliers' QA is qualified by the utility before a contract is awarded. Ontario Hydro has chosen to drop "Quality Assurance" in favor of "Quality Engineering" (QE). The Quality Engineering Manual was produced in 1975 and issued formally in 1978. Quality Engineering is defined as a planned and systematic application of scientific and technical skills and management activities to achieve the required level of quality and to provide assurance that this is being done effectively and efficiently.

The Quality Engineering Program is administered by Level 3 managers (divisional directors). Specific responsibilities in each of the areas of design, procurement, construction, commissioning and operation are defined consistent with the line responsibilities for engineering activities in each area. The Quality Engineering Department in the Design and Development Division is responsible for providing the secretariat, including necessary staff support, to the Quality Engineering Policy Committee.

The goal of the Quality Engineering Policy Committee is to promote a coordinated approach to quality engineering in the Operations Group and provide to the executive vice-president, Operations, recommendations on QE policies, objectives and strategies for all areas of design, procurement, construction commissioning and operation; to provide advice to the committee chairman with respect to the suitability of the QE procedures (for adherence to policies, support of objectives, etc.); and to keep members mutually informed on QE matters.

For each project, the project engineering and construction departments under a project manager are assembled within the Generation Projects Division. These departments perform the detailed design, procurement and construction processes for that particular project. The project manager is responsible for designing the project to the requirements specified by the Design and Development Division. During this stage, the project manager is responsible for the overall quality engineering program, engineering manager for the part of the program related to quality engineering in design, including procurement, and the construction manager for the part of the program related to QE construction.

Prior to awarding a contract, equipment purchaser must ensure that a supplier can immediately perform in compliance with the relevant QA codes and regulatory requirements, or alternatively, be able to so perform prior to commencement of the work. Ontario Hydro is qualifying suppliers' quality programs either by a formal audit or by an evaluation by inspection. To minimize the number of formal audits being performed, several utilities, consulting engineering firms and regulatory agencies combine to carry out joint audits. CSA has now embarked on a program of qualifying suppliers' programs to the Z229 Standards. A supplier will then be subjected to a periodic audit by CSA.

Product Engineers holds post-award meetings with major suppliers during the life of the contract. Participation at these meetings might include other functions within Ontario Hydro, along with the suppliers' representation from Design, Project Management, Quality Assurance, Purchasing, Manufacturing, Production Control, Contract Administration or Management. The inspector assigned to the contract attends the meeting to provide input from day-to-day surveillance of the contract. The purpose of the meeting is to ensure that the supplier has planned for and carries out all aspects of the contract, including development work, qualifications, submission of manufacturing, welding, nondestructive testing and shipping procedures, and submission of inspection and test plans and history dockets.

D.4.6.2 Federal Republic of Germany

By early 1983, the Federal Republic of Germany (West Germany) had 15 nuclear power reactors in operation, with a total installed capacity of 9,800 MWe. At that time, there were 12 additional units under construction which are expected to add an additional 13,000 MWe installed capacity (Nuclear Engineering International 1983).

Organization. The legal base for QA/QC programs applying to the planning, construction and operation of nuclear power plants in West Germany rests in the Atomic Energy Act (ATG), last revised in 1976. However, the term "QA/QC" is not defined in the ATG. The ATG provides the legal framework for the licensing proceedings for nuclear power plants, details of which are prescribed in the License Procedure Ordinance (AVerfVO). This ordinance states explicitly that the applicant for a license provide a safety report, as well as

Factual statements enabling the examination of the reliability and expert knowledge of the persons responsible for the erection of the installation and the management and control of its operation, as well as factual statements enabling the examination of the requisite knowledge of all persons working on the installation.

These general requirements provide the basis of all ensuing QA/QC programs imposed by the regulatory authorities.

Nuclear power plants in West Germany are licensed by the individual federal states on behalf of the federal government under the supervision of the Department of the Interior (BMI). With respect to nuclear power plants, the BMI has three major advisory bodies: the Committee for Reactor Safety, the

Committee for Radiological Protection, and the Committee for Nuclear Safety Standards. There is also a State Committee for Nuclear Energy, which in 1982 issued a "standard set of information to be submitted to the licensing authority in the course of licensing proceedings." In 1980, the Committee for Nuclear Safety Standards (KTA) published the General Requirements for QA/QC. These rules have the force of regulations.

Quality Program Requirements. The Committee for Reactor Safety (RSK) is a consulting body set up by the BMI. RSK's findings are limited in that they only have the force of recommendations (to the BMI, and, via BMI's supervisory role, to the state authorities). The actual importance of these findings cannot be overstated, however. The RSK has recommended guidelines for pressurized water reactors ("RSK-Leitlinein fur Druckwasserreaktoren," 3rd edition, October 1981) which constitute the framework of safety-related standards that must be adhered to by an applicant. The RSK Guidelines consolidate a wealth of BMI regulations and KTA rules supplemented by the RSK and its subcommittees. In various instances, the extent, methods, and even specifications of QC test procedures are detailed by the RSK guidelines. The RSK has set for itself the duty of regularly revising and updating the Guidelines to keep them abreast of "the up-to-date scientific and technical knowledge."

The Committee for Radiological Protection (SSK), is an important advisory body to BMI, but has little direct involvement with the AOQ program.

The Committee for Nuclear Safety Standards (KTA) also reports to the BMI. Its task is to establish safety-related standards and to further their adoption in all sectors of nuclear technology. The KTA provides a highly collaborative approach to the development of the rules. Its membership includes representatives from many sources: suppliers, vendors, utilities operating nuclear power plants, Department of Interior, state licensing authorities, expert institutions, other governmental departments, national nuclear laboratories, trade unions, insurance companies and the Commission for Industrial Standards.

In the context of licensing and surveillance of nuclear power plants, TUV (Technische Uberwachungs-Vereine) and GRS (Gesellschaft fur Reaktorsicherheit) organizations are of utmost importance. Historically, the TUV organizations have been set up by industry as self-financing, independent agencies to act as "watch dogs" on technical hazards in and through large industrial plants. They have built an excellent reputation for technical scientific ability and trustworthiness. They inspect and test all kinds of technical installations (pressure vessels, lifting equipment, bridges, motor vehicles, etc.) or materials on behalf of government authorities or act as supervisory or inspecting agencies for industrial customers. Seven of the 11 TUV organizations have established nuclear departments that work exclusively on inspections, controls, and audits of nuclear power plants on behalf of the licensing authorities. The government licensing authorities do not perform significant inspection activities at the construction sites.

TUV organizations, being independent expert institutions, are also called upon frequently by buyers of complex industrial projects to act as auditors/QC agents. This is also the case with regard to various nuclear power plant projects. However, though not explicitly excluded by law, a situation in which one TUV organization would perform duties on behalf of both the licensee and the licensing authority on the same project is avoided as a matter of principle.

The GRS is a semi-governmental, limited corporation (jointly owned by the TUVs, the federal government and two state administrations). It is also active in the field of licensing proceedings, either directly for the authorities or in a supporting role to one of the TUV organizations. Except for questions regarding prevention of human threat (sabotage or terrorist attack), the GRS has little direct involvement with QA/QC matters.

Quality Program Implementation. Light water reactor plants are constructed under a turnkey arrangement, with a single corporation responsible for plant design, procurement, construction-management and construction. This general contractor and the suppliers of parts, material and components are required to establish their own QA/QC procedural systems. These QA/QC procedure systems are considered the vendors' proprietary material and are not published. They are, however, reviewed and approved by the authorities.

The licensing authority holds the licensee, a private sector utility, totally responsible for the nuclear power plant. However, the general contractor is responsible to the utility to conform to all regulatory requirements and provides a warranty for the plant as a "vendor liability."

One characteristic of the regulatory QA/QC system in West Germany is that control measures are predominantly hardware-oriented. There are no "supplier certificates" or "N stamps." The burden of proof for adequate quality of every item rests on the supplier; who must satisfy inspection authorities that standards are met.

The required QC measures apply to all materials, parts, components and systems deemed safety-related through all steps of assembly and erection.

- Pre-construction Audits: Audits of design and specifications according to fixed criteria and standards.

- Inspection and Tests during Production Phase: Materials, production, acceptance tests and functional testing on all assembly phases, documented and certified by authorized inspectors.

- Commissioning: Functional testing and acceptance testing supervised by authorized inspectors who have to release every system for operational (hot) commissioning.

An important aspect of the principal philosophy of liability in the West German nuclear industry is the fact that the (licensing) authorities perform their control duties on behalf of the populace, which in turn can have the administrative courts (three stages of appeal) control every administrative act.

The obligation of general contractors and suppliers of parts, materials or components to establish in-house QA/QC systems is caused by the regulatory requirements, which mandate a QA level satisfying all relevant safety issues; and the warranty issues in context with the vendor's liability. In essence, the regulatory QA/QC system in West Germany can be characterized by three elements:

- Control measures are predominantly object-oriented. There are no "suppliers certificates" or "N" stamps. The burden of proof for adequate quality of every item rests on the supplier, who must satisfy the inspection authorities that standards are met.

- The inspecting and controlling agency is not an administrative (governmental) body; rather, it is the independent institution, TUV, which has a long record in inspecting services in conventional fields and is accepted as highly competent and trustworthy by all interested parties.

- Basic contractual arrangements are supportive to clear-cut responsibilities and facilitate controls: one licensee/applicant, one general contractor who sells the complete plant on a turnkey basis and who bears full vendor's liability.

D.4.6.3 France

As of January 1, 1983, 32 reactor units were in operation in France. Nuclear power accounts for approximately 40% of Frances electrical production in 1981. Also, at the beginning of 1983, there were 25 additional power reactors under construction in France, one of which is a liquid metal fast breeder reactor (NEI 1983).

All of the light water reactor plants are designed and built under a turnkey arrangement with Framatome, a government-owned corporation, for operation by Electricité de France (EDF), the government-owned utility. With few exceptions,[a] the applicant for a construction and operating license of a nuclear power plant in France is EdF. The EdF is the French monopolistic electric utility that is organized and run like a public company, although it is 100% state-owned.

A series of three standard pressurized water reactor plant designs have been developed. The EDF has developed and utilizes an information system to collect information on the operating experience in all of its plants. This information is used as a basis for improvements in designs and in the overall program. The designs are also modified as necessary to meet site-specific needs.

(a) These exceptions include plants jointly owned and operated by EdF and foreign utilities in locations near the French border, and the Phenix and Super-Phenix fast breeder reactors.

Organization. There is no fundamental French law for regulation of nuclear matters in an encompassing way like the U.S. Atomic Energy Act of 1954. When necessary, matters are settled on an ad-hoc basis via legally binding governmental decrees (decrets) that do not require parliamentary support. Thus, the construction permit for an individual plant is granted by a special decree that is signed by the Prime Minister.

Nearly all matters concerning nuclear activities in France are regulated through government ordinances (degrets or arretes). The only legal act providing explicit legislative approval for nuclear matters is the Bill on Protection and Control of Nuclear Materials of May 25, 1980. This act does not, however, provide a general legal base for nuclear power plant regulation. Governmental ordinances applicable to nuclear plants can be divided into two categories:

- ordinances concerning safety of nuclear installations in general or safety in handling nuclear material in general

- special ordinances concerning individual installations (e.g., construction or operating licenses of named units).

Construction permits and operating licenses are, as a rule, granted by governmental ordinances. General requirements prerequisite to a construction permit are defined in a decree of December 11, 1963, as amended February 26, 1974, and December 6, 1974. These amending decrees address specifically the issues related to pressurized water reactors.

The license to build and operate a nuclear power plant is granted by the Department for Industry, which also acts as a supervisory authority for operating plants. The licensing proceedings take place in the national capitol. Regionally, a public inquiry is held at the proposed location of a new plant. This inquiry is headed by the regional administrator, and deals only with site-specific aspects, e.g., water consumption or environmental matters.

Due to this centralized organization, responsibilities in nuclear matters are organized vertically. Supervision and licensing of nuclear power plants fall within the jurisdiction of the Department of Industry, which has a special organizational unit, the Central Service for Safety of Nuclear Installations (SCSIN). The SCSIN has two consulting bodies consisting of senior administrators and technical experts that report directly to it:

- The Section Permanente Nucleaire (SPN) has the task of developing the rules and regulations concerning nuclear power plants.

- The Commission Centrale des Appareils a Pression (CCAP) has the task of further developing rules and regulations regarding pressurized systems in general.

Because most regulatory matters deal with nuclear power plant safety and the complete range of handling fissionable materials (fabrication, transport, marine propulsion, refabrication, etc.), an Interdepartmental Committee (CISN) was established in 1975 to coordinate all governmental actions "to protect people and property against dangers of any nature resulting from construction, operation or dismantling of (stationary or mobile) nuclear installations as well as all stages of handling of fissionable materials or radioactive wastes."

Of somewhat lower hierarchical rank, but still of eminent importance, is the Atomic Energy Commission (CEA). Wholly state-owned, the CEA has become not only the expert authority on all nuclear matters, but also the major economic entity controlling a sizeable sector of all parts of the French nuclear market. Rules and regulations developed by the CEA or one of its organizational units are adopted and made official through the decrees, Official interpretations of decrees come from the Department of Industry. The various supervisory boards of the CEA encompass state representatives, public interest groups, e.g., (trade unions), the EdF and major banking houses.

Quality Program Requirements. There is only one governmental ordinance and its official interpretation (circulaire) that explicitly addresses QA/QC matters at nuclear installations: Both exist in draft form as of September 1983. The draft papers are as follows:

- Directive Regarding Quality Design, Construction and Operation of Nuclear Installations

- Circular Regarding the Application of Regulations on Quality of Design, Construction and Operation of Nuclear Installations.

The directive and circular define a general provision for the regulatory authority to intervene in any particular case if there is a suspected shortcoming in safety or quality-related matters.

The directive places responsibility for quality assurance at a nuclear installation on the applicant/licensee for all phases of design, construction and operation. For each system or component, the level of quality to be guaranteed is correlated with its safety importance. The applicant/licensee must install a QA system that takes into account:

- definition of safety requirements and quality standards system-by-system, taking into account all applicable regulations and standards

- design of a QA system

- implementation of a QA system

- installation of a special organizational unit for quality assurance

- documentation of all measures taken

- provision for the required number of adequately trained personnel for QA/QC activities

- provision for adequate technical resources

- updating of the QA system itself in step with advancing technical knowledge

- explicit and definitive procedures to be followed in case of off-normal events

- complete and readily accessible documentation of all steps taken.

Portions of the outlined QA procedure may be performed by suppliers of parts and subsystems on behalf of the applicant/licensee. The applicant/licensee has then the duty to supervise and control the suppliers' QA/QC activities. In any case, the applicant is required to submit to the licensing and supervisory authority (in this case, the Control Service for Safety of Nuclear Installations, SCSIN) a (provisional) safety report on the installation. The QA/QC system of the applicant and the ways and means of supervision of the suppliers' activities are defined in this report.

The chairman of the SCSIN may require additional measures to be taken by the applicant/licensee and may control adherence to these measures. In case of dispute, the applicant may appeal to the Minister of State for Research and Industry.

In addition to the (provisional) safety report, the licensee is required to assemble a QA manual defining all QA/QC measures (technical, organizational and personnel) taken, including surveillance measures over suppliers.

The directive is officially interpreted by an accompanying circular. Of special interest in the circular are the following points:

The applicant/licensee is in principle responsible for safety of the installation through all stages of design, construction and operation. He may delegate the responsibility for safety and quality of subsystems or parts to suppliers, but ultimate responsibility remains with the applicant/licensee. The licensee must be sure that suppliers who perform QA duties on his behalf strictly adhere to the approved procedures.

The applicant/licensee has some flexibility in defining the range and extent of "safety-related activities" (including their application to activities of subcontractors). The regulatory authorities do, however, reserve judgment on the applicant's views through approval/disapproval when the provisional safety report is submitted.

The applicant/licensee is assigned an important role in supervising the activities of his suppliers. The supplier has to prove to the licensee's satisfaction that he has an adequate QA system. The applicant/licensee may transfer his duty of surveillance of suppliers' activities onto third-party independent experts or expert institutions. Authorized experts/expert institutions are required to be independent of contractual or economic ties with suppliers they are to control.

The supplier of subsystems or parts may define adequate levels of safety and quality for their products; however, the applicant/licensee having ultimate responsibility for the safety of the installation must approve them.

Quality Program Implementation. The licensee is required to submit safety reports that correspond to defined stages of the project:

- A preliminary report that gives an overview of general design criteria aimed at safe operation of the plant

- A provisional safety report that gives detailed design information (including safety and quality standards), demonstrates adherence to applicable norms and regulations, and gives preliminary information towards an operating license

- A final safety report that includes documentation of QA/QC during the construction phase and commissioning test.

Again, the high level of standardization for "series type" PWR nuclear power plants causes a high level of standardization in the safety reports and the licensing proceedings.

During construction, an onsite resident government inspector overviews the QA/QC activities. A major onsite quality control inspection effort is provided by "authorized experts." These inspectors are individuals or staff from small associations under contract to the utility who have been examined and certified by the government's Nuclear Safety Inspectorate. In addition to the certification, the individual inspectors take an oath of office and essentially function as government deputies. The utility also performs onsite quality control inspections.

The Design and Construction Standards for nuclear power plants are planned to be a comprehensive, self-contained set of standards. The AFCEN, an organization encompassing representatives from industry and the EdF, set up the RCC-codes, a consistent system of rules and standards applying to all safety and reliability aspects of nuclear installations. The RCC-Code is published through the Association Francaise de Normalization (AFNOR), which is comparable to the ANSI organization in the United States.

The RCC-Code refers to the nuclear island. Many rules of the RCC-Code are similar to parts of the ASME code, which may be explained by the fact that Framatome developed its standard PWR from a Westinghouse design. One example of the similarity between RCC and ASME codes is that the RCC-M code divides the components of the nuclear island into three classes according to their safety importance. Since the complete RCC-Code is not finalized, existing standards from other technical fields and from the ASME Code are referenced for convenience.

Like the ANSI in the U.S., the AFNOR in France defines general technical standards and codes of practice and keeps them updated. The licensee is free in principle to define the systems according to safety requirements as perceived by him. However, the regulatory authority, in this case primarily the SCSIN, has to concur.

Overall, the French QA/QC system is characterized by three elements:

° Regulatory control measures are predominantly organization-oriented. They consist primarily of auditing and approving QA/QC systems implemented by the applicant or by suppliers.

° With regard to PWRs, the high degree of standardization of plants facilitates regulatory tasks. There are only two basic plant designs (900 and 1300 MWe), one vendor (Framatome), and one applicant/licensee (EdF). This assures maximum nuclear experience in all groups involved.

° The central role of the CEA (and all its organizational affiliates) ensures that maximum experience on all technical or organizational aspects is available.

D.4.6.4 Japan

Japan began using nuclear power plants to generate electricity in 1966. As of March 1984, 25 reactors were in operation (18,277 MWe), 12 (11,804 MWe) were under construction and 7 (6,053 MWe) were in planning. The annual capacity factor has improved in recent years, to 71.5% for 1983. This is a significant load factor considering the obligated three-month downtime for in-service inspection. Nuclear power plants currently produce 20% of the electricity generated in Japan.

The German system of Civil Law was introduced into Japan almost a century ago. Over time, this system was developed and modified to fit Japanese customs. After World War II, this system was exposed to a great amount of information from the U.S. In technical and administrative areas, where governmental influence was not significant, many aspects of the U.S. system were implemented, and today many of the Japanese codes and standards refer to the technical requirements of U.S. codes and standards. Administrative areas in Japan's heavy industries have not been so strongly influenced by the U.S. system.

In the nuclear industry, Japan's QA program was introduced through U.S. companies, such as General Electric and Westinghouse, which contracted with the Japanese utilities to construct nuclear power plants. For the initial construction projects, the regulatory authority performed its duties similar to practice with conventional power plants. QA practices were passed on to sub-tiered contractors through Japanese vendor-vendee relationships. These QA practices emphasized inspections and records rather than system design and performance.

Organization. The regulation of nuclear power plants in Japan is conducted in accordance with the Electric Utility Industry Law (EUIL) and the Law for the Regulation of Nuclear Source Material, Nuclear Fuel Material and Reactors (LRNR). The LRNR was established in 1957. Until 1978, the Japanese Atomic Energy Commission (AEC) had the responsibility for both nuclear development and nuclear safety. The Law for Revision of the Atomic Energy Law enacted in 1978, established the Nuclear Safety Commission (NSC) to control nuclear safety. These laws do not include requirements for quality assurance: however, the Ministry of International Trade and Industry (MITI) has imposed administrative guidelines requiring a QA program.

MITI has the authority to issue licenses for the construction and operation of commercial nuclear power plants in Japan. The Prime Minister and Minister of Transportation has the authority to issue licenses for the construction and operation of research reactors and nuclear vessels respectively. Under the EUIL, applicants for a license to construct a commercial nuclear power plant (research and ship reactors are covered by different organizations) must submit siting data for environmental impact review by the MITI. The MITI reviews the data with consultation from the Committee on Environmental Matters, and then holds public hearings where local governments and citizens participate. Once site approval has been obtained from the local governments, the MITI submits the application to the Electric Power Resources Development Coordination Council for its approval. Before issuing a license for construction, the MITI also consults with the AEC and the NSC about the reactor design.

The Japanes nuclear power program has a large number of participants compared to many other countries. Nine private utilities have nuclear power plants in Japan.

The Ministry of International Trade and Industry (MITI) also includes the Agency of Natural Resources and Energy (ANRE) and advisory committees on both environmental matters and nuclear power technology. MITI is responsible for commercial nuclear plant licensing and safety regulations on all the construction, maintenance and operation stage. The MITI provides technical reviews for the licensing of commercial reactors and conducts safety reviews of their installation. The MITI inspects operating reactors. The MITI currently has about 50 people on its staff who perform technical safety and licensing reviews of commercial nuclear power plants in support of the STA's administrative responsibility for licensing review. The MITI has approximately 100 people qualified to perform inspections.

The Science and Technology Agency (STA) is an administrative body attached to the Prime Minister's office. The STA has both management and technical review responsibility for research reactors and reactor systems still under development.

The Atomic Energy Commission (AEC), which is responsible for nuclear development, is made up of five Commissioners appointed by the Prime Minister with the consent of both houses of the Diet. The AEC is an advisory body to the Prime Minister concerning the development and use of nuclear power.

The Nuclear Safety Commission is also made up of five Commissioners and is under the authority of the Prime Minister. This commission is an advisory body to the Prime Minister concerning the safe use and regulatory requirements of nuclear energy.

The Japanese Institute for Nuclear Safety (JINS) was recently formed within the Nuclear Power Engineering Test Center (NUPEC) as a joint venture of the MITI and the STA. The JINS assists the MITI and the STA in the technical regulation, licensing, and standards development for nuclear power plants.

Quality Program Requirements. The laws governing construction and operation of commercial nuclear power plants in Japan do not specifically include quality assurance. Administrative guidelines imposed by the MITI on licensees do, however, include some requirements for QA, but these guidelines are not as specific as those in 10 CFR 50 Appendix B. Instead, the licensee's QA program is expected to include those QA/QC elements that have evolved during the development of Japan's nuclear power program. Certain inspections are required by the EUIL and LRNR laws, including inspection of components and structures during manufacture, installation and construction, and inspection of welds.

In 1972, the first QA standard for construction of nuclear power plants was published by the Japan Electric Association. The Nuclear Safety Standards (NUSS) program established by the International Atomic Energy Agency (IAEA) in 1974 was another impact on both governmental agencies and utilities. The Japan Electric Association revised the QA standard (JEAG-4101) according to IAEA QA Code of Practice in 1981, and has been preparing additional QA guidances corresponding to IAEA safety guides. For domestic nuclear contracts, JEAG-4101-1981 is referred to in procurement documents and is used as the criterion to survey, audit and quality the vendors.

ASME Stamp Association, starting from 1973, had a strong impact on Japanese heavy industries, especially for nuclear installations. ASME survey teams have taught QA concepts and importance of QA program maintenance. Now, in Japan, many factories hold ASME Stamps and most of the sub-tiered contractors have QA manuals similar to the ASME QA manual.

In concert with the NUSS program of the IAEA, two programs have emerged. The first program established a QA Investigation Committee under MITI, which is the responsible regulatory and enforcement agency for construction and operation of nuclear power plants. Established in 1980, this committee 1) analyzes nonconformities reported from utilities, 2) identifies QA problems with suppliers, and 3) investigates QA practices in the U.S. and Europe. The committee recommended the introduction of QA programs recognized in the U.S. and in European countries with some modifications suitable to Japanese industries.

The second program established the Committee for Nuclear Accreditation organized under the Japan Power Plant Inspection Institute (JPPII), which is authorized to inspect the nuclear power plant components on behalf of MITI. MITI has the procedures and criteria to qualify the manufacturers of nuclear power plant components concerning welding, but it is most concerned with the capability of facilities and personnel, not with the details of the QA programs. The committee has discussed the introduction of a system similar to the ASME "N" Stamp Accreditation system, and is also considering establishing a third-party agency to conduct surveys and audits.

Quality Program Implementation. The initial phase of the inspection program takes place to following issuance of the license and prior to authorizing construction. This plan includes details of the design (technical specifications), methods of construction, and a general description of the QA program. As part of the construction plan approval, the licensee must convince the MITI that the QA program meets the MITI administrative guidelines for quality assurance. QA program review at this stage is normally limited to the general description with a limited review of procedures. While it may appear that, in the absence of QA criteria, it would be difficult to meet the guidelines, the system apparently works well due to a limited number of licensees, most of them having prior nuclear experience.

The licensee has primary responsibility for QA inspections, and the MITI performs inspections on an oversight basis as necessary to meet their legal responsibilities. MITI inspections are normally a review of documentation, with limited hardware inspection except for those specifically required by law, e.g., reactor vessel, reactor cooling system, containment, etc. In addition to the inspection of specific documentation, the MITI also performs audits of the licensee's QA program to verify compliance with commitments made in the construction plan. If problems are discovered during the audit, the MITI may choose to perform a more detailed inspection of documentation and hardware.

While the MITI has approximately 100 inspectors, the level of effort expended on direct inspection activities is limited. Inspection activities at construction sites consume about 200 man-days for each nuclear power unit being constructed. The inspections are scheduled when required, depending on construction activities being performed. At present, the MITI is trying to relieve inspectors at each of the nuclear power plants. The MITI staff is relatively fixed due to budgetary restraints, and the number of reactors is increasing. Future inspections will be less technical and more programmatic than current inspections, resulting in an inspection program that will be primarily an auditing activity.

The Electric Utility Industry Law requires inspection of welds in vessels that contain radioactive fluids or that fulfill a safety-related function. The MITI is responsible for inspection of such welds. The actual inspection is performed by the JPPII, a non-profit organization established in 1970. The JPPII performs inspection of welds and administers tests for welding procedure qualification and welder qualification. It is funded by the users of its inspection services, who pay a fee for each inspection. The MITI has licensed the JPPII to perform weld inspections and almost all other hardware-type inspections of operating plants that the MITI is required by law to perform periodically. When the JPPII performs an inspection, no additional inspection is peformed by the MITI.

The JPPII has nore than 100 inspectors who perform required inspections for new as well as for operating nuclear plants. No estimate was available form the MITI as to the total JPPII effort expended per year for each nuclear power unit under construction.

To further the effectiveness of the audit program, in the past year the MI has instituted team inspections. These are audits by a team of three to fou inspectors, conducted during a period of three to four days. The team inspections are performed annually and, among other inspection activities, review the QA program in more detial than was done previously. It was reported that the team inspections (audits) have not uncovered any major problems.

The relationship between the MITI and licensees (and their contractors) is one of mutual trust and cooperation. All parties share a common goal, to build nuclear power plants that can be operated safely. The MITI stresses that their role is to oversee the licensee; the licensee is responsible for controlling activities of contractors.

Licensee requires submittal of detailed construction design approval including QA program. Requirements for inspections, other than MITI-required inspections, are the responsibility of the licensee. Therefore, the licensee establishes the inspections to be performed by the licensee and by contractors. Likewise, qualificaiton of licensee/contractor inspection personnel is the responsibility of the licensee. When the MITI requires specialized knowledge for an inspection, they normally expect the licensee/contractor inspectors to satisfy themselves that the technical requirements have been met.

D.4.6.5 Sweden

The Swedish nuclear power program began producing electricity in 1972, and by the end of 1982 had 10 operating reactors with a capacity of 7330 MWe. The average annual load factor in 1982 was 68.3%, and nuclear power produced 39% of the electricity in Sweden. Sweden currently has two power reactors under construction which will add 2110 MWe to the capacity. Three of the reactors were supplied by Westinghouse, and ASEA/ATOM developed the remainder. ASEA/ATOM, which is owned equally by ASEA and the Swedish government, designs and supplies BWR systems and fuel. ASEA/ATOM has had a technical exchange program with the General Electric Co. which has resulted in an American influence on Sweden's QA programs. ASEA/ATOM also functions as the architect-engineer and construction manager-contractor for the mechanical systems. By popular vote in 1980, a moratorium on nuclear power was approved which precludes the construction of additional units beyond the two currently being built. When these two are completed, nuclear power will provide nearly 50% of Sweden's total electrical energy (NEI 1983) .Organization. Nuclear installations in Sweden are governed primarily by the Atomic Energy Act of 1956. Other acts regulating nuclear power in Sweden are the Radiation Protection Act of 1958, the Emergency Preparedness Act of 1960, the Atomic Liability Act of 1968, and the act regulating special permission to load nuclear reactors (1977) (Stevenson and Thomas 1982). Another act relating to construction and considered applicable to nuclear power plants is the Building Act of 1947.

Licenses for nuclear power reactors are issued by the Swedish government according to the Atomic Energy Act of 1956. This Act places the responsibility for licensing with the Ministry of Industry. The Swedish Nuclear Power Inspectorate (SKI) administers the licensing process and reports to the Ministry of Industry. The granting of a license for a nuclear power plant by the Ministry of Industry is subject to approval by the Parliament. The SKI uses 10 CFR 50 Appendix B from the United States as a guideline for the scope of the QA program. The SKI has issued control procedures relevant to quality assurance of nuclear power plant construction.

There are six major participants in the Swedish nuclear power plant construction program:

- Ministry of Housing
- Swedish Nuclear Power Inspectorate (SKI)
- National Institute of Radiation Inspection (SSI)
- Swedish Plant Inspectors (SA)
- Utilities (SSPB, FKG, Sydkraft, and OKG)
- ASEA/ATOM.

If the proposed plant is to be constructed on a new site, the Building Act states that permission is required from the concerned municipalities before construction can began. The Building Act empowers local administrations to regulate construction in the vicinity of nuclear power plants, and also applies to establishing other types of industrial operations. The Licensing Board for Environmental Protection issues conditions and directives.

Two government agencies in Sweden are involved in the licensing and inspection of nuclear power plants. The SKI, under the Ministry of Industry, is responsible for technical and safety aspects of the nuclear power program. The other agency, the SSI, under the Ministry of Agriculture, reviews license applications and inspects facilities with respect to radiation protection and environmental impact of radioactive releases. Both agencies are relatively small. The SKI currently has about 80 employees, and 17 of the total 150-person staff of the SSI are assigned to nuclear powers matters. Funding for both agencies is provided by fees paid by the applicants or licensees. A third agency, the Labour Protection Board (KAS), provides assurance of the pressure circuits. The KAS regulates the design, manufacture, and construction of all industrial plants which present potential hazards other than radiation.

The SKI is the component of the Ministry of Industry that is responsible for administrating the licensing process for nuclear power plants. The SKI consists of five members appointed by the government, who are assisted by a staff and advisory committees. The advisory committees consist of chairmen and at least four members nominated by the government from the SKI and its staff.

The SKI's primary objective is to promote safety in nuclear power plants. It reviews safety assessments, inspects nuclear installations, and initiates research and development (R&D) within the field of nuclear safety. The SSI is the national authority for radiation protection for both occupational and environmental exposures. Its scope includes external and internal environments, and emergency planning. The SSI works with local authorities in preparing emergency plans.

The Swedish Plant Inspectorate (SA) is a nonprofit, government-owned company, formed in 1975, which performs third-party inspection and testing. It has a staff of 540 which is organized into pressure vessel engineering, machinery engineering, nuclear plant inspection, and four regional offices that perform inspections of shops, lifting devices, and pressure vessels. The SA is funded by fees for specific inspection and testing activities. Until 1975, there were two important nonprofit companies in Sweden which specialized in quality verification. These companies had shareholders, some of whom were involved in nuclear projects, and the government established SA based on the existing organization of these two companies to ensure independence of the project inspection agency.

Quality Program Requirements. The Swedish licensing procedure is basically a four-step system (Stevenson and Thomas 1982):

- The plant owner prepares a PSAR and applies for a construction permit. The SKI and the SSI grant permits for construction.

- The plant owner transmits data to the SKI demonstrating capability to meet conditions in approval license, and components and systems are tested as the plant is constructed.

- The plant owner submits a FSAR, and after SKI approval a fuel loading and reduced power permit is issued.

- The SSI reviews radiation protection and informs the SKI of its approval, who if satisfied, issues the operational license.

The owner is required to establish a QA program which meets the formal commitments for QA in the PSAR and is approved by the SKI. The SKI has utilized the criteria in 10 CFR 50 Appendix B as guidelines for quality assurance. The RKS has developed guidelines specific to the Swedish conditions, and, although not approved by the SKI, the utilities have been using these guidelines as a basis for internal QA work.

General inspection plans for safety class items were originally established by the utilities and approved by the SKI. These plans then developed into standardized inspection plans issued by the SKI. The general inspection plan identifies the documentation, inspections, tests, and examinations which are required for the various activities, and lists the responsibility for performing each. The SA has specified responsibilities to perform, review, verify, and report on certain of the tests, inspections and examinations. The SA inspects (including nondestructive testing) the pressure containment features in the power plants in addition to such inspection by the utility. The licensee may use independent inspection agencies, which includes SA, for its inspection activities.

Plant designs, inspection plans, and fabrication/work plans and procedures are all reviewed by the SA and approved before the licensee or vendor can proceed with specific activities. In essence, a "hold point" system is used and enforced by the SA. The SA supervisor designated for each nuclear power plant must be approved by the SKI. This supervisor must report to the SKI that the plant is satisfactorily completed before fuel loading and start-up operations can begin.

Quality Program Implementation. Four utilities operate or build nuclear power plants in Sweden. One is state owned, one is privately owned, and two are consortiums of local governments. These consortiums were established specifically to build and operate nuclear power plants. As owners, they have the overall responsibility for the design, construction, startup, and operation of nuclear power plants. The four owners have formed the Nuclear Safety Board of the Swedish Utilities (RKS), a joint body for collaboration in safety matters. The RKS collects, processes, and evaluates information on operational disturbances and incidents at Swedish and foreign nuclear power plants, and devises common policy and standards. Requirements for quality assurance were established in 1982 and then received trial use. In 1983, the RKS sent the QA requirements to the SKI and requested that they be designated as the reference for quality assurance of nuclear power plants in place of 10 CFR 50 Appendix B.

The largest utility is the state-owned Swedish State Power Board (SSPB), which provides approximately 45% of Sweden's capacity. The SSPB has five operating reactors plus one under construction. The Thermal Design Department has overall responsibility for quality assurance. It assigns QA functions to other components of the SSPB, and reviews and approves both internal and contractor procedures. The QC group is responsible for the quality of all mechanical equipment and assists in the preparation of specifications, reviews contractor proposals, prepares inspection plans and performs contractor surveillance for both manufacturer and site installations. The SSPB has a group to collect and analyze operations information with the objective of improving quality and reliability; arrangements have been made with the other Swedish utilities and some foreign utilities to exchange operating data and reports on failures, repairs, modifications and maintenance.

Sweden has used two systems for building nuclear power plants. The first two plants were obtained on a turnkey basis. For the remaining plants, the utilities used another system whereby the plants were divided into several large packages with the utility as overall coordinator: nuclear island, turbine-generators, and structure. The construction contract is normally a cost-plus system combined with economic incentives.

The nuclear steam supply system for nine of the twelve plants was provided by the same company, ASEA/ATOM. ASEA-ATOM's business is primarily the Swedish nuclear power plants, but it has also supplied two reactors to Finland and is supplying components to other countries in Western Europe. ASEA/ATOM's manager of design and development has overall responsibility for quality assurance and is directly responsible for design control. The production manager is responsible for compliance with the QA program. The program is set up with one individual responsible for each type of component or equipment. Design criteria are described in the design basis documentation which is included in a proposal and negotiated as part of a contract. The contractual design basis documentation is subject to formal change controls and customer approval. Quality verification is contracted to an independent inspection agency which prepares detailed procedures. While performing quality verification, the inspection agency compiles inspection reports on behalf of ASEA/ATOM.

For the remaining three plants, the steam supply system was supplied by Westinghouse. This arrangement has facilitated the use of replicated designs, with usually three or four units being built from the same basic design. ASEA/ATOM's first unit was supplied on a turnkey basis, and for the others, ASAE/ATOM functioned as the A-E for the mechanical systems that they supplied and as the construction manager/contractor for the mechanical systems that they subcontracted.

Most Swedish nuclear power plants have been built in a relatively short time (four to six years), and most of the activity is performed by experienced personnel who are with a job through completion. While there is a somewhat adverse relationship between the regulators/inspectors (SKI, SSI, SA) and the builders (the utilities and ASEA/ATOM), the limited resources of a small country (eight million people) and stability of the industry result in the interaction being less formal than in other countries.

The fabricator-installer provides "special" inspection plans, based on the general inspection plan, which cover the specific items being fabricated or installed. Special inspection plans are submitted to ASEA/ATOM for approval and for forwarding to the SA for approval. Following SA approval, the vendor can proceed with the specific activity covered by the plan.

Official third-party inspections are required by statute for certain activities and components and are performed by the SA. Further inspection and testing not required by statute may be prescribed by the owner, and the owner normally designates an independent agency to perform these inspections and tests. The SA designates a supervisor for each nuclear plant. This supervisor, who must be approved by the SKI, is responsible for ensuring that the plant meets codes, standards and requirements. The SKI has one inspector per unit. This inspector is not a resident but keeps frequent contact with construction, utility and other SKI personnel.

The program for nuclear power plant construction in Sweden has taken advantage of replicated designs and stability of personnel involved in the construction. The government regulatory agencies in Sweden have relatively small staffs and rely on independent-third party reviews and inspections. The utilities and the nuclear steam supply system supplier have taken an active role in formulating QA policy and in working with the regulators to adapt requirements to the Swedish environment.

D.4.6.6 The United Kingdom

In 1956, the United Kingdom (UK) initiated commercial nuclear generation of electricity when it began operating four 50-MWe reactors at Calder Hall. Until 1968, the UK had the largest nuclear power capacity in the world with 17 reactors producing electricity. Currently, the UK has 32 reactors producing 16% of the country's electricity. Forty-two reactors are projected for 1990 with a capacity of 12,514 MWe.

Unlike the United States, the UK regulatory agency does not prescribe the detailed methods for the compliance and implementation of the QA/QC requirements as part of the statutory regulation. The regulatory agency promulgates only the more general requirements in the form of guidelines for the safety and quality assurance for licensing. The licensee (utility is responsible for developing detailed requirements and for implementing safety and QA procedures that will satisfy the broad requirements of the regulatory agency.

Although the UK nuclear industry has had over two decades of QA/QC programs for nuclear plant design and construction, the Heysham II AGR plant (1978) is the first nuclear power plant in the UK with a license specification (1978) containing a formal QA requirement. The UK's QA/QC program is in a state of transition from gas-cooled reactor technology to PWR technology, and the British are taking steps to incorporate U.S. QA/QC requirements into their system. Therefore, the emerging UK QA/QC program will be a blend of U.S. requirements and British industry practices.

Organization. The main legislative acts governing commercial nuclear power plants in the UK are the Nuclear Installations Acts of 1965 and 1969 and the associated provisions of the Health and Safety at Work etc. Act of 1974. The Nuclear Installations Acts provide the regulatory framework for licensing of commercial nuclear power plants by the Health and Safety Executive (HSE). Under these Acts, no site may be used for the purpose of installing or operating a commercial nuclear installation unless a nuclear site license has been granted by the HSE. These Acts lay down only general requirements for the safety of nuclear power plants, and impose an absolute liability upon the licensee for any injury or damage caused by the release of radioactive material from its installations. The licensee is also responsible under the Health and Safety at Work Act for the safe design and operation of nuclear installations to ensure the health and safety of employees and other persons.

There are a limited number of major participants in the UK nuclear power program, and the character of those organizations has been changing in recent years. A description of their roles, internal organization and interrelationships is presented here. The four major participants in the UK nuclear power program are the HM Nuclear Installations Inspectorate (HMNII), the Central Electricity Generating Board (CEGB), the National Nuclear Corporation (NNC), and the major contractors (national and private). The Nuclear Installations Inspectorate (NII) was established in 1960 and became a part of the HSE when that organization was set up in 1975. The HSE brought together a number of existing inspectorates, including the Nuclear Installations Inspectorate, the Factor Inspectorate, the Alkali and Clean Air Inspectorate, and the Mines and

Quarries Inspectorate.

The HSE is a corporate body able to take independent action on safety enforcement, although it takes its general policy instructions from the Health and Safety Commission (HSC). The HSC then reports to the Secretary of State for Employment.

On nuclear safety matters, the HSC reports directly to the Secretary of State for Energy and to the Secretary of State for Scotland. While these Ministers have a limited power of directing, the NII operates independently of any government department.

The NII is organized under the Chief Inspector into four branches, each headed by a Deputy Chief Inspector. Three branches are responsible for the work on commercial nuclear power stations. Of these, one branch deals with future systems, and at present gives priority to the Inspectorate's assessment of the pressurized water reactor (PWR program). The fourth branch is responsible for the licensing of installations concerned with fuel fabrication and reprocessing, isotope separation and waste management. Inspectors in each section carry out such detailed work such as design safety assessments, quality assurance assessments, site inspections and other work connected with licensing. There are approximately 100 staff members, more or less evenly allocated among the four branches.

In addition to the NII, there is a further independent body, the Advisory Committee for the Safety of Nuclear Installations (ACSNI), which advises the HSC and the appropriate secretaries of State on major issues affecting the safety of nuclear installations that are referred to it or that it considers in need of attention. The ACSNI's function is to provide advice on policy matters rather than become involved in the regulatory process.

The CEGB is the government-owned utility that is responsible for design, construction and operation of nuclear power plants in England and Wales of the UK. An equivalent role is played by the South of Scotland Electricity Board (SSEB) for the regions of Scotland, but the scope and capacity of the SSEB is much smaller than the CEGB. Since the governing laws and regulatory requirements for nuclear power generation are the same for both England and Scotland, the SSEB practices are similar to these of the CEGB in licensing nuclear power plants. Consequently, this discussion is limited to the CEGB and its roles in the overall nuclear program of the UK.

The CEGB, as the owner and operator of commercial nuclear power plants, is responsible for the safety of its employees and the public from any nuclear hazard arising from its installations. This responsibility is formally defined in the Nuclear Installations Acts of 1965 and 1969. These Acts impose an absolute liability upon the CEGB, as licensee, for any injury or damage caused by the release of radioactive material from its installations. Recognizing this responsibility, the CEGB is committed to maintaining the highest nuclear safety standards to ensure the radiological protection of the employees and the public. The safety standards established by the CEGB are generally acceptable to the NII and are reviewed regularly in light of scientific and technical developments.

Among the departments and divisions of the CEGB, the following organizations have direct bearing on the safety of nuclear power plants:

- Health and Safety Department (HSD). The HSD is the primary interface for the CEGB with the NII. It is independent of all other parts of the CEGB organization, and its director reports directly to the Chairman and Executive of the CEGB. The HSD is responsible for assessing and monitoring the CEGB's activities to ensure a satisfactory standard of nuclear safety and compliance with regulatory requirements during all phases of the project and the subsequent operational lifetime and decommissioning. The HSD is also responsible for consultation and liaison with the NII on all licensing matters, including quality assurance, to obtain formal approvals under the nuclear site license.

- Generation Development and Construction Division (GDCD). The GDCD's major responsibility is designing and constructing nuclear power stations. Within the CEGB, the GDCD has total responsibility and authority for the design, procurement, manufacture and construction during the construction phase of a nuclear power station, responsibility including verification that QA/QC programs are satisfactorily implemented in the constituent phases. The GDCD is also responsible for developing and implementing relevant QA/QC program procedures from design through to commissioning and for establishing appropriate interface procedures for all principal participants.

- Technology Planning and Research Division (TPRD). The TPRD operates three CEGB laboratories involved in research work associated with nuclear technology, nuclear safety, fuel performance, materials science, thermal hydraulics, radiological protection and water chemistry.

- Nuclear Operations Support Group (NOSG). The NOSG administrates the CEGB's procedures to satisfy the conditions of the nuclear site licenses and coordinates the preparation of safety submissions to the Nuclear Safety Committee.

- Transmission and Technical Services Department (TTSD). The Engineering Services Department of TTSD is responsible for developing the CEGB's corporate policy on QA practices and for providing certain QA services to the GDCD.

- Nuclear Power Training Center (NPTC). The NPTC is used to train nuclear plant operators. Training is conducted primarily with simulators.

- Regions. The immediate delegated responsibility for operating a nuclear power station and for ensuring that QA practices are followed during plant operation rests with the Station Manager. This person is accountable through the CEGB's Regional line management to the Regional Director-General.

Quality Program Requirements. Although the NII does not issue standards or codes of practice for nuclear power plants, it does formulate and enforce the general requirements for the safety and the quality of the plant design, construction and operation. The NII's general requirements are set forth in a document entitled A Guide to the Quality Assurance Program for Nuclear Power Plants.

The NII is responsible for ensuring that the licensee develops and maintains appropriate standards that meet the general requirements of the NII and for monitoring and enforcing the quality of design, construction, and safe operation of the plant. This responsibility includes inspecting for compliance with the requirements at all stages of plant construction, operation, and decommissioning.

The applicant is granted a nuclear site license following a satisfactory outcome of the NII's review and assessment of the documents and preliminary plant design submitted by the applicant. Granting of a site license signifies that the NII is satisfied that the design intent, safety principles, and contract design description are such that construction can proceed with little risk of significant changes being subsequently required for safety reasons.

The NII maintains close surveillance, during construction, of the licensee's activities to ensure that the licensee follows appropriate QA/QC practices in construction. The NII's site inspector visits the site (average every two to four weeks) for inspection purposes, witnessing tests and examining test records. Where necessary, NII inspectors visit manufacturers' shops to monitor fabrication, witness tests and audit QA/QC procedures.

Some of the specific requirements and procedures that are followed during the construction phase to ensure quality of construction are listed below:

- The licensee must make arrangements for inspection and testing of major items of the plant both on-site and at manufacturer's shops. These activities may be carried out either by recognized independent inspecting agencies or by the licensees' own inspection organization, but the arrangement requires approval of the NII.

- The licensee must keep detailed case histories of the construction of important items such as pressure vessels, which must be retained throughout the life of the plant.

- The licensee must formulate appropriate QA/QC procedures that must be approved by the NII.

- The licensee must update the PCSR by a Station Safety Report (equivalent to U.S. FSAR) as the design and construction approach completion, which forms the basis of the NII's acceptance of the station for commercial operation.

- The licensee must obtain the NII's consent to proceed further at various major steps in the construction phase.

The QA/QC practices of the UK nuclear industry are based on two primary documents. These are BS-5882, Total Quality Assurance Programme for Nuclear Power Plants (1980); and NII/R/38/78/Issue 2, Guide to the Quality Assurance Programme for Nuclear Power Plants (1980).

BS-5882 closely parallels Appendix B to 10 CFR 50 and ANSI N45.2, and establishes the QA principles for the UK nuclear industry. The implementation methods of the principles are set out in another standard called BS-5750, Quality Systems, that consists of six parts delineating specific procedures for implementing the BS-5882 principles. NII/R/38/78 is the Guidelines document issued by the NII and is the Assessment Criteria Document for NII inspectors.

Quality Program Implementation. In implementing the QA program requirements of BS-5882 for design, procurement and construction of nuclear power plants, a graded categorization system is applied to various items. The QA category of an item or service is assigned according to its safety importance or operational reliability and performance importance. Factors considered in assigning the level of QA requirements for an item are as follows:

- the consequence of malfunction or failure of an item

- the design and fabrication complexity or novel features of an item

- the need for special controls and surveillance over processes and equipment

- the degree to which functional compliance of an item can be demonstrated by inspection or test

- the quality history and degree of standardization of an item

- the difficulty of repairing or replacing an item, or its accessibility for in-service inspection.

Essentially there are two QA category levels assigned to various plant items and services at a typical nuclear power station. The "Q" category is assigned to items and services of safety class, and the "N" category is assigned to non-safety class items and services. However, in practice the UK uses four grades of QA requirements to categorize control and verification requirements at nuclear power stations. The various grades of QA requirements for plant items and services are as follows:

- "Q". Items and services categorized as "Q" are subjected to the highest grade of control and verification requirements. These include all safety class plant items and services of a nuclear power plant.

- "N/S," "N/O." "N/S" includes items and services "Important to Safety." "N/O" includes items and services "Important to Operational Reliability."

- "N/E." The "N/E" grade is assigned to those items and services that are non-safety, non-operationally significant, but require significant design engineering by the contractor or manufacturer.

- "N/-." This grade is assigned to non-safety, non-operationally significant, off-the-shelf items that are mass-produced by a routine production process.

The "Q" graded items and services are required to satisfy BS-5882 QA program requirements. Items designated as "N" grade are not required to satisfy the BS-5882 requirement but must meet standards of quality assurance appropriate to the contract specifications.

An important aspect of the UK QA/QC practices is the so-called "hierarchical system," in which the extent of responsibility and authority and the line of communication channels are clearly defined in a descending order starting from the licensee at the top, through the main contractor and finally to the smallest supplier at the bottom. Although a higher-order organization may audit the QA/QC practices of any lower-order organization, an organization is only accountable to the organization immediately above it in the hierarchy. Under this hierarchical system, the CEGB interfaces with the National Nuclear Corporation (NNC) on all matters concerning quality assurance within the UK PWR program. The NNC was incorporated in 1974 as a partially government-funded (35%) private nuclear engineering company, and is the only such company in the UK. The NNC is a contractor to the CEGB, and is responsible for its own QA/QC practices as well as for those of other suppliers. A Joint Project Team (JPT) is formed, primarily of NNC and CEGB staff, and is responsible for developing, coordinating, and monitoring the implementation of the project QA program at all project stages. Contractors are responsible to the JPT for the quality of the products and services they supply to the CEGB. Each purchaser, including the NNC acting as the CEGB's agent, is responsible in turn for ensuring that each supplier has acceptable QA/QC programs and procedures and for verifying that the performance of each supplier against these procedures is appropriate.

An Independent Third Party Inspection Authority (ITPIA) is employed by the GDCD to provide independent services involving all items procured to the intent of Section III of the ASME Boiler and Pressure Vessel Code. The ITPIA undertakes the tasks ascribed to an "Authorized Inspection Agency" in Section III of the ASME Code as adapted by the CEGB and NNC for use in the UK.

The underlying characteristic of regulatory practices in the UK is that the regulatory agency emphasis is on the actual accomplishment of the licensee in the safe design, quality construction, and safe operation of nuclear power plants rather than on documentation requirements.

Another point of interest about regulatory practices in the UK is that the responsibility for safety is placed on the licensee (utility), requiring it to formulate the design safety criteria and standards, QA standards and implementing procedures. The UK approach to safety does not accept the premise that designers and operators ensure safety by meeting a prescribed standard or guidance set by the regulatory agency.

D.5 REFERENCES

Adkins, B. 1973. "Public Understanding and Acceptance of Nuclear Energy in Japan," Nuclear Engineering International.

Atomic Energy Control Board. Annual Report 1982-1983. Canada.

Atomic Energy Mission - White Paper. 1980. Ministry for Science and Technology, Japan.

Boeing Commercial Airplane Company. 1982. Quality Control Manual. D6-1979-4-C2-4203-8200. 747-767 Division, Seattle, Washington.

Booz-Allen and Hamilton, Inc. 1981. Analysis of Foreign Maritime Research. PB-81-176364, U.S. Maritime Administration, Washington, D.C.

British Standards Institution. 1980. Total Quality Assurance Programme for Nuclear Power Plants. BS-5882, United Kingdom.

British Standards Institution. 1981. Quality Systems, 1979-1981, Part 1-Part 6. BS-5750, United Kingdom.

Bundesanzeiger. 1974. Bekanntmachung über die Bildung eines kerntechnischen Ausschesses. Bonn, Federal Republic of Germany.

Central Electricity Generating Board (CEGB). 1982. Design Safety Criteria for CEGB Nuclear Power Stations. United Kingdom.

Commissariate a l'Energique Atomique. 1980. Principaux Textes Législatifs et Réglementaires Relatifs a l'Energie Nucleaire ou au C.E.A.. Paris, France.

Ernst and Whinney. 1979. Cost Impact of U.S. Government Regulations on U.S. Flag Ocean Carriers. Report NO. PB-80-151392. Prepared for the U.S. Maritime Administration, Washington, D.C.

Federal Aviation Administration (FAA). 1967. Designated Engineering Representatives. Washington, D.C.

Fischerhof, H. 1979. Deutsches Atomgesetz und Strahlenschutzrecht, Kommentar. 2nd edition. Baden-Baden, Federal Republic of Germany.

Fremling, A. G. 1980. "Quality Assurance." DOE Order 5700.1 Memorandum to Contractors, Richland, Washington.

Gumley, P. N.D. Use of Fault Tree/Event Sequence Analyses in a Safety Review of CANDU Plants, AECL-7373, Canada.

Hanford Engineering Development Laboratory (HEDL). 1982. Instructions for Submittal and Control of FFTF Principal Design Documentation. WHAN-M-17, Revision 11, Richland, Washington.

Hurst, D. G., and F. C. Boyd. N.D. Reactor Licensing and Safety Requirements. 72-CNA-102, Canada.

International Atomic Energy Agency (IAEA). 1978. Safety Series, IAEA Safety Standards. Vienna, Austria.

International Atomic Energy Agency (IAEA). 1982. Quality Assurance for Nuclear Power Plants. Proceedings of IAEA Symposium, Paris, May 1981. Vienna, Austria.

Japan Atomic Energy Commission. 1980. Guidelines in Reactor Site Evaluation. Japan

Japan Electric Association. 1981. Japanese Quality Assurance Criteria. JEAG-4101-1981. Japan.

Jennekens, J. H. N.D. Nuclear Energy--Some Regulatory Aspects. INFO-0002, Atomic Energy Control Board, Canada.

Lisanby, J. W., and J. Haas. 1981. "Use of Commercial Specifications in the Shipbuilding Process." Naval Engineers Journal, 93(1).

Maritime Administration. 1982. Relative Cost of Shipbuilding. Department of Transportation, Washington, D.C.

National Aeronautics and Space Administration (NASA). 1969. Quality Program Provisions for Aeronautical and Space System Contractors. NHB 5300.4(1B), Washington, D.C.

National Aeronautics and Space Administration (NASA). 1970. Quality Program Provisions for Aeronautical and Space System Contractors. NHB 5300.4(1A), Washington, D.C.

National Aeronautics and Space Administration (NASA). 1981. Basic Policy and Responsibilities for Reliability and Quality Assurance. NASA Management Instruction NMI 5300.7B, Washington, D.C.

National Nuclear Corporation (NNC). 1983. Joint Project Team Quality Assurance Program. United Kingdom.

Naval Sea Systems Command (NAVSEA). 1975. Naval Shipyard Quality Program Manual. NAVSEA 0900-LP-083-0010, Department of the Navy, Washington, D.C.

Naval Sea Systems Command (NAVSEA). 1976. Naval Shipyard Quality Program Manual. NAVSEA INST 4355.3, Department of the Navy, Washington, D.C.

Naval Sea Systems Command (NAVSEA). 1982. Naval Shipyard Quality Program Manual. NAVSEA INST 4855.5A, Department of the Navy, Washington, D.C.

NEI. 1983. "World Survey of Nuclear Power." Nuclear Engineering International, 28(341).

NEI. 1983. "Power Reactors 1983." Nuclear Engineering International, 28(344).

Nuclear Installations Inspectorate (NII). 1979. Safety Assessment Principles for Nuclear Power Reactors. United Kingdom.

Nuclear Installations Inspectorate (NII). 1980. A Guide to the Quality Assurance Programme for Nuclear Power Plants. NII/R/38/78/Issue 2, United Kingdom.

Nuclear Safety Analysis Center (NASC). 1981. Space and Missile Reliability and Safety Programs. NSAC 31, Pickering Research Corporation, Pasadena, California.

Sandia National Laboratories. 1977. A Study of the Nuclear Regulatory Commission's Quality Assurance program. NUREG-0321, National Technical Information Service, Springfield, Virginia.

Stevenson, J. D., and F. A. Thomas. 1982. Selected Review of Foreign Licensing Practices for Nuclear Power Plants. NUREG/CR-2664, National Technical Information Service, Springfield, Virginia.

Swedish Nuclear Power Inspectorate. 1980. SKI Control Procedures. Stockholm, Sweden.

U.S. Department of Energy (DOE). 1980. Application of Space and Aviation Technology to improve the Safety and Reliability of Nuclear Power Plant Operations. DOE/TIC-11143, International Energy Associates, LTD, Washington, D.C.

U.S. Department of Energy (DOE). 1981. Quality Assurance. DOE Order 5700.6A, Washington, D.C.

U.S. Department of Energy (DOE). 1982. Quality Assurance - ORO Site Implementation Plan. OR 5700.6, Oak Ridge Operations Office, Oak Ridge, Tennessee.

U.S. Energy Research and Development Administration (ERDA). 1973. RDT Standard Quality Assurance Program Requirements. RDT F2-2, Washington, D.C.

U.S. Energy Research and Development Administration (ERDA). 1974. RDT Standard Quality Verification Program Requirements. RDT F2-4, Washington, D.C.

U.S. General Accounting Office. 1978. The Nuclear Regulatory Commission Needs to Aggressively Monitor and Independently Evaluate Nuclear Power Plant Construction. EMD-78-80, Washington, D.C.

U.S. Nuclear Regulatory Commission (NRC). 1975. "Quality Assurance," Chapter 17 in Standard Review Plan for the Review of Safety Analysis Reports for Nuclear Power Plants, LWR Edition. NUREG-75, Washington, D.C.

U.S. Nuclear Regulatory Commission (NRC). 1983. Inspection and Enforcement Manual, Washington, D.C.

Westinghouse Hanford Corporation (WHC). 1983. Quality Assurance Program for Compliance with the American Society of Mechanical Engineers Boiler and Pressure Vessel Code. MG-23, Richland, Washington.

Wood, K. and N. H. Harding. 1974. Quality Assurance Applied to Nuclear Power Construction Project in Western Europe. ANS Report No. 61, Associated Nuclear Services, London, England.

APPENDIX E

<u>GLOSSARY OF ACRONYMS/ABBREVIATIONS</u>

GLOSSARY OF ACRONYMS/ABBREVIATIONS

ACRS	Advisory Committee on Reactor Safeguards
A/E	architect-engineer
A&E	architectural and engineering
AEC	Atomic Energy Commission
ANSI	American National Standards Institute
ASLB	Atomic Safety and Licensing Board
ASME	American Society of Mechanical Engineers
ASQC	American Society for Quality Control
AWS	American Welding Society
CAT	Construction Appraisal Team
CM	construction manager
CP	construction permit
CPE	Construction Project Evaluation
CRGR	Committee to Review Generic Requirements
DOE	Department of Energy
DR	designated representative
FAA	Federal Aviation Administration
FSAR	Final Safety Analysis Report
HARC	Human Affairs Research Center
IDCVP	Independent Design and Construction Verification Program
IDI	integrated design inspection
IDVP	Independent Design Verification Program
IE	Office of Inspection and Enforcement
IEEE	Institute of Electrical and Electronics Engineers
INPO	Institute of Nuclear Power Operations
MarAd	Maritime Administration

NASA	National Aeronautics and Space Administration
NB	The National Board of Boiler and Pressure Vessel Inspectors
NRC	Nuclear Regulatory Commission
NRR	Office of Nuclear Reactor Regulation
NSSS	nuclear steam supply system manufacturer
OTA	Office of Technology Assessment
PAT	Performance Appraisal Team
PNL	Pacific Northwest Laboratory
PSAR	Preliminary Safety Analysis Report
PUC	Public Utility Commission
QA	quality assurance
QC	quality control
SALP	Systematic Assessment of Licensee Performance
SRP	Standard Review Plan
TPT	Torrey Pines Technology
USN	U.S. Navy

☆ U.S. GOVERNMENT PRINTING OFFICE: 1984-421-297:3941

NRC FORM 335
(11-81)

U.S. NUCLEAR REGULATORY COMMISSION

BIBLIOGRAPHIC DATA SHEET

1. REPORT NUMBER *(Assigned by DDC)*	NUREG-1055

4. TITLE AND SUBTITLE *(Add Volume No., if appropriate)*

Improving Quality and the Assurance of Quality in the Design and Construction of Commercial Nuclear Power Plants - A Report To Congress

2. *(Leave blank)*

3. RECIPIENT'S ACCESSION NO.

7. AUTHOR(S)

W. Altman, T. Ankrum, W. Brach

5. DATE REPORT COMPLETED

MONTH	YEAR
April	1984

9. PERFORMING ORGANIZATION NAME AND MAILING ADDRESS *(Include Zip Code)*

Quality Assurance Branch
Division of Quality Assurance, Safeguards, and
   Inspection Programs, IE
Washington, D.C. 20555

DATE REPORT ISSUED

MONTH	YEAR
May	1984

6. *(Leave blank)*

8. *(Leave blank)*

12. SPONSORING ORGANIZATION NAME AND MAILING ADDRESS *(Include Zip Code)*

Quality Assurance Branch
Division of Quality Assurance, Safeguards, and
  Inspection Programs, IE
Washington, D.C. 20555

10. PROJECT/TASK/WORK UNIT NO.

11. FIN NO.

13. TYPE OF REPORT

Report to Congress

PERIOD COVERED *(Inclusive dates)*

15. SUPPLEMENTARY NOTES

14. *(Leave blank)*

16. ABSTRACT *(200 words or less)*

At the request of Congress, NRC conducted a study of existing and alternative programs for improving quality and the assurance of quality in the design and construction of commercial nuclear power plants. A primary focus of the study was to determine the underlying causes of major quality-related problems in the construction of some nuclear power plants and the untimely detection and correction of these problems. The study concluded that the root cause for major quality-related problems was the failure or inability of some utility managements to effectively implement a management system that ensured adequate control over all aspects of the project. These management shortcomings arose in part from inexperience on the part of some project teams in the construction of nuclear power plants. As a corrollary, NRC's past licensing and inspection practices did not adequately screen construction permit applicants for overall capability to manage or provide effective management oversight over the construction project. The study recommends a number of improvements in industry and NRC programs.

17. KEY WORDS AND DOCUMENT ANALYSIS

Quality
Quality Assurance
Quality Control
Design
Construction
Commercial Nuclear Power Plants
Ford Amendment

17a. DESCRIPTORS

Management Capability
Prior Experience
Architect-Engineer
Construction Manager
Constructor
Utility

b. IDENTIFIERS/OPEN-ENDED TERMS

18. AVAILABILITY STATEMENT

Unlimited

19. SECURITY CLASS *(This report)*	21. NO. OF PAGES
20. SECURITY CLASS *(This page)*	22. PRICE S

www.ingramcontent.com/pod-product-compliance
Lightning Source LLC
Chambersburg PA
CBHW081427170526
45166CB00008B/2115